Udo Paulitz

HANOMAG
Traktoren

Hanomag R 55 mit 5 t-Seilwinde, Baujahr 1957

 2002
Verlag Podszun-Motorbücher GmbH
Elisabethstraße 23-25, D-59929 Brilon
Herstellung Druckhaus Cramer, Greven
ISBN 3-86133-294-9

Udo Paulitz

Hanomag Traktoren

Inhalt

Von Maschinen und Lokomotiven zu Tragpflügen	5
Die Entwicklung der WD-Radschlepper	7
Die Hanomag-Dieselschlepper der 30er Jahre	15
Die schweren Diesel-Radschlepper R 40 und deren Folgemuster	28
Die Hanomag-Schlepper in den ersten Nachkriegsjahren	65
Die Ära der Zweitakt-Dieselmotoren	94
Bewährte Technik unter moderner Verkleidung	108
Die 60er Jahre – Wirbelkammer-Dieselmotoren und kantige Hauben	122
Granit, Brillant und Robust – die letzten Hanomag-Schlepperkonstruktionen	133

Udo Paulitz

Geboren am 14. Dezember 1946 in Geesthacht, Kreis Lauenburg/Elbe, gelernter Sortimentsbuchhändler und begeisterter Fotoamateur, entdeckte bereits als Heranwachsender seine Vorliebe für alle Bereiche der historischen Technik. In den 70er Jahren war er über viele Jahre lang mit Fotoapparat und Filmkamera den letzten Dampflokomotiven Mitteleuropas auf der Spur, bis er ab etwa 1980 sein fotografisches Hauptinteresse vor allem auf historische Feuerwehrfahrzeuge und Lastwagen und deren Historie verlegte. Vor einigen Jahren wandte er sich zusätzlich den älteren Traktoren und Ackerschleppern zu, die er in bekannter Manier stets in einem ausgesuchten Umfeld abzubilden versuchte. Seit 1987 hat er viele Beiträge in Hobby- und Fachzeitschriften sowie zahlreiche Bild/Textbände und Kalender zu den oben genannten Themenkreisen – auch in unserem Verlag – veröffentlicht.

Zu diesem Buch

Genau 31 Jahre ist es her, seit die letzten Hanomag-Traktoren die Fertigungsstraßen des traditionsreichen Herstellers in Hannover verließen. Damit endete eine wechselvolle, von Höhen und Tiefen geprägte, fast sechs Jahrzehnte dauernde Periode der Schlepperherstellung an diesem Standort, und sowohl der deutsche als auch der internationale Markt war um einen bedeutenden Anbieter dieser Branche ärmer geworden.

Mit diesem neuen Buch über Hanomag-Traktoren soll der Versuch unternommen werden, der bisher vorliegenden eher umfangreichen Hanomag-Literatur ein weiteres Werk zur Seite zur stellen, welches die rund 60-jährige Geschichte der Traktoren dieses Herstellers – in Wort und Bild – in kompakter und informativer Form umreißen soll. Auf die Darstellung der vielen Kettenschlepper und Zugmaschinen wurde bewusst verzichtet.

Den Schwerpunkt dieses Buches bildet zweifelsohne das vom Autor in den letzten Jahren selbst aufgenommene Fotomaterial, welches durch ausgesuchte historische Faksimiledrucke sinnvoll ergänzt wird. Zusätzlich werden die wichtigsten technischen Daten der gezeigten Traktorenmodelle anhand von kurzen Tabellen aufgezeigt.

Die vielen brillanten, fast ausschließlich unveröffentlichten Farbaufnahmen derart vieler restaurierter Traktoren sind gleichzeitig ein Spiegelbild dessen, welcher Wertschätzung sich die renommierte Schlepper-Marke Hanomag in Oldtimerkreisen auch heute noch erfreut, denn viele hundert originalgetreu wieder hergerichteter Traktoren der gesamten Modellpalette befinden sich in Sammlerhand.

Herzlich danken möchte ich an dieser Stelle den vielen Traktorbesitzern, die mir ihre Fahrzeuge entgegenkommenderweise in die richtige Fotografierposition fuhren und Herrn Kurt Häfner vom KMH-Verlag sowie Herrn Klaus Vollmar vom WK-Verlag für die Bereitstellung von Unterlagen. Mein Dank richtet sich nicht zuletzt an meine Frau Halena, die mich auf vielen dieser Fotoexkursionen begleitete und meine Fotoleidenschaft sowie die regelmäßige Abwesenheit an den Wochenenden der Sommermonate nahezu geduldig ertrug.

Udo Paulitz, Braunsberger Weg 69, 47279 Duisburg

Von Maschinen und Lokomotiven zu Tragpflügen

Die Geschichte des Hauses Hanomag geht auf das Jahr 1835 zurück, als Georg Egestorff im Dorf Linden bei Hannover eine Metall-, Gusswaren- und Maschinenfabrik gründete, die ein breit gefächertes Programm von Guss- und Eisenwaren – von Kochtöpfen über Grabkreuze bis zu Zahnrädern – fertigte. Schon ein Jahr später wurde – zunächst mit Unterstützung englischer Fachkräfte, da es in Deutschland derzeit nur wenig ausgebildete Facharbeiter dieser Sparten gab – mit dem Bau von Dampfmaschinen und Dampfkesseln begonnen und zwei Jahre darauf die Fertigung von Schiffmaschinen aufgenommen. Ab 1846 kam als weiterer Zweig der Lokomotivbau hinzu. So konnte am 14. Juni 1846 mit der "Ernst August" die erste Egestorffsche Lokomotive an die Hannoversche Staatsbahn übergeben werden. Aus diesen noch bescheidenen Anfängen entwickelte sich eine der größten und berühmtesten Lokomotivfabriken Europas.

Im Jahr 1871, drei Jahre nach dem Tod des Firmengründers, wurde das rund 400 Belegschaftsmitglieder zählende Unternehmen in eine Aktiengesellschaft umgewandelt. Bereits im Juni 1873 gelangte die 1000ste Lokomotive zur Ablieferung. Die gut florierende und trotz mehrerer Wirtschaftskrisen schnell expandierende Fabrik, die sich auf nahezu allen Gebieten der Eisen- und Metallfertigung einen Namen gemacht hatte, firmierte seither als "Hannoversche Maschinenbau Aktiengesellschaft, vormals Georg Egestorff". 1894 waren annähernd 2000 Beschäftigte in dem Werk tätig und ab ca. 1905 lautete die aus dem Telegrammnamen hervorgegangene Kurzbezeichnung des Unternehmens "Hanomag". Dieses war der Name, unter dem das auch im Exportgeschäft ungeheuer vielseitige Unternehmen einen derart großen Bekanntheitsgrad erlangte, dass dieser auch über 30 Jahre nach der Produktionseinstellung unvergessen geblieben ist. Im Jahre 1914, zu Beginn des Ersten Weltkrieges, zählte diese Firma etwa 4400 Arbeitnehmer und gehörte damit zu den größten Industriebetrieben Norddeutschlands.

Unmittelbar vor Ausbruch des Krieges wurde mit der Fertigung von Verbrennungsmotoren und dem Bau von Tragpflügen die Angebotspalette um zwei zusätzliche Zweige erweitert, mit denen das Werk in der Folgezeit sowohl für den Export als auch auf dem Inlandsmarkt große Erfolge erzielen sollte.

Mit dem in Anlehnung an die Motorpflüge von Stock nach den Ideen und Vorstellungen des Maschinenbauingenieurs Ernst Wendeler und des in der Uckermark ansässigen Gutbesitzers Boguslav Dohrn entwickelten und bei Hanomag gebauten WD-Großpfluges wurde im Oktober 1912 der Einstieg in die Landtechnik unternommen. Dieser Großpflug, für dessen Vertrieb eigens die Deutsche Kraftpflug-Gesellschaft mbH in Berlin gegründet wurde, als dessen Direktor Ernst Wendeler fungierte, war ein gewaltiges Unikum. Der WD stand auf einer Achse mit übermannshohen Treibrädern (Durchmesser 2,20 m) und einem am Ende eines Auslegers angebrachten lenkbaren Hinter- und Stützrad. Er war mit einem beweglichen Pflugrahmen versehen und brachte rund sieben Tonnen auf die Waage. Angetrieben wurde er zunächst

WD-Großpflug mit 80 PS Hanomag-Motor

von einem vor der Antriebsachse liegenden 50-PS-Vergaseraggregat der Berliner Motorenfabrik Kämper. Dem Fahrer war es möglich, mittels eines unterhalb des Lenkrades angeordneten Handrades, die Pflugtiefe stufenlos einzustellen. Schon bald aber erwies sich die Motorstärke als zu gering, so dass erst 65-, später 80-PS-Motoren – letztere wurden von Hanomag gefertigt – zum Einbau gelangten. Der mit zwei Vorwärts- und einem Rückwärtsgang ausgerüstete Tragpflug besaß eiserne Räder mit für Straßenfahrten umklappbaren Greifern und rief schon bald das Interesse von Großbauern und Besitzern großer Güter hervor, die lange eine einfache und robuste Kraftmaschine für die Verwendung in der Landwirtschaft gesucht hatten. Diese schweren, als Vorläufer der späteren Traktoren anzusehenden, Geräte stellten im Vergleich zu den bisherigen Methoden der Feldbearbeitung einen großen Fortschritt dar, und trotz der recht eingeschränkten Verwendbarkeit dieser gewaltigen Maschinen – sie waren in der Anschaffung recht kostspielig und daher eigentlich nur auf den großen Gütern im Osten Deutschlands wirtschaftlich einsetzbar – konnte im Jahr 1918 die Deutsche Kraftpflug-Gesellschaft die Auslieferung des 500sten Exemplars vermelden. Dafür zeichneten nicht zuletzt die Kriegsereignisse verantwortlich, denn die Maschinen wurden durch spezielle Motorpflugkommandos, in denen bis zu zehn Tragpflüge zusammengefasst waren, auch auf den großen Flächen besetzter Gebiete beispielsweise in Nordfrankreich, Polen, Rumänien und Bulgarien zur Bodenkultivierung und zur Bearbeitung von Ödlandgebieten im Osten und Südosten Europas eingesetzt.

	WD-Großpflug	WD-Kleinpflug
Produktionszeitraum	1912-1919	1921-1924
Motorleistung (PS)	50/65/80	35
Hubraum (cm³)	14137/15095	5702
Anzahl der Zylinder	4	4
Anzahl der Gänge	2+1	2+1
Eigengewicht (kg)	6000	3000
Höchstgeschwindigkeit	5,5 km/h	5,5 km/h

Im Jahr 1919 wurde der Bau des WD-Großpfluges eingestellt, und zwei Jahre später folgte ein wesentlich kleinerer, 35 PS starker WD-Tragpflug, der im Prinzip seinem größeren Vorgänger entsprach, aber mit einer Riemenscheibe zum Antrieb stationärer Maschinen ausgerüstet war. Damit erfolgte die werksseitige Anpassung an die neuen wirtschaftlichen Gegebenheiten. Da der aufwendige und teure Tragpflug nur bei Betriebsgrößen von mindestens 50 ha rentabel war und sich als Arbeitsgerät nur zum Pflügen des Ackers eignete, erwies er sich gegenüber den vielseitigeren, wesentlich kleineren und preiswerteren Motorschleppern als unterlegen. Daher wurde die Fertigung von Tragpflügen im Jahr 1924 gänzlich aufgegeben. Insgesamt wurden mehr als 1000 WD-Tragpflüge bei den Hanomag-Werken gebaut. Eine größere Stückzahl davon ging in den Export.

Darüber hinaus wurde ab 1919 mit dem Bau eines vom Ingenieur Josef Vollmer konstruierten 25-PS-Kettenschleppers den nach dem Tragpflugprinzip arbeitenden Maschinen ein Raupenfahrzeug zur Seite gestellt, das nicht nur in der Landwirtschaft, sondern auch in der Forstwirtschaft, im Bauwesen und in der Industrie sinnvolle Verwendungsbereiche finden konnte. Hieraus entwickelte Hanomag das Modell Z 50, eine äußerst gelungene 50-PS-Raupe, deren Bau in ihren Grundzügen bis weit in die 50er Jahre hinein fortgeführt wurde.

Die Entwicklung der WD-Radschlepper

Die Hanomag begann 1924 mit dem Automobilbau – das erste Produkt dieser Art war ein sehr erfolgreicher 10-PS-Kleinwagen mit dem treffenden Spitznamen "Kommissbrot" – und mit dem WD-Radschlepper R 26 A den erfolgreichen Vorstoß in den Traktorenmarkt. Diese auf der 1924 in Hamburg stattfindenden Deutschen Landwirtschaftsgesellschaft (DLG)-Ausstellung erstmals der Öffentlichkeit vorgestellte Konstruktion – aus der Hand des Hanomag-Konstrukteurs Josef Vollmer – entstand vor allem unter dem Eindruck des seit 1917 in den USA mittels Fließband in Serie gefertigten sehr fortschrittlichen, universell einsetzbaren und gleichzeitig preisgünstigen Fordson-Traktors, der die deutsche Industrie einem Zugzwang aussetzte. Darüber hinaus bestand ein in erster Linie durch den Krieg verursachter großer Fehlbestand an Zugtieren, verbunden mit dem Mangel an einfachen und preiswerten motorisierten Arbeitsgeräten, in der unter den Altlasten des verlorenen Krieges leidenden Landwirtschaft des Deutschen Reiches.

Mit dem neuen, überaus gut gelungenen WD konnten sich die Hanomag-Werke zweifelsohne zu den Pionieren des deutschen Schlepperbaus zählen. Die Konstruktion dieses soliden Schleppers orientierte sich mit seiner rahmenlos gegossenen, gewichtssparenden und gleichzeitig preiswerteren Blockbauweise, bei der Motor, Getriebe und Hinterachse eine Einheit bildeten, weitgehend an dem amerikanischen Vorbild. Er besaß einen wassergekühlten Reihen-Vierzylinder-Viertakt-Doppelvergaser-Benzol- oder Petroleummotor mit 4252 ccm Rauminhalt, mit dem eine Leistung von 26 PS bei 1100 U/min erzielt werden konnte. Die besondere Vielseitigkeit des universell einsetzbaren WD-Schleppers war die duale Verwendbarkeit sowohl auf dem Acker, als auch für Straßentransportzwecke. Daher war dieser in einer Ackerausführung mit Eisenbereifung und als Straßenzugmaschine mit gummibereiften Vollscheibenrädern, der sogenannten Elastikbereifung, und einer Anhängekupplung für Fuhrunternehmen erhältlich. Die Eigengewichte betrugen – je nach Ausführung und Ausrüstung – zwischen 1950 kg als Ackerschlepper und 3500 kg mit schweren, für eine angemessene Zugkraft sorgenden, Gussfelgen bei

Restaurierter WD-R-26-Straßenschlepper, Baujahr 1925

Zwei WD-R-28-Traktoren, links die Ackerausführung mit Eisenrädern, rechts als Straßenzugmaschine, beide Baujahr 1928

der Straßenverwendung. Während in der eisenbereiften Ackerausführung Geschwindigkeiten zwischen 3-8 km/h möglich waren, erreichte der WD-Schlepper in der elastikbereiften ab 1926 lieferbaren Variante 15 km/h. Beide Ausführungen waren mit drei Vorwärtsgängen und einem Rückwärtsgang bestückt. Dieser mittelschwere und sehr solide wirkende überaus zuverlässige Schlepper mit seinem charakteristischen, vor dem Lenkrad angebrachten, trommelförmigen Kraftstofftank, besaß eine unter dem Motorblock angebrachte, gefederte vordere Pendelachse, eine Bosch-Magnetzündung, Druckumlaufschmierung und Ölpumpe, sowie eine Umlaufkühlung mit Pumpe. Mit einem Gewicht von nahezu zwei Tonnen war er gegenüber dem 1240 kg wiegenden Fordson allerdings deutlich schwerer, wohl aber auch um rund ein Drittel leistungsfähiger geraten. Eine Sandstreueinrichtung erhöhte Anzugs- und Bremsvermögen des Fahrzeugs und der Wendekreis war mit neun Metern recht günstig. Die Ausrüstung mit Riemenscheiben-Antrieb machten ihn auch als stationäre Kraftquelle vielseitig einsetzbar.

Um den Landwirten die Anschaffung des 4800 Reichsmark teuren Schleppers zu erleichtern, erhielt eine unter der Leitung des Reichsernährungsministeriums geschaffene Finanzierungsgesellschaft die Aufgabe, zinsgünstige Kredite für potentielle Erwerber bereitzustellen. Sein Preis lag damit allerdings immer noch um etwa 1000 Reichsmark höher gegenüber dem wohl leistungsmäßig und in technischer Hinsicht unterlegenen Fordson-Traktor. In der ab 1927 lieferbaren Ausführung R 28/32 wurde die Leistung des Schleppers angehoben sowie mehrere technische Verbesserungen vorgenommen. Die Motorleistung bei Verwendung von Benzin oder Benzol anstelle des Petroleums stieg dann auf 30 beziehungsweise 32 PS. Bereits seit 1925 besaßen die Kolben vier Ringe gegenüber früher drei; die Kurbelwelle wurde einem besonderen Härtungsverfahren unterzogen und neben einer Erhöhung der Ölpumpenleistung gelangten größere Kolbenbolzen zum Einbau. Auch war der WD-Schlepper dieser überarbeiteten Ausführung mit Luftbereifung lieferbar, wodurch eine universelle Verwendbarkeit – auf der Straße und auch auf dem Acker – möglich wurde. Auch war das gegen Aufpreis erhältliche Sonderzubehör recht umfangreich; es reichte von der elektrischen Beleuchtung über eine Druckluftanlage, Seilwinde und Wetterverdeck bis zu Radverbreiterungen für Moorböden und anstelle der Hinterräder anbaubaren Halbraupen. Um ortsfeste Maschinen und Geräte wie Mischmaschinen, Sägen, Dreschkästen und Steinbrecher antreiben zu können, war eine seitliche Riemenscheibe vorgesehen.

	WD R 26	WD R 28/32
Produktionszeitraum	1924-1925	1926-1931
Motorleistung (PS)	26	28-32
Hubraum (cm^3)	4252	4252
Anzahl der Zylinder	4	4
Anzahl der Gänge	3+1	3+1
Eigengewicht (kg)	1950	2100
Höchstgeschwindigkeit	8,0 km/h	15,0 km/h

Zunächst blieb auch für die WD-Schlepper die Deutsche Kraftpflug-Gesellschaft in Berlin für die Belange des Vertriebs zuständig. Mit steigenden Verkaufserfolgen aber versuchte die Hanomag ihren Namen aus verständlichen Gründen mehr und mehr in den Vordergrund zu schieben. Nach der Auflösung der Deutschen Kraftpflug-Gesellschaft im Jahr 1928 prangte auf allen bei Hanomag in Hannover produzierten Schleppern auch deren Name.

Der Hanomag WD 28/32 PS wurde mit vielen Sonderausrüstungen geliefert, zum Beispiel mit einer Sandstreuanlage, die auf schlüpfrigem Boden das Gleiten der Triebräder verhindern sollte, oder mit besonders schweren Triebrädern mit verstellbaren Greifern. Eine elektrische Lichtanlage gehörte damals auch noch zur Sonderausstattung ebenso wie das Klappverdeck.

Bild 1

Bild 2

Bild 3

Bild 4

Bild 5

WD-Radschle

Betrieb mit Benzol, Pe

vom Reichsernährungsministerium **preisgekrönt**

Aus dem Prüfungsbericht: „Der WD-Radschlepper i
Arbeitsleistung auf dem Acker ist **einwandfr**
Gummibereifung auf den schweren Straßenräder

Für vielseitige Verwendung eingerichtet, billig im
Dauerleistungen, ist der WD-Radschlepper geeignet, sow
setzen, wie in Zeiten der Arbeitshäufung als hochwertige
pausen wie das Zugtier und kann bei Bedarf mit Ablösu
wird der Betrieb intensiver gestaltet und gleichzeitig ver

Betriebsgewicht: Als Ackermaschine ca. 1700 kg
(Als Straßenzugmaschine ca. 3200 kg)
Geschwindigkeiten: 2 km; 4,5 km; 8 km/Stunde
Rückwärtsgang: 1,6 km/Stunde
Riemenscheibe: Drehzahl 745, Durchmesser: 400 mm
bzw. 365 mm
Pflugleistung bei 25–30 cm Tiefe: auf mittlerem Boden
ca. 1 Morgen/Stunde. Schälen auf Durchschnitts-
boden mit 10 Scharen
Ausrüstung: paarweise gleich große Greiferräder,
leichte Laufreifen für kurze Straßenfahrten
Motor: bei 1100 Umdrehungen/Min. effektive Leistung
mit Petroleum, Zugolin 28 PS, mit Benzol 32 PS

Der Viertakt-Vierzylindermotor ist in einem Block
gegossen, der abnehmbare Zylinderkopf enthält die
hängenden Ventile, die sich über den Kolben befinden
(daher günstiger Kompressionsraum, gute Wärmevertei-
lung und geringer Brennstoffverbrauch). Sämtliche Lager-
stellen werden von einer Ölpumpe ständig geschmiert.
Der Motor ist öl- und staubdicht gekapselt und gegen
die Vorderachse abgefedert; auf Zugänglichkeit von allen
Seiten ist besonderer Wert gelegt.
Der Vergaser ist ein Doppelvergaser neuester Konstruktion (anlassen mit Benzol, dann umstellen auf

Bild 6

Bild 7

pper / 28 PS

roleum, Zugolin usw.

egen seiner guten Pflug- und Zugleistungen.

die **billigste** und leichteste aller Maschinen ... Die ... Das Lastenziehen gelang mit Hilfe doppelter sehr gut." (Geheimrat Fischer im Landmaschinenmarkt Pößneck 1925, No. 92, S. 4.)

triebe, stets dienstbereit, unermüdlich auch bei größten teuere menschliche und tierische Arbeitskraft zu erusatzarbeitskraft zu dienen. Er braucht keine Betriebssbedienung unbegrenzt ausgenutzt werden. Dadurch ligt.

Sonderausrüstungen:

urchenräder größeren Durchmessers für Tiefkultur (Bild 7.)

adverbreiterungen und Greifererhöhungen für Sandboden (Bild 5.)

adverbreiterungen für Moor (Bild 6.)

prossenräder für Wiesenkultur

chwere eiserne Laufreifen zum Ziehen kleiner Lasten.

ollscheibenräder mit Gummibereifung für Zugleistung bis zu 20 t (Bild 1.)

bere Zugvorrichtung.

Die Räder sind leicht auswechselbar.

hweren Brennstoff). Die Vorwärmung des Brennstoffemisches ist regulierbar.

Die Triebkraft wird über eine Lamellenkupplung und n sperrbares Differential auf die Hinterräder übertragen.

Die federnde Mittenzugvorrichtung kann sich seitlich nstellen bzw. in beliebiger Stellung durch Vorsteckbolzen stgehalten werden. Sie schont Anhängegerät und hlepper beim Anfahren und beim Auftreffen auf Steine. ie Anpassungsmöglichkeit und Elastizität dieser Zugrrichtung hat eine gute Lenkbarkeit des Schleppers nd einen ruhigen Gang des Pfluges zur Folge.

Bild 8

Bild 9

Bild 10

Technische Vollendung
bürgt für größte Lebensdauer und höchste Leistung.

1. Aufklappbare **Motorhaube**.
2. Abnehmbarer **Zylinderkopf**.
3. Regelbare **Luft- u. Vergaservorwärmung**.
4. **Ölumlauf i. Kontrollglas** dauernd sichtbar.
5. Staubentfernung durch **Luftfilter**.
6. Feststellbare **Handbremse**, nachstellbar.
7. **Automobillenkung**, sehr enger Wendekreis.
8. Einfache **Kugelschaltung** 3 Gänge vorwärts, 1 Gang rückwärts.

Rahmenlose Bauart; alle umlaufenden Teile staubdicht gekapselt

9. Bequemer, weich **gefederter Sitz**, kein Ermüden des Fahrers.
10. **Schutzbleche** an den Triebrädern.
11. Hochelastisch. **Kissenreifen**, beste Federwirkung.
12. **Trittbrett** mit Werkzeugkasten.
13. **Ausgleichgetriebesperre**, kein einseitiges Gleiten oder Wühlen der Räder.
14. Geriffelter **Kupplungs-Fußhebel**, kein Abrutschen der Füße.
15. Kräftig wirkende, nachstellbare **Fußbremse**.
16. **Kühlwasserregler**.
17. **Konische Rollenlager** in den Vorderradnaben.
18. **Hochdruck-Preßölschmierung** durch Ölpumpe.
19. Doppelte **Ölreinigung** durch Saugsieb- und Druckfilter.
20. Doppelte **Blattfederung** zwischen Motor und Vorderachse.
21. **Boschmagnet** mit Abschnappkupplung. Leichtes Anspringen.
22. Zwangsläufige **Umlaufkühlung** durch Wasserpumpe.

Alle Getriebewellen laufen in Kugel- und Rollenlagern

23. **Drehzahlregler**, keine Überbeanspruchung d. Motors, stets gleiche Drehzahl b. ortsfestem Antrieb.
24. **Beleuchtung** auf gefederten Stützen.
25. **Schweröldoppelvergaser**, geringe Brennstoffkosten.
26. Großbemessener **Röhrenkühler** in fester Schutzhaube.

Außer vorgenannten Einrichtungen wird noch das erforderliche Werkzeug kostenlos mitgeliefert.

Technische Einzelheiten:

Länge	mm	3130
Breite	"	1600
Höhe	"	1500
Achsabstand	"	1692
Spurbreite, vorn	"	1210
" hinten	"	1344
Spur zwischen den Hinterrädern	"	1000
Raddurchmesser, vorn	"	700
" hinten	"	1065
Bodenfreiheit	"	250
Wenderadius	cm	450
Betriebsgewicht, normal	kg	3200
" schwer	"	3500
Motor, Zylinderzahl		4
Bohrung	mm	95
Hub	"	150
Tourenzahl in der Minute		1100
Hubvolumen	cm³	4252
PS mit Gasöl, mit Petroleum		28
PS mit Benzin		30
PS mit Benzol, Monopolin		32

Betriebsstoffverbrauch bei Voll-Last:

Benzol, Benzin, Monopolin	gr/PS/st	275
Petroleum, Gasöl	"	290

Öl- und Fettverbrauch 5—10 % des Betriebsstoffverbrauches.

Fahrgeschwindigkeiten:

1. Gang	km/st	3,7
2. Gang	"	7,5
3. Gang	"	15
Rückwärtsgang	"	2,7
Riemenscheibe, Breite	mm	150
" Durchmesser	"	365
" Drehzahl	Min.	770
Zapfwelle, Drehzahl	"	600
Haupttank, Inhalt	Liter	62
Hilfstank, Inhalt	"	22
Verdichtung des Petrolmotors		1:4,6
" " Benzolmotors		1:5,1

Zündung: Boschmagnet mit Abschnappkupplung.
Vergaser: Graetzin.
Schmierung: Hochdruckumlaufschmierung mit Öldruckkontrolle und Ölfilteranlage.
Luftreinigung: Ölfilter.

Wenn Sie einen Schlepper kaufen,
denken Sie bitte an die hochwertige Ausstattung des Hanomag.

Die Hanomag-Dieselschlepper der 30er Jahre

Erst relativ spät, gegen Ende der 20er Jahre und zur Zeit der großen Weltwirtschaftskrise, die auch den Hanomag-Werken schwer zu schaffen machte, und in dessen Zuge man im Sommer 1931 die Lokomotivfertigung und den Bau von Dampfkesseln aufgab, begann Hanomag, nachdem bereits einige Mitbewerber, im Jahr 1926 auch Deutz, Schlepper mit Fahrzeug-Dieselmotoren herausgebracht hatten, ein nach dem Vorkammerverfahren arbeitendes Dieselantriebsaggregat für den Schlepperbau zu entwickeln.

Die Hanomag-Werke zählten damals nur noch 1260 Beschäftigte, das war kaum mehr als ein Fünftel der Belegschaftszahl des Jahres 1926. Ab 1932 gehörte das Unternehmen dem Bochumer Verein für Bergbau- und Gussstahlfabrikation an. Im Jahr 1931 – nach einer Entwicklungszeit von etwa zwei Jahren – war der unter der Leitung von Dipl.-Ingenieur Lazar Schargorodsky, einem aus Odessa stammenden und seit 1924 für die Hanomag tätigen überaus fähigen Konstrukteur, entworfene und nach dem Vorkammerprinzip arbeitende Dieselmotor des Vierzylinder-Viertakt-Modells D 52 produktionsreif. Dieser stehende Motor mit 5195 ccm Hubvolumen, der eine Leistung von 36 PS bei 1100 U/min abgab, trat in konstruktiver Hinsicht durch eine ebenfalls neuentwickelte, im Ölbad laufende, linksseitig angebrachte Schrägnocken-Einspritzpumpe besonders hervor und wurde im gleichen Jahr erstmals in den Radschlepper RD 36 sowie in den neuen Kettenschlepper K 35/40 eingebaut. Damit konnten auch Hanomag-Schlepper das gegenüber anderen Kraftstoffen wesentlich billigere Gasöl verwenden und die Kraftstoffkosten um rund 80 Prozent senken. Die gesamte Motorkonstruktion war derart fortschrittlich, dass sie länger als 20 Jahre nichts an ihrer Aktualität verlieren sollte. In den folgenden Jahren konnte deren Leistung kontinuierlich bis auf 50 PS gesteigert werden.

Bei dem neuen Radschlepper RD 36 griff man auf die beim R 28/32 bereits bewährten Komponenten wie Getriebe, Hinterachse und Einscheiben-Trocken-

Hanomag-Dieselschlepper RD 36 mit 36 PS

HANOMAG-DIESEL-ZUGMASCHINE

Magdeburg, den 6. Januar 1931.

HANOMAG-Zweigstelle MAGDEBURG Bärickestr. Nr. 38
Telefon: 42389

Firma
Fritz Wöhler,
Kohlenhandlung,
<u>Benzingerode,</u>
Krs. Blankenburg.

Wir haben das Vergnügen, Ihnen ein Angebot auf unsere

<u>HANOMAG-DIESEL-ZUGMASCHINE</u>

zu überreichen und bitten Sie, dasselbe einer geneigten Durchsicht zu unterziehen.

Bei der Prüfung werden Sie zunächst feststellen können, daß unsere Straßen-Zugmaschine, gemessen an der Nutzleistung, preislich am günstigsten liegt.

Aber nicht allein der niedrige Preis ist ein besonderer Vorzug der Hanomag-Diesel-Zugmaschine, sondern mehr noch die außerordentlich niedrigen Betriebskosten, die bestbewährte Konstruktion, die Herstellung aus nur erstklassigem Material, die technische Überlegenheit, der ruhige und störungsfreie Lauf der Maschine in allen Belastungsstufen, die ständige Betriebssicherheit.

Ferner bieten wir Ihnen auf Wunsch weitgehende Zahlungserleichterungen und als Hanomag-Kunden einen gut organisierten Überwachungsdienst, in Verbindung mit den über ganz Deutschland verteilten, prompt arbeitenden Ersatzteillagern.

Sollten Sie noch weitere Auskünfte zu erhalten wünschen, bitten wir um gefl. Benachrichtigung; einer unserer sachverständigen Herren steht zu Beratungszwecken an Ort und Stelle jederzeit ohne Verbindlichkeit zu Ihrer Verfügung.

Wir empfehlen uns Ihnen und zeichnen

hochachtungsvoll

Hannoversche Maschinenbau-Actien-Gesellschaft
vormals Georg Egestorff (Hanomag) Hannover-Linden
Schlepper-Vertrieb Magdeburg.

Der Hanomag-Dieselschlepper RD 36 verfügt über eine Zugleistung von bis zu 600 Zentner

kupplung zurück. Das Gewicht war auf 2750 kg in der Ackerausführung und auf 3700 kg beim Straßenschlepper angewachsen. Fertigungstechnisch und auch vom Aussehen her änderte sich gegenüber dem Vormodell durch die Verwendung vieler baugleicher Teile nicht allzu viel. Infolge des größeren und schwereren Motors hatte der RD 36 gegenüber seinem Vorgänger an Länge und Gewicht erheblich zugenommen. Es standen für den Betrieb auf Straßen wahlweise Ackerluftreifen ab Mitte der 30er Jahre zur Verfügung. Besonders beim Einsatz als Verkehrsschlepper im schweren Nahverkehr, im Speditionsgewerbe, im Kohlen- und Baustoffhandel, sowie im Baugewerbe oder bei Schaustellern waren diese unverwüstlichen und bis zu 25 km/h schnellen, mit Druckluftanlagen bestückten Traktoren von Hanomag sehr beliebt. Die Werbung des Unternehmens bezeichnete diesen und auch die nachfolgenden Typen der mit umfangreichem Zubehör lieferbaren Modellreihe mit Fug und Recht als einfach, betriebssicher und unverwüstlich. Seinem größten Konkurrenten gegenüber, dem Lanz-Bulldog 15/30 PS des Typs HR 5, erwies sich das neue und stärkere Hanomag-Modell mehr als ebenbürtig.

Zwischen 1933 und 1938 hatte Hanomag mit den Modellen GR/AR beziehungsweise AGR 50 dem RD 36 einen 50 PS starken Radschlepper mit unterschiedlicher Bereifung zur Seite gestellt. Dessen auf 50 PS leistungsgesteigertes D 52- Dieselaggregat stammte samt Motorblock, Dreiganggetriebe und Hinterachse vom Kettenschlepper K 50 ab. Die 50-PS-Modelle waren optisch vor allem durch den eckigen, der Motorhaubenform der Raupe angepassten Kraftstofftank vom RD 36 zu unterscheiden. Infolge der von der Raupe stammenden, in Portalbauweise konstruierten Hinterachse erhielten diese Schlepper eine sehr hohe Bodenfreiheit,

	RD 36	AR 38/AGR 38	AGR 38/45	SR 38/45	SR 45	RL 20
Produktionszeitraum	1931-1936	1936-1942	1936-1942	1936-1942	1936-1942	1937-1949
Motorleistung (PS)	36	38	38/45	38/45	45	20
Hubraum (cm³)	5195	5195	5195	5195	5195	1911
Anzahl der Zylinder	4	4	4	4	4	4
Anzahl der Gänge	3+1	3+1	3+1	3+1	3+1	4+1
Eigengewicht (kg)	2750	3200	3200	3600	3900	1680
Höchstgeschwindigkeit	18,0 km/h	13,7 km/h	16,0 km/h	22,0 km/h	28,6 km/h	24,0 km/h

und auch die Vorderachse musste entsprechend angepasst werden. Dieser schwere Schlepper war wahlweise auch mit einem Sechsganggetriebe erhältlich.

Auch die Hanomag-Werke konnten an der nach Hitlers Machtübernahme im Jahr 1933 einsetzenden raschen wirtschaftlichen Belebung nachhaltig partizipieren. In schneller Folge kamen nun immer neue, ständig verbesserte Typen auf den Markt. Die verschiedenen, mit starken Dieselmotoren bestückten Schlepper entsprachen dem verbreitet vorgetragenen Wunsch nach leistungsfähigen Maschinen für die Feldarbeit und konnten hinfort eine kraftvolle Position am Markt einnehmen.

Die Landwirte erkannten rasch die vielen Vorzüge der Luftbereifung, und daher begann diese Bereifungsart sich gegenüber den eisenbereiften Varianten immer stärker durchzusetzen.

Im Jahr 1936 löste das weiterentwickelte und leistungsgesteigerte Modell AR 38 den RD 36 in der Ackerausführung ab. Die mit Ackerluftreifen und einer schnelleren Getriebeübersetzung als Geländeschlepper AGR bezeichnete Modellvariante gab durch Drehzahlsteigerung sogar 45 PS in der Ausführung AGR 38/45 ab. Das Eigengewicht dieser Modelle lag zwischen 2750 kg bei Verwendung von Eisenrädern und 3200 kg mit Luftbereifung. Das eingebaute Dreiganggetriebe ließ beim eisenbereiften AR 38 Geschwindigkeiten bis zu 8 km/h und solche von 13,7 beim AGR 38 mit Luftreifen zu. Der AGR 38/45 konnte sogar 16 km/h als größtmögliche Geschwindigkeit erreichen. Diese Modelle gehörten zu jener Zeit zu den wichtigsten schweren Schleppern in Deutschland, die auf größeren landwirtschaftlichen Betrieben zum Einsatz kamen.

Hanomag AGR 38 mit Ackerluftbereifung und schnellerem Getriebe, Baujahr 1938

Somit befanden sich Mitte der 30er Jahre im Hanomag-Schlepperprogramm Grundtypen mit 38, 45 beziehungsweise 50 PS Leistung im Angebot.

Als SR 38/45 gab es ab 1936 auch eine Dieselschlepperausführung mit Elastik- (Vollgummi) Bereifung und mit Zwillings-Hinterrad-Luftbereifung. Die schnellste Ausführung dieses Straßenschleppers wurde als SR 45 bezeichnet und sie befand sich von 1936 bis 1942 in der Produktion. Sie verfügte über die sogenannten Riesenluftreifen mit besonders guten Federungseigenschaften. Der mit einem Dreiganggetriebe bestückte bis zu 3900 kg wiegende Schlepper erreichte im dritten Gang Geschwindigkeiten von bis zu 28,6 km/h und die Fußbremse wirkte nicht wie bei den langsameren Varianten auf das Getriebe, sondern direkt auf die Hinterräder. Für den Anhängerbetrieb – 25 t Last waren bequem zu bewältigen – war wahlweise eine Druckluftbremsanlage nachrüstbar.

Weitere Ausstattungsmerkmale waren Seilwinde und Verdeck, welche das Gesamtgewicht weiter erhöhten.

Um auch für mittlere und kleinere bäuerliche Betriebe ein Fahrzeug im Angebot zu haben wurde 1937, ein Jahr nachdem Deutz seinen berühmten 11-PS-Bauernschlepper vorgestellt hatte, auch bei Hanomag das Modell eines Bauernschleppers in das Verkaufsprogramm aufgenommen. Nicht zuletzt die Reichsregierung hatte die Traktorenhersteller dazu gedrängt, mit speziell für bäuerliche Kleinbetriebe konzipierten Konstruktionen aufzuwarten. Der als RL 20 bezeichnete Typ verkörperte mit seiner Pkw-ähnlichen Motorhaube und dessen bewusst einfach gehaltener Bauweise eine völlig neue Designrichtung im Schlepperbau. Er war dazu ausersehen, das Angebot nach unten hin zu erweitern. Mit einem Abgabepreis von 3950 Reichsmark kam das Werk der meist dünnen Kapitaldecke kleinerer Bauernhöfe sehr entgegen. Die in der Firmenwerbung herausgestellten positiven Eigenschaften wie geringe Betriebskosten, hohe Leistung, einfache Bedienung und Wartung hatten durchaus ihre Berechtigung. Der Bauernschlepper war abgeleitet von der ebenfalls mit Komponenten aus dem Pkw-Bau ähnlich gestalteten Straßenzugmaschine SS 20 und wurde ohne deren Fahrerhaus mit einem einfachen offenen Schleppersitz geliefert. Ein durch Quertraversen verstärkter U-Profilrahmen bildete das Rückgrat des 1680 kg schweren Schleppers, der mit dem auch im firmeneigenen Pkw-Diesel-Modell "Rekord" verwendeten, gleichfalls vom Leiter der Hanomag-Motorenkonstruktion Dipl.-Ingenieur Lazar Schargorodsky entwickelten vierzylindrigen D 19-Motor mit 1911 ccm Hubraum bestückt, dessen Leistung auf 20 PS bei 2000 U/min reduziert worden war. Der Erwerber konnte zwischen einem Drei- oder Vierganggetriebe der Zahnradfabrik

Friedrichshafen auswählen. Während die erstere Ausführung, da auf 13 km/h Geschwindigkeit begrenzt, noch ohne Führerschein gefahren werden durfte, erreichte die Viergangvariante, für die eine Fahrerlaubnis erforderlich war, 24 km/h. Sowohl die als Schwingachse ausgebildete Vorderachse als auch die gegen den U-Profilrahmen abgestützte und abgefederte Hinterachse und weitere Teile stammten aus der Pkw-Fertigung. Der mit vier Rädern gleicher Größe ausgestattete RL 20 besaß außerdem einen elektrischen Anlasser, Kotflügel vorn und hinten, elektrische Beleuchtung und eine Anhängevorrichtung. Als Zusatzausstattung waren Zapfwelle, Mähwerk, Seilspill, Windschutzscheibe und Verdeck gegen Aufpreis erhältlich. Die Zubehörteile ermöglichten dem Verwender dieses erfolgreichen Kompromissproduktes beachtliche Anwendungsmöglichkeiten, was sich nicht zuletzt in den Verkaufszahlen niederschlug. Zwischen 1937 und 1948 konnte insgesamt die beachtliche Zahl von 4320 Einheiten verkauft werden.

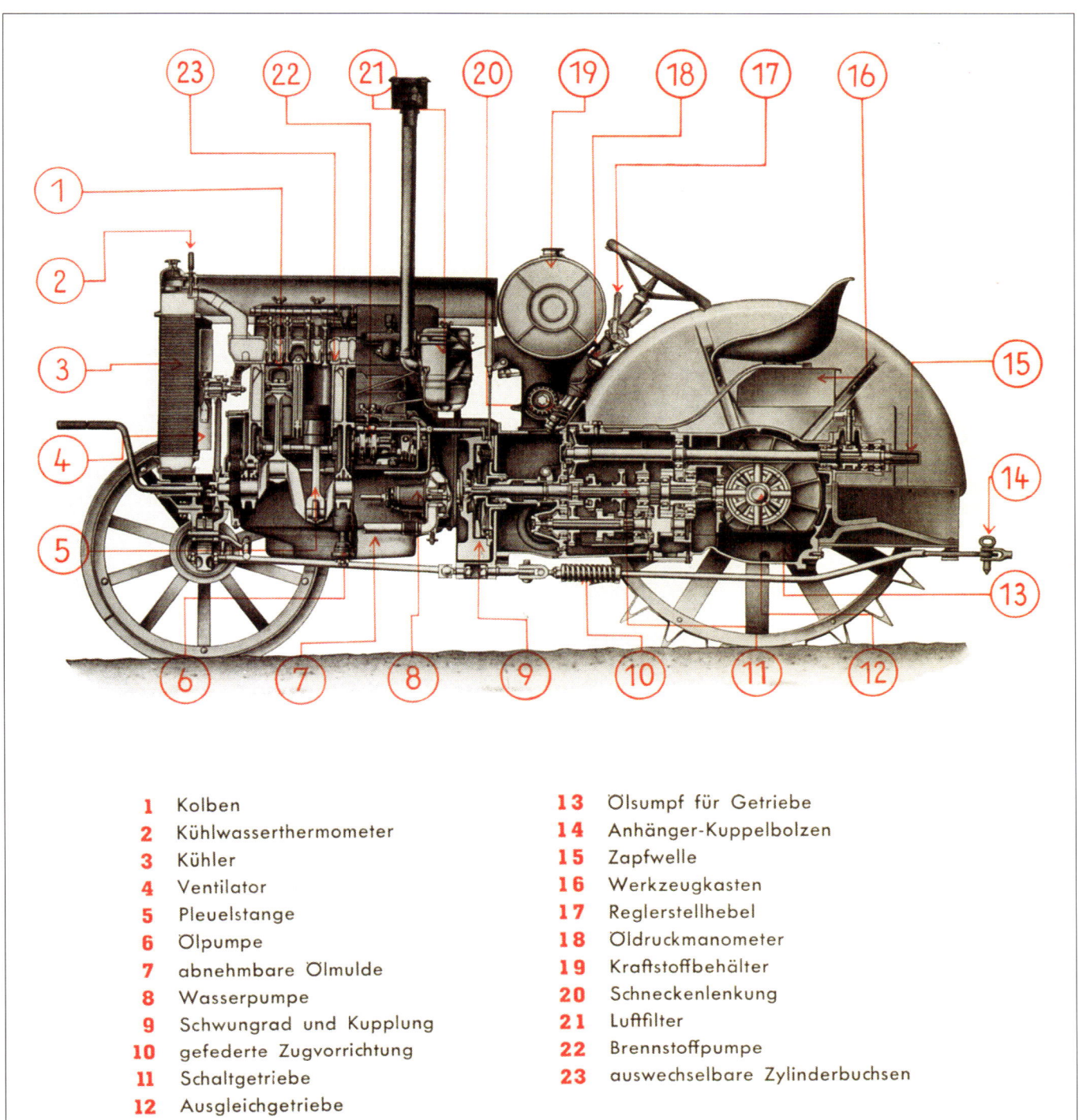

1 Kolben
2 Kühlwasserthermometer
3 Kühler
4 Ventilator
5 Pleuelstange
6 Ölpumpe
7 abnehmbare Ölmulde
8 Wasserpumpe
9 Schwungrad und Kupplung
10 gefederte Zugvorrichtung
11 Schaltgetriebe
12 Ausgleichgetriebe
13 Ölsumpf für Getriebe
14 Anhänger-Kuppelbolzen
15 Zapfwelle
16 Werkzeugkasten
17 Reglerstellhebel
18 Öldruckmanometer
19 Kraftstoffbehälter
20 Schneckenlenkung
21 Luftfilter
22 Brennstoffpumpe
23 auswechselbare Zylinderbuchsen

Das Innenleben des Dieselschleppers 38 PS

Restaurierter Hanomag AGR 38 in roter Exportausführung, Baujahr 1939

Die Hanomag-Dieselschlepper der 30er Jahre **21**

Hanomag-Dieselschlepper 38 PS, unten mit Mähdrescher und Spreuwagen

Hanomag-Dieselschlepper 38 PS, oben mit Dreischar-Tiefpflug

Hanomag SR 45 als Straßenschlepper, Baujahr 1937

24 Die Hanomag-Dieselschlepper der 30er Jahre

HANOMAG-DIESEL
20 PS BAUERNSCHLEPPER

Der Hanomag-Bauernschlepper wird geliefert:

A) <u>für 13 km/Std.</u> Höchstgeschwindigkeit (führerscheinfrei),

B) <u>für 24 km/Std.</u> Höchstgeschwindigkeit (führerscheinpflichtig).

Im Zeichen des Aufbaues, der Erfassung auch der letzten Reserve steht heute die Landwirtschaft. Hohe Ziele wurden gesteckt und müssen erreicht werden. Der Staat gab den Bauern durch seine Gesetzgebung sicheren Rückhalt. Er forderte die Industrie auf, dem Bauernstand nunmehr ihrerseits das nötige Rüstzeug zur Verfügung zu stellen, das ihm helfen soll, die großen Aufgaben zu lösen. Unter diesen Bedingungen entstand der **Hanomag-Diesel-Bauernschlepper.**

RL 20 Bauernschlepper mit 20 PS bei der Feldarbeit

Hanomag RL 20 Bauernschlepper, Baujahr 1940

Die schweren Diesel-Radschlepper R 40 und deren Folgemuster

Im Zuge des sogenannten Schell-Planes, der auch im Schlepperbau des Deutschen Reiches eine weitgehende Typenbegrenzung herbeiführen sollte, wurden der Hanomag Entwicklung und Fertigung unter anderen von Radschleppern mit 30 und 40 PS Leistung zugestanden. Der Hanomag wurde sowohl für den 40-PS-Radschlepper als auch für einen 60-PS-Kettenschlepper die Kooperation mit den Breslauer FAMO-Werken, früher Linke-Hofmann-Busch (LHB), auferlegt mit dem Ziel, die wichtigsten Teile beider Fabrikate austauschbar zu gestalten. Infolge der Kriegsereignisse aber gelangten diese Maßnahmen nur in einem geringen Umfang zu greifbaren Erfolgen. Im Jahr 1940 kamen die ersten Versuchsexemplare der Neukonstruktion des schweren 40-PS-Verkehrsschleppers R 40 von Hanomag auf den Markt und lösten damit die bisherigen Typen R 38 und SR 45 ab. Als ein wichtiges Bauteil wurde der bewährte D 52-Dieselmotor, welchem eine Leistungssteigerung auf 40 PS bei 1200 U/min verpasst wurde, von den Vormodellen übernommen. Dieser war entweder mit einer Benzin-Anlassvorrichtung oder mit vorzuglühendem elektrischem Anlasser erhältlich. Gänzlich neu war das mittels eines Hebels zu schaltende, mit fünf Vorwärts- und einem Rückwärtsgang ausgerüstete, aus der Zusammenarbeit mit FAMO entstandene Getriebe, das für Geschwindigkeiten von 4,3 bis 18,7 km/h ausgelegt war und sich durch nur geringe Kraftverluste auszeichnete. Gleichfalls neu entwickelt und gestaltet wurden neben dem mit einer Doppelsitzbank versehenen Fahrerplatz auch die Vorderachse – diese war als blattgefederte Pendelachse ausgebildet – und die Motorverkleidung. Die ungefederte Hinterachse erhielt eine abschaltbare Differentialsperre, und die Kraftersparnis durch die installierte ZF-Roßlenkung ließ den R 40 auch für Frauen und Jugendliche, die in der Kriegszeit ja immer häufiger als Schlepperführer eingesetzt werden mussten, in Frage kommen. Der Motor konnte sowohl bei der Acker- als auch bei der Universalausführung entweder mit elektrischem Anlasser zum Vorglühen oder mit einer Benzin-Anlassvorrichtung geliefert werden.

Hanomag R 40 mit...

...Imbert-Gaserzeugungsanlage, Baujahr 1943

Der R 40 war in den Versionen als luft- oder eisenbereifter Ackerschlepper oder als luftbereifter Universal- und Verkehrsschlepper erhältlich. In der Ausführung mit Eisenrädern war die größtmögliche Geschwindigkeit auf 8 km/h begrenzt, da vierter und fünfter Getriebegang gesperrt waren. Zur Standardausrüstung gehörten Riemenscheibe und Zapfwelle, während vordere Kotflügel und Fahrerdach nur beim Universalschlepper zum Ablieferungszustand gehörten. Ein festes Fahrerhaus, Seilwinde und die für Straßentransporte unverzichtbare Druckluftbremsanlage hingegen waren nur gegen Aufpreis auf Kundenwunsch lieferbar.

Nachdem das 30-PS-Modell nicht realisiert wurde, beschränkten sich die Hanomag-Werke bei der Radschlepperfertigung hinfort auf das Modell R 40. Ab Juli 1940 gab es Ackerschlepper nur noch auf Bezugschein, und von Mitte des Jahres 1942 an musste der Bau von Dieselschleppern gemäß einer Anordnung der Reichsregierung zugunsten solcher mit Holzgasantrieb mittels Einheitsgenerator eingestellt werden. Durch eine geringfügige Änderung der Zylinderbohrung, welche eine Vergrößerung des Hubraums auf 5702 ccm Rauminhalt bewirkte, sowie durch Austausch des bisherigen Zylinderkopfes und Wegfalls der Einspritzanlage war die Anpassung auf die Imbert-Gaserzeugungsanlage ohne größere Schwierigkeiten zu bewerkstelligen. Bis in die ersten Nachkriegsjahre entstanden daher zahlreiche Gasschlepper des Typs R 40 HG mit seitlich angebautem Generator und darauf abgestimmtem Vergasermotor. Infolge der Knappheit an flüssigen Kraftstoffen wurden viele der bisherigen Dieselschlepper auf Holzgasgeneratorantrieb umgerüstet. So war es möglich, die Produktion landwirtschaftlicher Erzeugnisse auch in den letzten Kriegsjahren aufrechtzuerhalten.

Die Fertigung des R 40 konnte während des Krieges und auch bereits kurz nach Kriegsende, bis 1951 mit einer Gesamtstückzahl von etwa 12000 Einheiten weiter geführt werden. Bereits ab Juli 1945 nahm man auf Zugeständnis der britischen Militärregierung bei Hanomag den Schlepperbau, trotz der stark durch den Bombenkrieg in Mitleidenschaft gezogenen Produktionsanlagen und der Materialkontingentierungen, teilweise unter freiem Himmel und unter Verwendung von Lagerbeständen, wieder auf. Mit einem im Dezember des Jahres 1947 ausgeführten Auftrag nach Belgien, konnte das Unternehmen erstmals nach dem Krieg erfolgreich an die bestehenden Exporttraditionen anknüpfen und mit der Währungsreform im Jahr 1948 kam mit dem Modell R 40 C eine 25 km/h schnelle Ausführung hinzu, die sich aufgrund ihrer höheren Geschwindigkeit recht gut für den Straßenbetrieb eignete und daher oft mit Druckluftbremsanlage, festem Fahrerhaus und gepolstertem Sitz an die Kundschaft geliefert wurde. Deren Zugkraft betrug im 5. Gang auf ebener Straße 22 Tonnen.

Bevor auf die Entwicklung der vielen übrigen, leichteren Hanomag-Nachkriegsschleppertypen eingegangen werden soll, sei in den nachfolgenden Ausführungen zunächst der Blick auf die Nachfolgemodelle der R-40-Baureihe, die bereits zu Lebzeiten zur Legende wurden und zweifelsohne zu den erfolgreichsten Radschleppern aller Zeiten zählten, gerichtet.

1951 folgte dem erfolgreichen R 40 das mit leistungsgesteigertem, vergrößertem Motor sowie neugestalteter Motorverkleidung ausgerüstete und erstmals auf der DLG-Ausstellung in Frankfurt vorgestellte Modell R 45. Es war damit der stärk-

Hanomag R 40 A, Baujahr 1945

ste Radschlepper im damaligen Hanomag-Verkaufsprogramm, das bereits Modelle mit 16, 22, 28 und 45 PS aufweisen konnte. Dessen aus dem D 52 weiterentwickeltes, nun als D-57 bezeichnetes Aggregat – es war vom D-52-Motor an dem rechts außen am Motorblock angebrachten Ölfilter zu erkennen – konnte aus 5702 ccm Hubraum 45 PS Leistung bei 1200 U/min erzeugen. Das Aggregat allein brachte etwa 800 kg auf die Waage. Dieses neue, bis 1964 (zuletzt unter der neuen Bezeichnung R 460) gebaute Modell war in den bereits vom R 40 her bekannten Varianten erhältlich, wobei die Ausführung R 45 A als gummibereifter Ackerschlepper mit Kotflügeln vorn und hinten für den Inlandsmarkt, die Variante B mit Eisenrädern und Dreiganggetriebe in erster Linie Exportzwecken vorbehalten war.

Neu war hingegen war die Ausstattung des Fahrzeugs mit hydraulischem Kraftheber und Fronthydraulik, welche die Bestückung mit dem Wittenburg-Frontlader zuließ, wodurch der R 45 auch für die Bauwirtschaft interessant wurde. Ab 1953 rüstete man den R 45 mit einer neuen Kühlermaske aus. Wie beim R 40-Programm war es erneut die Version C, die – teilweise ausgerüstet mit einem von der Firma Benze gelieferten festen Fahrerhaus mit Rundumverglasung und links angebrachter Einstiegtür – als Straßenzugmaschine mit Schnellgang sowie hinteren Gussfelgen und Druckluftbremsanlage zu einiger Bedeutung gelangte. Die Ausführungen A und C waren auf Wunsch auch mit einem Wittenburg-Frontlader mit 1,5 t Hubkraft erhältlich.

Der R 45 war überall dort anzutreffen, wo hohe Zugleistungen benötigt wurden, sei es vor schweren Anhängepflügen auf dem Acker, zum Antrieb von Mähdreschern oder auf der Straße mit zwei beladenen Anhängern im Schlepp. Die im Heck einbaubare 3,5-t-Seilwinde machte den R 45 auch für Holzrückarbeiten in der Wald- und Forstwirtschaft uneingeschränkt verwendbar. Auch bei Spediteuren und Schaustellern erfreute sich dieses Modell – im übrigen sein Vorgänger R 40 und die Nachfolger auch – größter Beliebtheit.

Im Jahr 1955 entsprachen die Hanomag-Werke dem Kundenwunsch nach einem schweren Radschlepper mit angehobener Motorleistung und brachten demzufolge das Modell R 55 als stärkere Abwandlung des R 45 auf den Markt, wobei man die höhere Leistung durch Drehzahlsteigerung erzielt hatte. Optisch unterschied sich das neue, ursprünglich speziell für Exportzwecke vorgesehene Modell nur durch die größere 15-30-AS-Bereifung von seinem Vorgänger. Es stand ein in zwei Schaltgruppen unterteiltes, von Hanomag entwickeltes Zehnganggetriebe mit zwei Rückwärtsgängen und Differentialsperre zur Verfügung. Mit diesem war bei der Wahl des 5. Ganges in der 2. Schaltgruppe eine größtmögliche Geschwindigkeit von 19,91 km/h zu erreichen. In der ebenfalls lieferbaren schnelleren Straßenausführung R 55 C waren es sogar 26,8 km/h. Das Eigengewicht des R 55 betrug – je nach Ausführung – zwischen 3270 und 3470 kg. Die umfangreiche Zubehörpalette war mit der des R 45 identisch.

Hanomag R 40 A, Baujahr 1945 und – ausgeführt ohne Kotflügel vorn und mit Muschelkotflügeln hinten – von 1947 (unten)

Seit 1957 gab es für die Hanomag-Schlepper neue, aus einer dreistelligen Zahlengruppe bestehende Verkaufsbezeichnungen. Dabei gab die erste Ziffer die Anzahl der Zylinder an, während die beiden letzten Ziffern die Motorleistung in PS ausdrückten. So wurde aus dem weiterhin gefertigten Modell R 45 der R 445 und aus dem R 55 der R 455, der bis 1958 in der Produktion blieb.

Das Modell R 445 wurde ab 1958 durch den auf 50 PS leistungsgesteigerten, ansonsten weitgehend unveränderten Typ R 450 ersetzt, während das ehemalige 55 PS starke Modell R 455, nicht ganz der Hanomag-Typenterminologie entsprechend, als R 450 E bezeichnet wurde. Neu war bei beiden Typen ein auf Wunsch erhältlicher Kriechgang und ein wie schon beim R 55 lieferbarer längerer Radstand. Auch hier war eine schnellere Getriebeausführung erhältlich. Beide Modelle wurden bis 1961 gebaut.

Im Jahr 1960 kam als letztes Glied dieser erfolgreichen Schlepperreihe das bis 1964 gefertigte Modell R 460 hinzu, das sich von den vorgenannten Typen durch die eine höhere Motorleistung von 60 PS und die Klauenschaltung für den 4. und 5. Gang unterschied. Dieses Fahrzeug war ausschließlich mit dem längerem Radstand zu erwerben. Auch in diesem Fall gab es eine schnelle Straßenausführung, mit der im 5. Gang 30 km/h Höchstgeschwindigkeit erreichbar waren.

Zum Abschluss dieses Kapitels soll die als R 460 BS bezeichnete und auf der Basis des R 460 angebotene Mehrzweckbaumaschine erwähnt werden, die das aus Kettenschleppern und Radladern bestehende Baumaschinenprogramm der Hanomag ergänzen sollte. Dieses Fahrzeug besaß ein zwei Meter breites Planierschild, einen um fast 360° schwenkbaren, heckseitig arretierten Hydraulikbagger des Fabrikats Weyhausen-Atlas mit einer Tragkraft von 1000 kg sowie als Sonderausstattung einen hydraulischen Frontlader, der über 1100 kg Hubkraft verfügte.

Bereits 1952 stellte Hanomag den mit 55 PS sehr leistungsstarken Spezialschlepper R 55 ATK vor, der das angebotene Radschlepperprogramm durch eine Variante für große, in Straßeneinsätzen ruckfrei zu bewegende Lasten ergänzen sollte. Das neue Modell erregte insbesondere ob seiner für diesen Zweck zusätzlich zwischen Schwungscheibe des Motor und Schaltkupplung installierten ölhydraulischen Voith-Turbo- oder Strömungskupplung vom Typ 464 TD in der Fachwelt großes Aufsehen. Durch diese Kupplung ließ sich das Drehmoment des Antriebsaggregats stufenlos und völlig ruckfrei auch unter Last auf das Fünfganggetriebe übertragen. Die Kupplung eröffnete weiterhin die Möglichkeit, in jedem beliebigen Gang anfahren zu können. Diese Eigenschaft machte den ATK zum Ziehen schwerer Lasten, wie zum Beispiel als Schwerlastschlepper für Flugzeuge auf Zivil- und Militärflughäfen besonders geeignet. So setzte die niederländische Luftfahrtgesellschaft KLM, im übrigen der erste Kunde dieser Spezialschlepper, die mit der Turbokupplung versehene ATK-Baureihe erfolgreich ein. Häufig verwendet wurde dieses Modell aber auch im industriellen Verschiebedienst, beim Verfahren schwerer Schiffsbauelemente auf Werftgeländen oder beim Verschub von Waggons auf Werksanschlussgleisen. Versuche hatten ergeben, dass der ATK Lasten von bis zu 50 t bewegen konnte. Er wurde auch bei verschiedenen Nato-Armeen in olivgrüner Lackierung eingesetzt. Durch seine verhältnismäßig hohe Höchstgeschwindigkeit von 32,5 km/h war der Radschlepper auch hervorragend als Straßenzugmaschine verwendbar. Auf Wunsch gab es neben dem normalen Fahrerdach auch ein von der Firma Benze hergestelltes geschlossenes Fahrerhaus mit abnehmbaren Plexiglasdach, das für eine gute Rundumsicht sorgte. Gleichfalls auf Wunsch konnte in dem je nach Ausstattung bis zu 5200 kg schweren Schlepper heckseitig eine 6 t-Seilwinde sowie eine Fronthydraulik und eine Vorbauplatte für Schiebearbeiten sowie ein Schneeräumer installiert werden. Als Sonderausrüstung war eine dreistufige, durch das Stromnetz zu betreibende Kühlwasservorwärmung, eingebaut. Dadurch war es möglich, eine ständige Betriebsbereitschaft der ATK-Schlepper auch bei Minustemperaturen aufrechterhalten zu können.

Hanomag R 40 A als Straßenzugmaschine mit festem Fahrerhaus und Druckluftbremsanlage, Baujahr 1942

Im Zuge der 1957 erfolgten Änderungen in den Typenbezeichnungen wurde das ohne nennenswerte Änderungen weitergebaute Modell nun als ATK 455 bezeichnet. Von 1960 bis 1964 hingegen befand sich als Pendant zum Modell R 460 der auf die gleiche Leistung von 60 PS angehobene, im Wesentlichen aber unveränderte Typ ATK 460 im Verkaufsprogramm der Hanomag-Werke.

Zu den markantesten Radschleppern seiner Zeit gehörte das erstmals im Jahr 1964 angebotene Modell Robust 800, welches gleichzeitig nach fast vier Jahrzehnten den Schlusspunkt dieser im Grundkonzept noch auf den legendären R 40 mit sei-

Hanomag R 40 A mit Druckluftbremsanlage, Baujahr 1946

nem langhubigen Vorkammer-Dieselmotor basierenden Traktors darstellte und den Typ R 460 ablöste. Dieses Modell verkörperte gleichzeitig den letzten klassischen schweren Hanomag-Schlepper. Während beispielsweise Getriebe und Vorderachse noch vom Vormodell R 460 stammten, war der vierzylindrige, ursprünglich für Bauraupen geschaffene Motor des Typs D 941 R mit 6786 ccm Hubvolumen und einer Leistung von 75 PS neu. Ferner besaß dieser konstruktiv überarbeitete Motor nunmehr fünf anstelle von drei Kugelwellenlagern. Die Hydraulikanlage mit einer Hubkraft von 2200 kg wurde gleichfalls neu dimensioniert und als weitere Neuerung erstmals als Sonderausstattung eine hydraulische Lenkhilfe in einen Hanomag-Schlepper eingebaut. Als weiteres Sonderzubehör waren Riemenscheibe, Seilwinde, Verdeck sowie eine Druckluftbremsanlage für den Straßenbetrieb erhältlich. Die Federung des bequem gepolsterten Einzelsitzes war individuell auf das Gewicht des Fahrers einstellbar. Unabhängig von der Fahrkupplung und auch unter Last schaltbar war die Zapfwelle, was vor allem bei

Hanomag R 40, Baujahr 1947

Hanomag R 40 A, Baujahr 1944 und Hanomag R 40 A, Baujahr 1946 (unten)

der Arbeit mit gezogenen Mähdreschern oder Vollerntemaschinen bedeutsam war. Der 3420 kg schwere Robust 800 war das derzeit leistungsstärkste Hanomag-Modell, welches ein in zwei Schaltgruppen gegliedertes Fünfganggetriebe mit Vorgelege (also 10 Vorwärts- und 2 Rückwärtsgänge) besaß. Im langsamsten Gang konnte die Geschwindigkeit des Robust 800 auf 1,4 km/h reduziert werden; dagegen erreichte er im 5. Gang der Schaltgruppe II mit 25,2 km/h seine Maximalgeschwindigkeit. Obwohl dieser gewaltige Bolide infolge seiner hohen Motorleistung und günstigen Gewichtsverteilung eine enorme, auf maximal 7000 kg zu steigernde Höchstzugkraft besaß, konnte er in technischer Hinsicht mit der gewachsenen Konkurrenz allerdings nur noch bedingt mithalten. Erst 1969, rund zwei Jahre nach dem Erscheinen des mit einem Kurzhubmotor bestückten völlig neuen Robust 900, wurde die Produktion eingestellt.

Nicht unerwähnt bleiben sollen einige interessante, bereits ab 1964 auf Basis des Robust 800 gebaute Prototypen, in die der aus dem Radlader B 11 stammende sechszylindrige D 961 B-Dieselmotor mit 10170 ccm Hubraum und 110 PS Leistung eingebaut war. Angeblich sechs Exemplare wurden unter der Bezeichnung Robust 800/6 beziehungsweise Robust 1100 gebaut, von denen bis heute noch einige bei Sammlern überlebt haben.

	R 40 A	R 40 Holzgas	R 40 C	R 45 A/R 445 E	R 55/R 455 E	R 450	R 460	R 55/455/460 ATK	Robust 800
Produktionszeitraum	1942-1951	1942-1945	1948-1951	1950-1958	1955-1958	1958-1961	1960-1964	1952-1964	1964-1969
Motorleistung (PS)	40	40	40	45	55	50	58/60	55/60	75
Hubraum (cm³)	5195	5702	5195	5702	5702	5702	5702	5702	6786
Anzahl der Zylinder	4	4	4	4	4	4	4	4	4
Anzahl der Gänge	5+1	5+1	5+1	5+1	10+2	10+2	10+2	5+1	10+2
Eigengewicht (kg)	3200	3580	3260	3270	3470	3600	3600	5200	3420
Höchstgeschwindigkeit	18,7 km/h	21,8 km/h	25,0 km/h	17,3 km/h	19,9 km/h	18,5 km/h	30,0 km/h	32,5 km/h	25,2 km/h

Hanomag R 40 A als Straßenzugmaschine mit Druckluftbremsanlage, Baujahr 1944

HANOMAG

45 PS HANOMAG-Diesel-Straßenschlepper R 45

Wendig, robust und zugstark, das sind die großen Vorteile des 45 PS HANOMAG-Straßenschleppers R 45 C. Sei es im Baubetrieb, beim Holztransport, in der Spedition, im Werk- oder Pendelverkehr, stets erledigt der R 45 alle Schwer-Transporte schnell und wirtschaftlich. Der niedrigtourige 45 PS 4 Zylinder-Dieselmotor in Verbindung mit dem verlustarmen und richtig abgestuften Getriebe ermöglicht große Zugleistungen bei hoher Geschwindigkeit. Zwei beladene 8 t Anhänger werden vom R 45 ohne Überlastung **im 5. Gang** gezogen. Der Schlepper ist mit einer Luftdruckbremsanlage für Anhänger ausgestattet und wird dadurch allen Anforderungen, die Stadtverkehr und Transporte im bergigen Gelände stellen, gerecht. Die selbsttätige, gefederte Anhängerkupplung ermöglicht weiches Anfahren und Bremsen. Fahrer und Beifahrer finden bequem Platz auf zwei Polstersitzen. Das Ganzstahl-Fahrerhaus bietet Dank der großen Fenster freie Sicht nach allen Seiten.

Ist der R 45 mit hydraulischem Lader ausgerüstet, so wird die Einsatzmöglichkeit für den Baubetrieb und für alle Betriebe, die Schüttguter laden müssen, ganz erheblich erweitert. Ein Mann und ein Schlepper beladen die Anhänger in kurzer Zeit. 1000 bis 1500 kg können mit einem Hub 3,50 m hochgehoben werden — also weit höher als zum Beladen von Anhängern erforderlich ist. Die Schaufel des Laders kann schnell und einfach abgenommen und durch einen Erdhobel, einen Lasthaken oder einen Verlängerungsarm mit Lasthaken zum Heben in größere Höhen ersetzt werden. Zur Kippanhängerbetätigung kann die Hydraulik mit einer Abreißkupplung versehen werden.

Unsere Vertreter sind gern bereit, Ihnen den R 45 vorzuführen. Überzeugen Sie sich selbst von den Vorteilen, die Ihnen diese wirtschaftliche Zug- und Arbeitsmaschine bietet.

Technische Einzelheiten

Motor
- Arbeitsweise: Viertakt-Dieselmotor
- Zylinderanzahl: Vier
- Bohrung: 110 mm
- Hub: 150 mm
- Hubraum: 5702 ccm
- Drehzahl: 1200 Umdr./min.
- Motordauerleistung: 45 PS
- Kolbenwerkstoff: Leichtmetall
- Zylinderkopf: abnehmbar, wassergekühlt
- Ventilanordnung: hängend, von oben gesteuert
- Luftreinigung: Ölbad-Luftfilter mit automatischer Ölbenetzung und vorgeschaltetem Vorabscheider
- Brennstoffpumpe } HANOMAG
- Einspritzventile }
- Schmierung: Umlaufpreßölschmierung mit dreifacher Ölreinigung

- Kühlung: zwangsläufige Umlaufkühlung mit Wasserpumpe, Vierflügelventilator, Rollvorhang
- Drehzahlregler: Präzisionsfliehkraftregler
- Anlasseranlage: BOSCH

Fahrgestell
- Kupplung: Einscheiben-Trockenkupplung
- Gangzahl: 5 Vorwärtsgänge, 1 Rückwärtsgang
- Geschwindigkeiten:
  - 1. Gang: 3,5 oder 5,7 km/st
  - 2. Gang: 5,0 oder 7,5 km/st
  - 3. Gang: 6,6 oder 9,6 km/st
  - 4. Gang: 9,8 oder 14,4 km/st
  - 5. Gang: 17,3 oder 25,0 km/st
  - Rückwärtsgang: 2,7 oder 4,8 km/st
- Zugleistung auf ebener Straße: 22 t im 5. Gang
- Lenkung: ZF-Roßlenkung

- Fußbremse: mechanische Servobremse, auf Hinterräder wirkend, nachstellbar
- Druckluftbremse: für Anhänger
- Handbremse: mechanisch auf Getriebe wirkend, nachstellbar
- Zugvorrichtung: selbsttätige, gefederte Anhängerkupplung
- Räder: Stahl- oder Gußeisenräder
- Bereifung: vorn 6.50-20, e. H.D hinten 12.75-28
- Radstand: 2080 mm
- Vorderachse: Pendelachse mit Blattfederung
- Abmessungen: Länge: 3570 mm, Breite: 1780 mm, Höhe: 2310 mm
- Steuergewicht: 3260 kg
- Wenderadius (nach DIN 70020): 4,4 m
- Kraftstoffbehälter: 85 Liter Inhalt
- Fahrersitz: 2 bequeme Polstersitze

Auf Wunsch liefern wir: Seilwinde, Riemenscheibe, Lader.

Abbildungen, Angaben und technische Daten unverbindlich. Änderungen vorbehalten.

Hubvermögen 1000-1500 kg

Hubhöhe 3,50 m

HANOMAG HANNOVER
gegründet 1835

Überreicht durch:

Prospekt Nr. 202
Werbedruck Aug. Lönneker, Stadtoldendorf

Hanomag R 45

Hanomag R 45 A als Straßenzugmaschine mit Druckluftbremsanlage, Baujahr 1952

Hanomag R 45 A, Baujahr 1952

Hanomag R 45 C als Straßenzugmaschine mit Druckluftbremsanlage, 1978 noch im Einsatz

Hanomag R 45 C, Baujahr 1951

Hanomag R 45 A,
Baujahr 1952 (oben)
und R 45 C,
Baujahr 1957

Hanomag R 55 in roter Export-Ausführung, Baujahr 1955, vor einem Buschhoff-Dreschkasten

Hanomag R 55, Baujahr 1955

Hanomag R 55, ehemaliges Exportmodell mit hinteren Muschelkotflügeln, Baujahr 1957

Das Schaltgetriebe des Hanomag R 55

Die schweren Diesel-Radschlepper R 40 und deren Folgemuster

Der Motor

Dieser Viertakt-Motor hat dank seiner niedrigen, mittleren Kolbengeschwindigkeit eine lange Lebensdauer und besitzt die Kraft zur Bewältigung schwerster Arbeitsleistungen. Er arbeitet nach dem unempfindlichen Vorkammer-Verfahren, das saubere und wirtschaftliche Verbrennung des Kraftstoffes bei jeder Belastung garantiert. Ein robuster, 4 PS starker elektrischer Anlasser startet den Motor bei jeder Außentemperatur schnell und sicher. Die Leistung von 55 PS wird bei einer Drehzahl von 1300 U/min erreicht. Einfachheit, Zuverlässigkeit und Wirtschaftlichkeit sind die Hauptmerkmale dieses Motors. Zwei wesentliche Faktoren verringern die Betriebskosten und bringen bedeutende geldliche Vorteile: Das ist zunächst der billige Dieselkraftstoff und zum anderen der geringe Verbrauch an Kraftstoff und Motorenöl.

Aus einem Verkaufsprospekt...

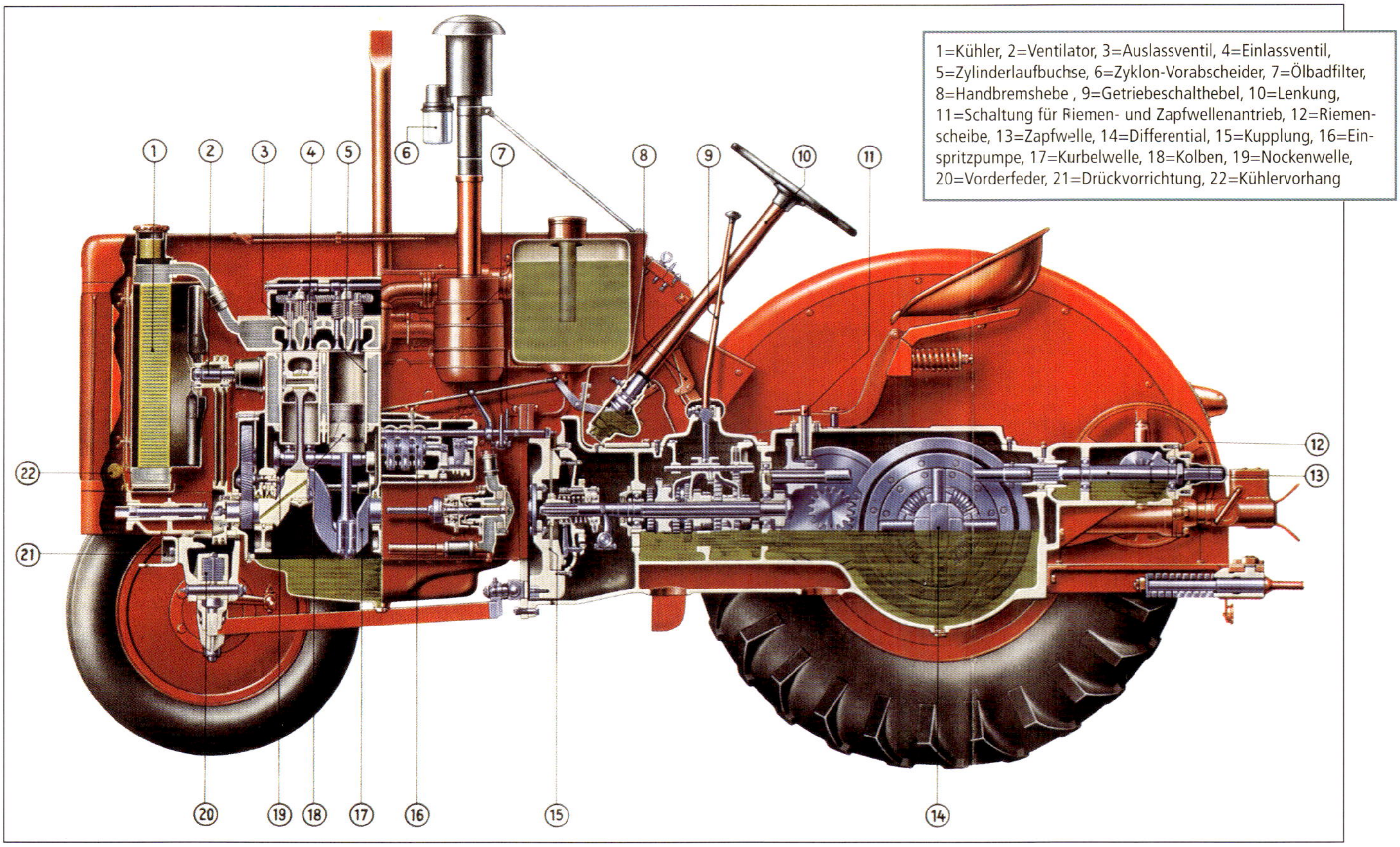

1=Kühler, 2=Ventilator, 3=Auslassventil, 4=Einlassventil, 5=Zylinderlaufbuchse, 6=Zyklon-Vorabscheider, 7=Ölbadfilter, 8=Handbremshebe, 9=Getriebeschalthebel, 10=Lenkung, 11=Schaltung für Riemen- und Zapfwellenantrieb, 12=Riemenscheibe, 13=Zapfwelle, 14=Differential, 15=Kupplung, 16=Einspritzpumpe, 17=Kurbelwelle, 18=Kolben, 19=Nockenwelle, 20=Vorderfeder, 21=Drückvorrichtung, 22=Kühlervorhang

...des Hanomag R 55

Scheiben-Beetpflug ROLLER I

Dieser besonders beliebte Scheiben-Beetpflug wird mit 5 und 6 Scheiben gefertigt, hat 4 Räder und eignet sich besonders für Arbeiten in hartgebranntem oder auch verwachsenem Boden sowie für den Umbruch von Neuland. Ein Vierkant-Stahlbalken als Rahmen verleiht dem Gerät eine außergewöhnliche Festigkeit. Die Scheiben sind am Rande geschärft, verschleißfest und bruchsicher. Verstellbare Abstreifer an den Scheiben verhindern auch bei schwierigen Bodenarten ein Verstopfen des Gerätes.

Anhänge-Beetpflug TBS 720

Aus einem Verkaufsprospekt des Hanomag R 55 von 1955: Auf der linken Seite oben wird der Scheibenpflug „SVS" mit vier Scheiben präsentiert, auf dieser Seite oben die Doppelscheibenegge „Salta" und unten der Anhänge-Saatpflug „Titan"

Hanomag R 450 E in roter Exportlackierung mit festem Fahrerhaus, Baujahr 1960 (oben) und Hanomag R 455 S als Straßenzugmaschine mit Druckluftbremsanlage, Baujahr 1961

Hanomag R 55, ehemaliges Exportmodell mit Muschelkotflügeln, Baujahr 1957

Hanomag R 450, Baujahr 1958

Hanomag R 450 E als Straßenzugmaschine mit Druckluftbremsanlage, Baujahr 1961

Hanomag 450 EL mit Riemenscheibe am Heck, Baujahr 1959

Hanomag R 545 (5=Viertakter mit D-57-Motor, 4=Zylinderzahl, 5=starke Viertaktausführung) in Exportausführung, oben Baujahr 1959, unten 1960

Hanomag R 460,
Baujahr 1962 (oben),
Baujahr 1960 (Mitte),
Baujahr 1963 (unten)

Hanomag Schwerlastschlepper R 55 ATK für Flughafeneinsatz

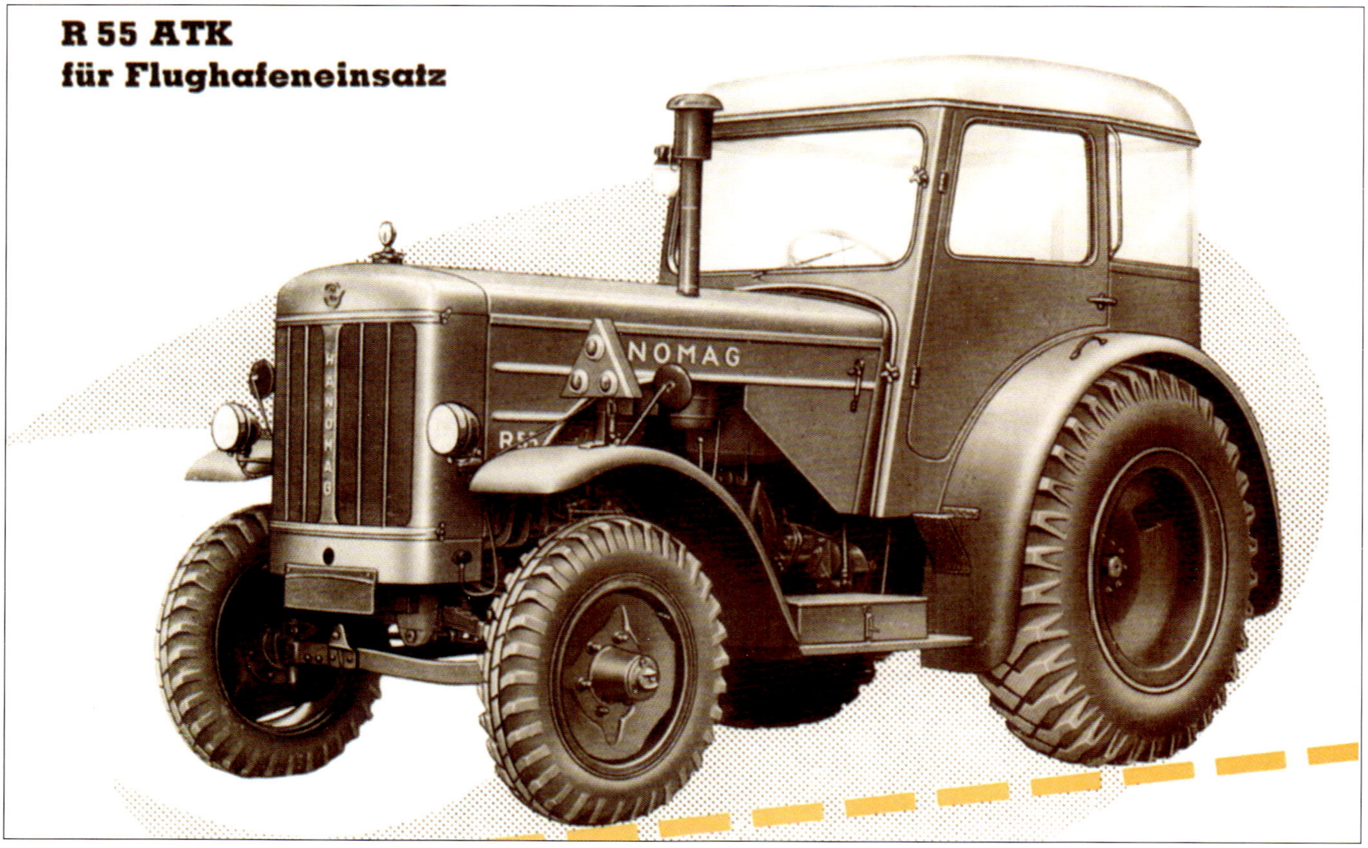
R 55 ATK
für Flughafeneinsatz
NOMAG

Hanomag R 55 ATK

HANOMAG
HANOMAG

Hanomag R 455 ATK, Baujahr 1961

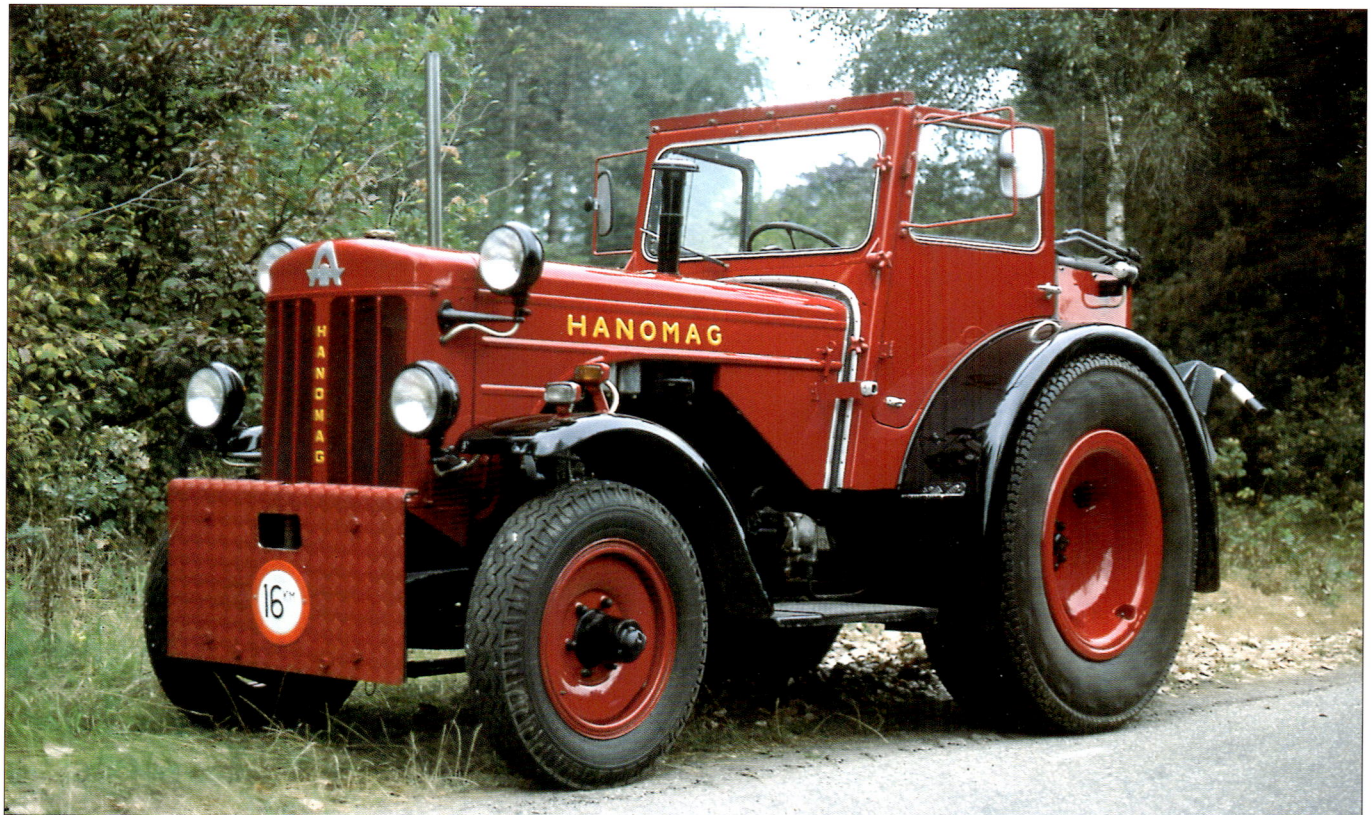

Hanomag R 455 ATK mit festem Fahrerhaus und 6-t-Seilwinde, Baujahr 1962

Hanomag R 455 ATK mit Klappverdeck und Seilwinde, Baujahr 1962

Hanomag
Robust 800
75 PS

Hanomag Robust 800, Baujahr 1966

Hanomag R 460 als Straßenzugmaschine mit Druckluftbremsanlage, Baujahr 1962

Hanomag Robust 800 mit festem Fahrerhaus, Baujahr 1965

Hanomag Robust 800 als Straßenzugmaschine mit Druckluftbremsanlage, Baujahr 1968

Hanomag Robust 800, Exportausführung Baujahr 1964 (oben), Hanomag Robust 800, Exportausführung, Baujahr 1965 (Mitte), Hanomag Robust 800, Exportausführung mit Dieteg-Fahrerkabine, Baujahr 1967 (unten)

Hanomag Robust 1100, Versuchsausführung, Baujahr 1968

Die Hanomag-Schlepper in den ersten Nachkriegsjahren

Drehen wir an dieser Stelle die Zeit um etwa 20 Jahre zurück, um die Entwicklung des übrigen Hanomag-Typenprogramms auf dem Schleppersektor näher zu betrachten.

Nachdem im Jahr 1947 die Produktion des Bauernschleppers RL 20 wieder angelaufen war, wurde dieser bereits zwei Jahre später durch den moderneren Nachfolger, den Allzweck-Diesel-Radschlepper R 25 ersetzt. Dieser Schlepper, dessen erste Entwicklungsansätze auf die Zeit etwa unmittelbar nach Ausbruch des Krieges zurückgingen und der zum Jahresende 1948 vorgestellt wurde, war das erste Glied einer Reihe mittelstarker Schlepper, die nahezu zwei Jahrzehnte als einer der Schwerpunkte im Verkaufsprogramm verbleiben sollten. Konstruktiv war der R 25 in der Halbrahmenbauweise, einer Kombination aus Rahmen- und Blockbauweise, gehalten. Hierbei wurde das neuentwickelte, mit fünf Vorwärtsgängen und einem Rückwärtsgang ausgebildete Getriebe wie üblich mit der Hinterachse zu einem Block vereint, an den sich zwei U-Profilträger anschlossen, auf welche die daran arretierte Vorderachskonsole aufgelegt war. Der hinten an das Getriebe angeflanschte Motor war vorn zusätzlich auf den Halbrahmen abgestützt. Diese Bauweise ermöglichte wartungstechnische Verbesserungen sowie die Anbaumöglichkeit von Geräten zwischen und seitlich an den Achsen. Zapfwelle und Riemenscheibe gehörten ebenfalls zum serienmäßigen Lieferumfang. Als Sonderzubehör konnten Mähwerk, Dach und hydraulischer Kraftheber bezogen werden.

Verwendet wurde bei den unterschiedlich bereiften lieferbaren Varianten, der Ausführung A (mit Speichenrädern und 9.00-40-Bereifung) und B (mit Scheibenrädern und 11,25-24-Reifen), zunächst der Vierzylinder-Dieselmotor D 19, der 20 PS

Hanomag R 25 C, Baujahr 1949

25 PS
2,8 Liter-Diesel-Motor

Der HANOMAG-Vierzylinder-Dieselmotor läuft ruhig und erschütterungsfrei. Die Verbrennung ist so vollkommen, daß sie rauch- und geruchlos erfolgt. Daraus erklärt sich auch der geringe Kraftstoffverbrauch. Irgendwelche besonderen Erfahrungen im Umgang mit Dieselmotoren sind nicht notwendig, weil der HANOMAG-Diesel unkompliziert und übersichtlich gebaut ist.

Technische Einzelheiten

4-Zylinder-Diesel-Motor
mit Vorkammer-System und 5mal gelagerter Kurbelwelle
- Bohrung . . . . 90 mm
- Hub . . . . 110 mm
- Hubraum . . . . 2800 cm³
- Drehzahl . . . . 1500 Umdreh./Min.
- Leistung . . . . 25 PS
- Zylinderbuchsen . . . . auswechselbar
- Schmiersystem: Umlaufpreßölschmierung
- Ölreinigung: Durch bewährten Spaltfilter, der vom Kupplungspedal zwangsläufig betätigt wird
- Einspritzpumpe: Fabrikat Bosch oder Deckel
- Kühlung: Umlaufkühlung mit Wasserpumpe
- Kühlwasser-Betriebstemperatur 70°—80°, Regelung automatisch

5-Gang-Getriebe
- Ganganzahl . . . . . . 5 Vorwärtsgänge und 1 Rückwärtsgang
- Geschwindigkeiten
  - 1. Gang 3,7 km/Std   3. Gang 6,3 km/Std
  - 2. Gang 5,0 km/Std   4. Gang 11,6 km/Std
  - 5. Gang 18,3 km/Std

Kupplung: Einscheiben-Trockenkupplung

Handbremse: Mechanisch mit Duplexwirkung

Fußbremse: Mechanisch mit Servowirkung

Schutz gegen einseitiges Räderrutschen: Einzelradabbremsung

Zapfwelle
überträgt fast die ganze Motorleistung
- Genormtes Zapfwellenprofil
- Anordnung . . . . hintenliegend
- Drehzahl . . . . . . 560 Umdrehungen/min

Abmessungen
- Größte Länge . . . . 2850 mm
  - bzw. 2990 mm*
- Größte Breite . . . . . . normal 1580 mm
  - bei max. Spur 1750 mm
- Größte Höhe
  - Oberkante Lenkrad 1580 mm
  - bzw. 1760 mm*
- Radstand . . . . 1800 mm

* bei Bereifung 9,00—40

Bodenfreiheit:
- Durchgehend . . . . 300 mm
  - bzw. 450 mm*
- Bereifung Vorn . . . . 5,50—16
  - bzw. 6,50—20
- Hinten . . . . . . 11,25—24
  - bzw. 9,00—40

Spurweite vorn:
1300-1400-1500 mm . . . . . . verstellbar
hinten
1300-1400-1500 mm . . . . . . verstellbar

Wenderadius:
2500 mm ohne Einzelradabbremsung
1800 mm mit Einzelradabbremsung
(Innen)

Leergewicht . . . . 1720 kg
bzw. 1940 kg*

Zugvorrichtung
- unten . . . . gelochte Anhängeschiene für Ackergeräte
- oben . . . . Zugmaul, für Straßenzug seitlich verstellbar

Sonderausrüstungen
- Kraftheber   Vordere Kotflügel
- Mähbalken   Polstersitz
- Allwetterdach   gefed. Zugvorrichtg.
- Riementrieb   Anbaugeräte aller Art

Änderungen vorbehalten

HANOMAG-HANNOVER

Hanomag R 25 D mit Bandsäge, Baujahr 1950 – noch 1991 im Einsatz

Dauerleistung und 25 PS Höchstleistung aus 1911 ccm Hubraum entwickelte und sich bald für die Fahrzeuge als zu schwach erweisen sollte.

Daher wurde der bereits 1950 bei Hanomag bereitstehende, und vor allem für die leichten Lastwagen der L-28-Reihe neuentwickelte Vierzylinder-2,8-Liter-Dieselmotor des Modells D 28 auch zum Einbau in den R-25-Ackerschlepper (nunmehr als R 25 D bezeichnet) verwendet. Dieser stellte seine Höchstleistung von 25 PS wesentlich müheloser zur Verfügung. Erstmals präsentierte Hanomag die mit diesem Aggregat bestückten Nachfolgevarianten R 25 C und D im Juni 1950 auf der 40. DLG-Ausstellung in Frankfurt dem Publikum. Besitzern der alten Ausführung wurde werksseitig kulanterweise sogar eine kostenlose Umrüstung auf den neuen Motor angeboten.

Den R 25 gab es sowohl als Allzweckschlepper mit hoher schmaler Speichenradbereifung als auch als Standardschlepper mit Scheibenrädern und niedrigen, breiten Reifen. Die dreifachen Spurverstellmöglichkeiten zwischen 1300 und 1500 mm ließen ihn auch für Arbeiten in Reihenkulturen auf Hackfruchtfeldern, wie zum Beispiel bei Rüben und Kartoffeln als sehr geeignet erscheinen.

Nachdem rund 2700 Einheiten gebaut worden waren, wurde der R 25 bereits im Jahr 1951 durch das stärkere Modell R 28 ersetzt. Das mit dem leistungsgesteigerten, ansonsten aber baugleichen, Vierzylinder-28-PS-Dieseltriebwerk D 28 S ausgerüstete Fahrzeug zählte aufgrund seiner Stärke bereits zur Kategorie der größeren Schlepper. Neu an diesem Modell war das Fünfganggetriebe, das mit der gleichen Anzahl als Sonderausstattung erhältlicher Kriechgänge für Fahrgeschwindigkeiten zwischen 1,05 und 18,3 km/h bestückt war. Den R 28 gab es in der Variante A als Allzweckschlepper mit 9-42-Hinterradbereifung und in der Normalausführung B mit 10-28 oder 11.25-24-Hinterreifen. Als damals noch seltenes Zubehör zur Verwendung von Anbaugeräten galt der hydraulische Drei- oder Vierpunkt-Kraftheber. Außerdem waren auf Kundenwunsch Eisenräder, Moorbereifung, Frontlader sowie eine 3 t-Seilwinde zu bekommen. Die mit einem Schnellganggetriebe bestückte Ausführung R 28 S kam vor allem als Straßenzugmaschine in Frage und die Variante R 28 N war aufgrund ihrer niedrigen Bauhöhe besonders zum Einsatz in Plantagen geeignet. Aber schon 1953 wurde der R 28 zugunsten des stärkeren R 35 aus dem Programm genommen. Noch im gleichen Jahr erhielten alle Hanomag-Schlepper eine neue Kühlermaske, die aus fünf senkrechten Stäben bestand.

Zu jener Zeit waren insgesamt etwa 6300 Mitarbeiter bei Hanomag beschäftigt und für notwendige Investitionen suchte man einen finanzkräftigen Partner, der im Oktober 1952 in der Rheinstahl-Union gefunden wurde, zu deren größten Tochterfirma die Hanomag fortan wurde.

Das zwischen 1953 und 1957 gebaute Modell R 35 hatte man ebenfalls in Halbrahmenbauweise konstruiert. Durch Erhöhung der Drehzahl und Änderung der Vorkammer war aus dem Hubraum des vierzylindrigen 2,8-Liter-Motors eine

Hanomag R 25 mit Fahrerdach, Baujahr 1950

Mehrleistung von sieben PS herausgeholt worden. Wie beim Vorgänger war das in zwei Schaltgruppen unterteilte Fünfganggetriebe mit zusätzlichen Kriechgängen ausrüstbar. Die Geschwindigkeitsbandbreite reichte von 0,85 km/h bis zu 19,8 km/h im 5. Gang. Gleichwohl konnte die Spurweite des Schleppers dreifach verstellt werden. Die werksseitig erhältliche Zusatz- und Sonderausstattung war sehr umfangreich. Hierzu zählte auch ein überdachter Fahrerplatz aus Stahlblech mit Windschutzscheibe. Auch hier war wiederum eine Straßenzugmaschinen-Ausführung, die Variante R 35 S, mit schnellem Getriebe, hinteren Gussfelgen und Druckluftanlage erhältlich. Die Zugleistung dieser vor allem infolge der Gussfelge mit 2260 kg wesentlich schwereren Variante betrug auf ebener Straße im 5. Gang 15 Tonnen.

Zwischen 1955 und 1957 baute Hanomag das Modell R 35/45, einen schweren Schlepper für schwere Zapfwellengeräte und speziell für den Mähdrescherbetrieb, bei dem mittels eines vom Fahrersitz zu betätigenden Handhebels ein Roots-Gebläse durch Erhöhung der Kraftstoffeinspritzmenge zugeschaltet werden konnte, das die Motorleistung von 35 auf 45 PS steigerte. Durch eine installierte Doppelkupplung war das Schalten und Anhalten bei laufender Zapfwelle (der so genannten Motorzapfwelle) möglich. In Anbetracht seiner hohen Leistung war dieses Modell mit lediglich 2100 kg Eigengewicht als ausgesprochen leicht zu bezeichnen.

Um auch den damals noch beachtlich großen Markt der vielen Kleinlandwirte abdecken zu können, wurden auf der 1951 in Hamburg stattfindenden DLG-Ausstellung die beiden neuen Hanomag-Traktoren R 16 und R 22 vorgestellt. Erstmals wurden als Antrieb bei diesen Hanomag-Schleppern keine Vierzylindermotoren eingebaut. Der in rahmenloser Blockbauweise konstruierte und 1170 kg wiegende R 16 war das kleinere Modell, das mit dem nach dem Vorkammerverfahren arbeitenden, einer neuentwickelten Motorenbaureihe entstammenden wassergekühlten 16 PS starken Zweizylinder-Viertakt-Dieselmotor des Typs D 14 mit 1399 ccm Rauminhalt und dreifach gelagerter Kurbelwelle und Bosch-Einspritzpumpe ausgerüstet war. Für dieses für leichtere Feldarbeiten oder als Zweitschlepper universell verwendbare Modell stand ein in zwei Schaltgruppen aufgeteiltes Getriebe mit fünf Vorwärtsgängen und einem Rückwärtsgang sowie drei zusätzlichen Kriechgängen zur Verfügung, das Geschwindigkeiten zwischen 0,85 und 19,1 km/h ermöglichte. Das bedeutete eine Mindestgeschwindigkeit bei maximaler Motordrehzahl von 800 m pro Stunde, was bei Pflanz- und Hackarbeiten in Verbindung mit der dreifach verstellbaren Spurweite und der Bodenfreiheit von 40 cm ein großer Vorteil war. Während die Zapfwelle serienmäßig war, konnten Riemenscheibe, gefederte Vorderachse, Mähwerk, Hydraulik, Frontlader, Seilwinde und anderes Zubehör gegen Aufpreis bezogen werden. Der R 16, den es in den Ausführungen A als Allzweckschlepper und B als Standardausführung gab, war – in Details laufend verbessert – bis 1957 lieferbar.

Zu Beginn des Jahres 1953 wurde der R 16 durch den auf diesem basierenden Typ R 19 leistungsmäßig nach oben hin ergänzt. Ebenso wie der R 16 blieb dieses Modell bis 1957 in der Produktion. Auch in diesem Fall konnte durch Erhöhung der Nenndrehzahl und Änderung der Vorkammer die um drei PS höhere Leistung aus dem bewährten D-14-Triebwerk herausgeholt werden. Ein weiteres Unterscheidungsmerkmal zu seinem Vorgänger war der bis zur Motormitte zurückversetzte, links seitlich befestigte Ölbadluftfilter. Auch hier gelangte ein Fünfganggetriebe mit den auf Wunsch erhältlichen drei zusätzlichen Kriechgängen zum Einbau. Während der Geschwindigkeitsbereich der Kriechgänge von 0,76 bis 1,32 km/h reichte, lagen die Fahrtstufen der Normalgänge zwischen 3,8 und 19,6 km/h. Darüber hin-

Hanomag R 16,
Baujahr 1953 (oben),
Hanomag R 16 B,
Baujahr 1952 (unten)

	R 25 A+B	R 25 C+D	R 28	R 28 S	R 35	R 35/45	R 16	R 19	R 22	R 27
Produktionszeitraum	1949-1950	1950-1951	1951-1953	1951-1953	1953-1957	1955-1957	1951-1957	1953-1957	1951-1957	1953-1957
Motorleistung (PS)	20/25	25	28	28	35	35/45	16	19	22	27
Hubraum (cm³)	1911	2799	2799	2799	2799	2799	1400	1400	2099	2099
Anzahl der Zylinder	4	4	4	4	4	4	2	2	3	3
Anzahl der Gänge	5+1	5+1	10+2	10+2	10+2	10+2	8+2	8+2	10+2	10+2
Eigengewicht (kg)	1860	1940	1860	2260	1860	2122	1170	1250	1520	1585
Höchstgeschwindigkeit	18,3 km/h	18,3km/h	18,3 km/h	25,2 km/h	19,8 km/h	18,0 km/h	19,1 km/h	19,6 km/h	18,3 km/h	19,5 km/h

aus war der R 19 in vier unterschiedlich großen Hinterradbereifungen erhältlich.

Der zwischen 1951 und 1957 ebenfalls in Halbrahmenbauweise gefertigte und bereits erwähnte R 22 war besonders für kleinere, aber auch für mittelgroße landwirtschaftliche Betriebe vorgesehen. Er fügte sich in die zwischen R 16 und R 28 bestehende Angebotslücke nahtlos ein und war nur als Allzweckschlepper mit großen Antriebsrädern lieferbar. Konzeptionell entsprach der 1520 kg wiegende mit dem wassergekühlten 22-PS-Dreizylinder-Triebwerk des Typs D 21 S mit 2099 ccm Rauminhalt und vierfach gelagerter Kurbelwelle bestückte R 22 dem stärkeren R 28. Das Fünfganggetriebe mit den wahlweise erhältlichen Kriechgängen entsprach hinsichtlich der Geschwindigkeitsabstufungen dem der übrigen Typen dieses Zeitabschnitts. Das Zubehör war zu jener Zeit als recht umfangreich zu bezeichnen und bestand unter anderem aus hydraulischem Kraftheber, Frontlader, Mähbalken, seitlichem Riemenscheibenantrieb, Zapfwelle, festem Fahrerdach mit Windschutzscheibe und einer Seilwinde mit 3 t Zugkraft. Ebenso stand eine große Auswahl unterschiedlicher Anbaugeräte zur Verfügung. Die Spur des Traktors war dreifach verstellbar und ab 1953 trug er – wie alle übrigen Traktoren von Hanomag – die neugestaltete Kühlermaske mit ihren fünf senkrechten Streben, von denen die mittlere, breiter ausgeführte Strebe den Schriftzug "Hanomag" trug.

Parallel hierzu entstand im Jahr 1953 durch Erhöhung der Drehzahl aus diesem Modell der ansonsten nahezu baugleiche 27 PS starke Hanomag-Schlepper R 27. Dieser Traktor gehörte aufgrund seiner Leistung bereits zur oberen Mittelklasse. Andererseits war sein Eigengewicht mit maximal 1585 kg für einen Schlepper dieser Leistung – das in etwa vergleichbare 28-PS-Modell von Lanz wog 2200 kg – außerordentlich gering. Der R 27 war mit dem bekannten Fünfganggetriebe für maximal 19,5 km/h Geschwindigkeit bestückt. Auf Wunsch konnten fünf zusätzliche Kriechgang-Übersetzungen für Geschwindigkeiten zwischen 0,86 und 4,20 km/h eingebaut werden.

Vom R 27 und R 35 wurde im gleichen Zeitraum jeweils noch eine Spezialausführung als Row-Crop-Schlepper für den Export nach Übersee gebaut. Anstelle der herkömmlichen Vorderachse war bei beiden Typen ein Doppelrad vor dem Rahmen angebracht, das allerdings ohne großen Zeitaufwand gegen eine spurverstellbare Spezial-Vorderachse ausgewechselt werden konnte. Die Spurweiten mit der zusätzlichen Bezeichnung "RC" waren zwischen 1250 und 2060 mm stufenlos verstellbar. Sie waren zum Einsatz in Reihenkulturen wie Baumwoll- Mais- und Zuckerrohrplantagen vorgesehen und verfügten ebenfalls über großes Zubehörpotential.

Daneben war die zwischen 1951 und 1953 gefertigte Ausführung "N" des R-28-Schleppers ebenfalls für den Einsatz auf Plantagen vorgesehen. Zu diesem Zweck war dieses Modell durch Verwendung kleiner Reifengrößen besonders niedrig ausgeführt, besaß einen Astabweiser über dem Lenkrad sowie einen nach unten abgeleiteten Auspuff und einen geänderten Luftfilter.

Hanomag R 16, Baujahr 1951

Hanomag R 16 mit neuer Kühlermaske, Baujahr 1954

Hanomag R 16 mit neuer Kühlermaske, Baujahr 1955

Hanomag R 16 A mit Fahrerdach, Baujahr 1956

Hanomag R 16 A, Baujahr 1955 (oben), Hanomag R 16 mit Fahrerdach, Baujahr 1956 (unten)

76 Die Hanomag-Schlepper in den ersten Nachkriegsjahren

Hanomag R 16 Combitrac aus einem Prospekt von 1956

Hanomag R 19 Combitrac

Die Hanomag-Schlepper in den ersten Nachkriegsjahren

HANOMAG R19 Combitrac

TECHNISCHE EINZELHEITEN

Art und Arbeitsweise des Motors	Zweizylinder-Viertakt-Dieselmotor
Motordauerleistung	19 PS
Hubraum	1399 cm³
Drehzahlbereich	630—1975 U/min.
Kühlsystem	moderne Thermostat-Zweikreiskühlung
Länge über alles	2680 mm
Breite über alles (schmale Spur)	1515 mm
Radstand	1600 mm
Spurweiten	1250 / 1375 / 1500 mm
Eigengewicht	1250 kg
Zapfwellendrehzahl	563 U/min.
Riemenscheibendrehzahl	1215 U/min.
Riemenscheiben-Ø	273 mm
Bereifung, vorn	4,50—16
Bereifung, hinten	8—32 AS

Serienmäßig Speichenräder, auf Wunsch Scheibenräder

Bodenfreiheit	380 mm

Geschwindigkeiten

1. Gang	3,8 km/st
2. Gang	4,9 km/st
3. Gang	6,6 km/st
4. Gang	11,9 km/st
5. Gang	19,6 km/st
Kriechganggeschwindigkeiten	380 m/st — 1330 m/st

Sonderausrüstungen

Hydraulischer Kraftheber; Belastungsgewichte, vorn; Scheibenräder, hinten; Gitterräder; Komb. Riemenscheiben- u. Mähwerkantrieb; Mähwerkantrieb, seitl.; Kotflügel, vorn; Fahrerdach mit Windschutzscheibe; Doppelsitz auf Kotflügel; Kriechgang-Untersetzung; Bereifung 10—28 AS.

Konstruktive Änderungen und Abweichungen vorbehalten.

Der HANOMAG R 19 wurde vom Schlepper-Prüffeld Marburg unter Test-Nr. 97 geprüft.

Hanomag R 19 in roter Exportausführung, Baujahr 1957

Hanomag R 19, Baujahr 1955

Hanomag R 19 mit Fahrerdach, Baujahr 1954

Hanomag R 19, Baujahr 1953 (oben) und Hanomag R 19 mit Fahrerdach und Riemenscheibe, Baujahr 1956 (unten)

Hanomag R 22, Baujahr 1954 (oben), Hanomag R 22, Baujahr 1952 (Mitte) und Hanomag R 22, Exportausführung mit alter Kühlermaske, Baujahr 1952 (unten)

Hanomag R 22, Exportausführung mit Fahrerdach und alter Kühlermaske, Baujahr 1952

Hanomag R 27 mit Fahrerdach, Baujahr 1956

Hanomag R 27 Exportausführung mit Mähbalken, unrestauriert, Baujahr 1954 (großes Bild) und Hanomag R 27, Baujahr 1955

Hanomag R 28, Baujahr 1952 (oben) und Hanomag R 28, Exportausführung mit Muschelkotflügeln, Baujahr 1953 (unten)

Hanomag R 28 A mit Fahrerdach, Baujahr 1951 (oben), Hanomag R 28 A, Baujahr 1953 (Mitte) und Hanomag R 28 B, Baujahr 1951 (unten)

Hanomag R 35, Exportausführung, Baujahr 1955

Hanomag R 35 mit Fahrerdach und Frontlader, Baujahr 1956

Hanomag R 35, Baujahr 1956 (oben), Hanomag R 35 A, Baujahr 1954 (Mitte) und Hanomag R 35 A mit Fahrerdach, Baujahr 1953 (unten)

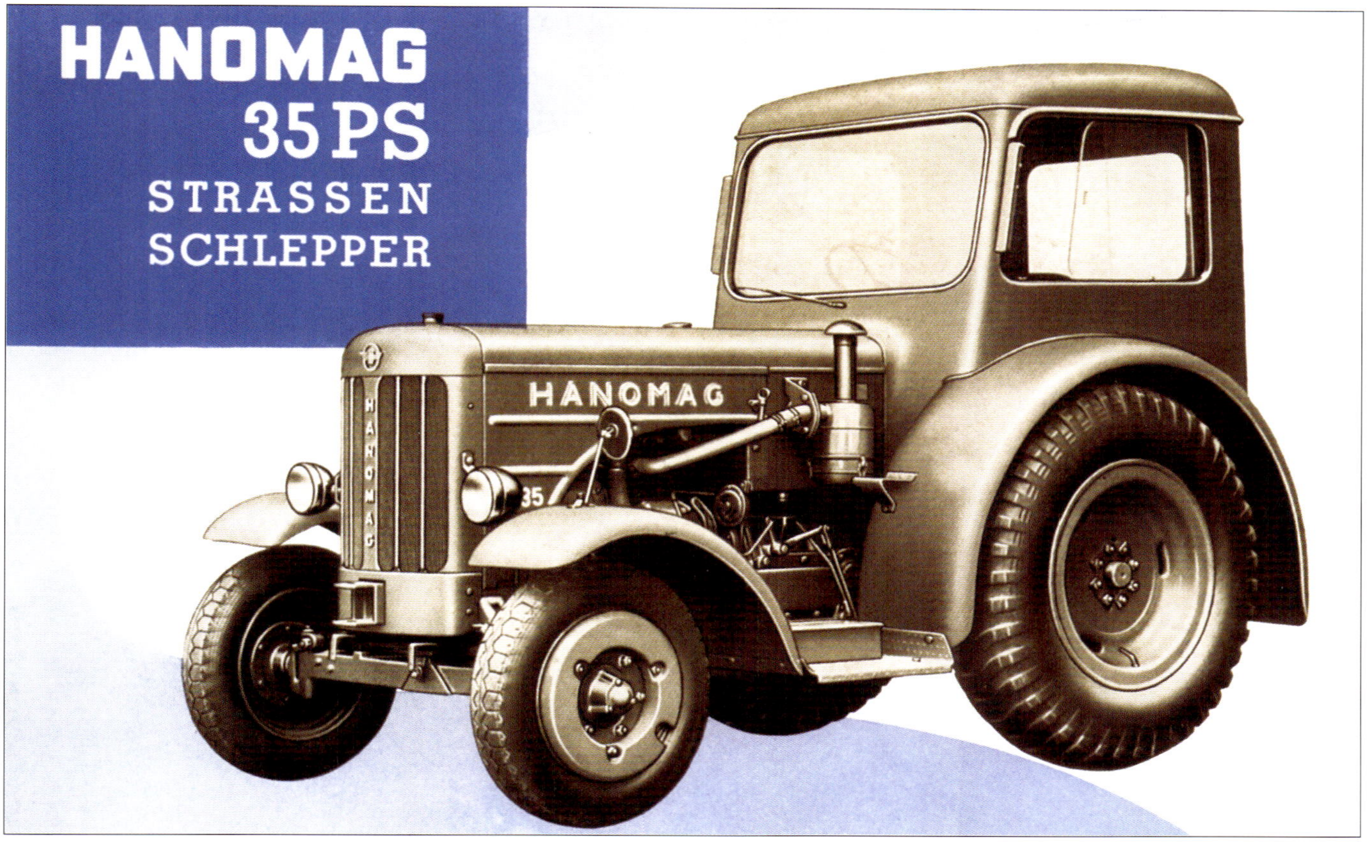
HANOMAG
35 PS
STRASSEN
SCHLEPPER
HANOMAG
35

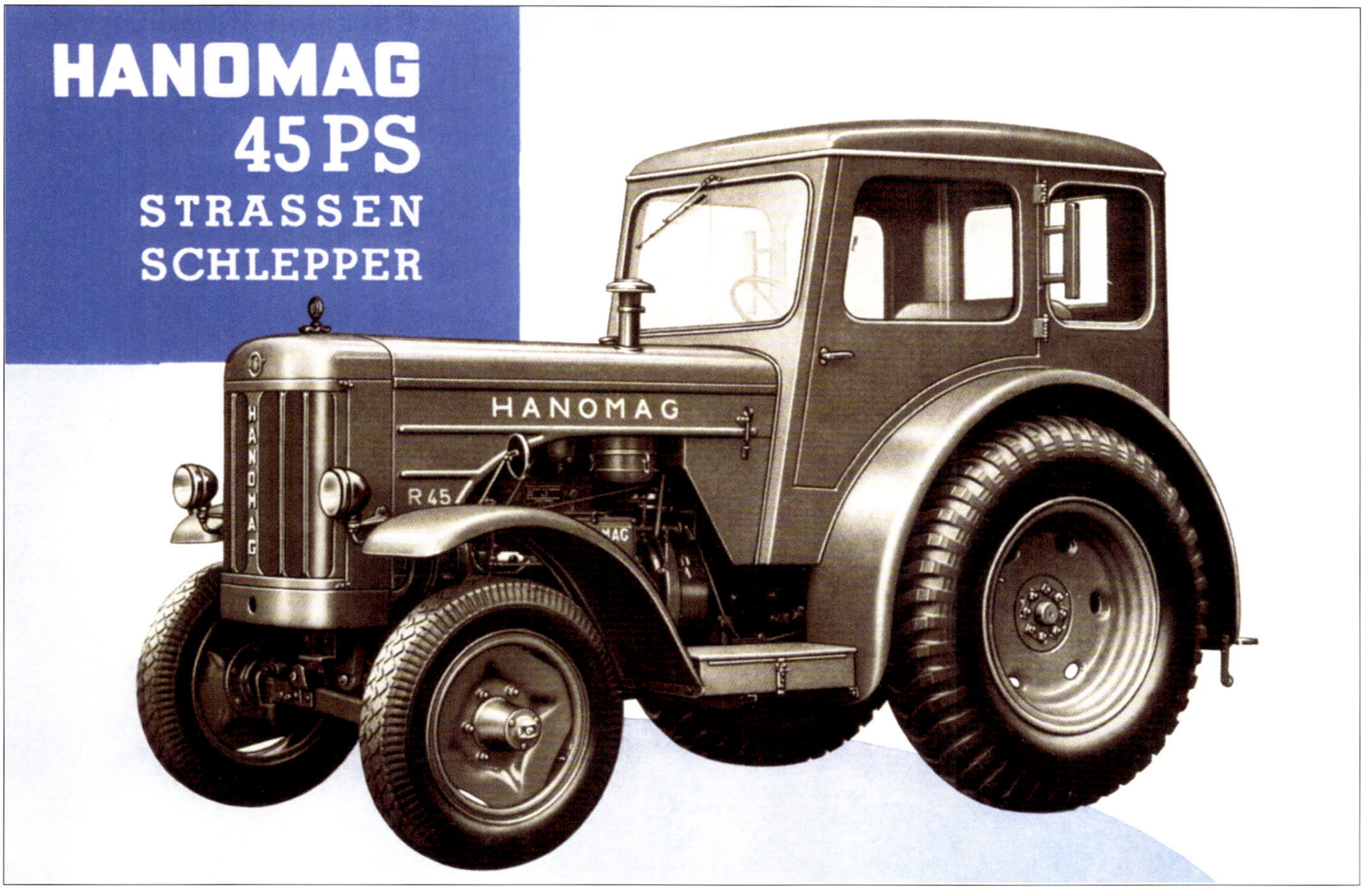
HANOMAG
45 PS
STRASSEN
SCHLEPPER
HANOMAG
R 45

Die Ära der Zweitakt-Dieselmotoren

Zur Abrundung der Angebotspalette im unteren Bereich präsentierten die Hanomag-Werke im Jahr 1953 das leichte und preiswerte Mehrzweckschleppermodell R 12, einen in einteiliger Rahmenbauweise aus Stahlprofilen und hoher Bodenfreiheit konzipierten Tragschlepper, der in der damals aktuellen Wespentaillenbauweise gefertigt wurde. Bereits seit 1951 war an dem Entwurf dieses Tragschleppers, welcher dazu vorgesehen war, als einfache und preisgünstige Alternative selbst kleinsten bäuerlichen Betrieben in der Größenklasse von unter 10 ha, die zu jener Zeit noch häufig auf Zugtiere angewiesen waren, den Einstieg in die Vollmotorisierung zu ermöglichen, gearbeitet worden. In die neugeschaffenen Produktionsanlagen für die Zweitakt-Dieselmotoren hatte die Hanomag die damals sehr beachtliche Summe von mehr als 20 Millionen DM investiert Besonderer Wert wurde bei diesem Entwurf auf die Möglichkeit der Einmann-Bedienung gelegt. Die Wespentaillenbauform mit ihrem langen Radstand und der erheblichen Bodenfreiheit von 425 mm kam dem Anbau von Zwischenachsgeräten am Rahmen sehr entgegen. Der Schlepper besaß eine Sitzbank für zwei Personen, wobei die rechts am Rahmen vorbeigeführte Lenkung den Fahrerplatz auf der rechten Seite vorschrieb. Das nur 800 kg betragende Eigengewicht des Schleppers, das unter anderem auf der nichttragenden und daher leichteren Bauweise des Motors – dessen Gewichtsanteil betrug lediglich 104 kg – beruhte, wurde von dem ventillosen, gebläsegespülten Einzylinder-Zweitakt-Dieselmotor D 611 mit Wasserkühlung und 511 ccm Hubvolumen bewegt, dessen Leistung von 12 PS bei einer Drehzahl von 2200 U/min erzeugt wurde. Im übrigen war dies der erste bei Hanomag nach dem Zweitaktverfahren

Hanomag R 12, Baujahr 1954, noch im Jahr 2002 im Einsatz

HANOMAG Combitrac R12

Der neue 12 PS Allzweck

Ein-Zylinder-Diesel-Zweitaktmotor, Kolbendurchmesser 85 mm, Kolbenhub 90 mm, Hubraum 510 cm³, Dauerleistung 12 PS, Wasserkühlung, Elektrischer Anlasser, Lichtmaschine, Ölbad-Luftfilter

Bereifung, hinten	7-30 AS	8-24 AS
Bereifung, vorn	4,00-16 AS Front	4,00-15 AS Front
Länge über alles	2730 mm	2730 mm
Breite über alles (schmale Spur)	1480 mm	1480 mm
Spurweiten		1250 - 1375 - 1500 mm
Radstand	1800 mm	1800 mm
Höhe bis Oberkante Sitz	1050 mm	1000 mm
Bodenfreiheit, vorn, Vorderachse	415 mm	400 mm
Bodenfreiheit, hinten	425 mm	375 mm
Schwerpunkthöhe über dem Boden	625 mm	515 mm
Ganganzahl	6 Vorwärts-, 2 Rückwärtsgänge	
Geschwindigkeiten	1,45 - 2,7 - 4,6 - 5,75	1,3 - 2,4 - 4,15 - 5,2
	10,85 - 19 km st	9,75-17,5 km/st
Gewicht (mit Wasser, Kraftstoff und Öl)		ca. 800 kg

SERIENMÄSSIGE AUSRÜSTUNG: 2 Polstersitze, Aufstiegbügel, elektrischer Anlasser, elektrische Beleuchtung, Zapfwelle (575 Umdr./min.) Differentialsperre.

SONDERAUSRÜSTUNG Riemenscheibe, Hydr. Kraftheber, gangabhängige Zapfwelle (144-1720 Umdr./min.).

Konstruktive Änderungen und Abweichungen vorbehalten!

Bilderklärungen

1) Schmale Bauweise gute Sicht auf Arbeitsgeräte 2) Dreipunkt-Hydraulik 3) Hauptkraftwelle in der Mitte des Schleppers angeordnet, ermöglicht hydraulisches Heben von Geräten hinten, in der Mitte und vorn 4) Zwei vollwertige Polstersitze für Fahrer und Beifahrer

Hanomag R 12 mit Mähbalken, Baujahr 1954

HANOMAG Combitrac R12

arbeitende Schleppermotor, der von dem neuen Hanomag-Chefkonstrukteur für den Motorenbau, dem aus Österreich stammenden Dr. Ing. Hans Kremser, dem Nachfolger Schargorodskys, entwickelt worden war. Dieses Aggregat zeichnete sich durch sehr kleine Abmessungen aus und arbeitete nach dem Prinzip der Umkehrspülung, wobei die unterstützende Tätigkeit eines Roots-Gebläses für die gute Leistungsausbeute des kleinen Motors mit verantwortlich war.

Aufgrund des im Vergleich zu den Modellen der Mitbewerber recht lauten und zumindest gewöhnungsbedürftigen Motorgeräuschs dieses schnelldrehenden Hochleistungstriebwerks wurde dem kleinen Schlepper schon bald der durchaus zutreffende Spitzname "Ackermoped" verliehen. Dieses helle, nicht kräftigte und damit vermeintlich keine Stärke vermittelnde Geräusch führte unter den eher konservativ eingestellten Landwirten allein schon zu einer subjektiven Ablehnung. Das Sechsganggetriebe mit zwei Rückwärtsgängen war in zwei Schaltgruppen unterteilt und ermöglichte Geschwindigkeiten von 1,45 bis 18,8 km/h. Gegen Aufpreis war eine Zentralhydraulik – in der damaligen Firmenwerbung sprach man wegen der vielen Geräteanbaumöglichkeiten vom Combitrac-System, welches schließlich auf die gesamte Angebotspalette übertragen wurde – mit einem an drei Punkten aufgehängtem Kraftheber sowie Mähwerk und Frontlader erhältlich. Die mittlerweile international genormte Dreipunkthydraulik ermöglichte den Anbau der dazu passenden Bodenbearbeitungsgeräte, wie beispielsweise Pflug, Grubber oder Eggenrahmen. Die Schlepperspurweite war mehrfach veränderbar.

Neben der großen Vielseitigkeit und der überraschend großen Zugkraft des neuen Zweitakt-Dieselschleppers hatten viele Landwirte aber Anlass zu Kritik. Zu Problemen führte der ständige Auswurf von unverbranntem schwarzen Schmieröl durch die Abgasanlage, mit dem neben einer erheblichen Verschmutzung des Traktors und der angebauten Geräte auch eine Geruchsbelästigung einherging. Dieser Mangel war dadurch begründet, weil die Abgastemperatur des Zweitakt-Diesels niedriger lag als beim Viertaktmotor, so dass das zwischen Kolben und Zylinderwand gedrückte Schmieröl nicht mitverbrannt wurde und deshalb mit den Abgasen aus dem Auspuff austrat. Dieses ließ sich zwar durch Einlegung eines höheren Gangs und damit verminderter Drehzahl umgehen, wurde aber in der Praxis – trotz werkseitig eingeleiteter Aufklärungskampagnen – viel zu wenig angewandt. Daher war es nicht weiter verwunderlich, dass dieses mit Kinderkrankheiten behaftete Modell bereits im folgenden Jahr durch eine überarbeitete Ausführung ersetzt wurde.

Das ab 1954 gebaute neue R-12-Modell mit der neuen verrundeten und aktualisierten Motorverkleidung war optisch wesentlich moderner und gefälliger gestaltet. Dieses neue Wespentaillenmodell mit dem hinsichtlich Hubraum und Leistungsabgabe unveränderten D-611-Motor, löste die bisherige Ausführung mit der kantigen Haube ab. Auch das Eigen-

Hanomag R 12 mit Mähbalken, Baujahr 1956

Hanomag R 12, Baujahr 1956

gewicht blieb mit 800 kg konstant und es wurde ein mit dem Vormodell identisches Getriebe eingebaut. Ab 1955 ersetzte man die Doppelsitzbank durch einen gefederten Muschelsitz und einen auf dem linken Kotflügel montierten zusätzlichen Sitzplatz. Eine weitere Neuerung des gleichen Jahres war der Antischlupf-Dreipunkt-Kraftheber, der das Gewicht eines Anbaugerätes zwecks Verbesserung der Zugkraft teilweise auf die Hinterachse des Schleppers verlagern konnte und dadurch eine um bis zu 210 kg erhöhte hintere Achslast erzielte. Ab 1957 wurde dieser 12-PS-Tragschlepper als C 112 bezeichnet und bis 1960 im Fertigungsprogramm belassen.

Erwähnt werden soll an dieser Stelle das Modell R 12 KB (kurze Bauweise), das vor allem auf jene Kunden zugeschnitten war, die keine Zwischenachsgeräte verwenden wollten. Durch die kurze Bauweise (2340 mm Gesamtlänge im Gegensatz zu 2810 mm bei der regulären Ausführung) wurde eine besonders gute Handlichkeit erreicht. 1957 wurde diese Sonderbauart aufgrund zu geringer Nachfrage aus dem Programm genommen.

Im übrigen gab es für Kunden, die keine Zwischenachsgeräte verwenden wollten, mit der Ausführung R 12 KB beziehungsweise R 112 ab 1954 eine um 30 cm gegenüber dem "normalen" R 12 kürzer gehaltene Bauform. Die Nachfrage nach diesem besonders handlichen Modell entsprach allerdings nicht den Vorstellungen, so dass diese Ausführung ab 1957 aus dem Angebot gestrichen wurde.

Der letzte Schlepper mit Einzylinder-Zweitaktmotor war der zwischen 1960 und 1962 gebaute Typ C 115 Greif, dessen Konstruktionsmerkmale sich von seinem Vorgänger wesentlich unterschieden. Dazu gehörte vor allem der Übergang von der Rahmen- zur Blockbauweise sowie das bis zu 20 km/h schnelle, in zwei Schaltgruppen aufgeteilte Sechsganggetriebe von ZF und außerdem der Kraftheber des gleichen Herstellers. Gleichzeitig war die Motorleistung dieses Modells auf 14 PS angehoben worden.

Es bleibt noch zu erwähnen, dass Hanomag zwar auf die Neukonstruktionen von Motoren nach dem Zweitaktprinzip fortan verzichtete allerdings die bereits fertig entwickelte und nach dem gleichen Verfahren arbeitende Zweizylinderausfüh-

Hanomag Combitrac R 12 KB

rung in Form von zwei neuen Modellen auf den Markt brachte. Dabei handelte es sich zunächst um das ab 1955 erhältliche Modell R 24, ein auf der motortechnischen Grundlage des R 12 basierender zweizylindriger Dieseltragschlepper mit Roots-Gebläse mit dem 24-PS-Motor vom Typ D 621. Die Typenbezeichnung des Schleppers wurde ab 1957, analog zu den neuen Hanomag-Modellbezeichnungen, in C 224 geändert. Zu dem ebenfalls recht lauten Motorgeräusch gesellte sich bei diesem Typ noch ein vom Kühlventilator herrührender hoher Pfeifton, was dem Modell den Spitznamen "Düsenjäger" bescherte. Des weiteren füllte seit 1956 der 18 PS starke Tragschlepper R 18 – ebenfalls mit Zweizylindermotor – die zwischen dem R 12 und dem R 24 bestehende Leistungslücke, welcher seit 1957 als C 218 bezeichnet wurde.

Sowohl der so genannte "Düsenjäger" als auch das "Ackermoped" trugen – trotz der innerhalb von zwei Jahren mit über 10000 gefertigten Einheiten durchaus guten Verkaufserfolgen – entscheidend zu dem Absatzrückgang und Kundenschwund der folgenden Jahre bei und schädigten den Markenruf des Unternehmens nachhaltig.

In den Jahren 1954 und 1955 hatte die Hanomag AG mit einem Marktanteil von 13,6 Prozent beziehungsweise 13,1 Prozent den ersten Platz (noch vor Deutz und Fendt) in der deutschen Zulassungsstatistik für Schlepper und Traktoren erobert und damit ihren Höhepunkt erreicht. Ab Mitte der 50er Jahre setzte eine allgemein stark rückläufige Nachfragetendenz ein, die sich vor allem auf das Segment der kleinen Schleppermodelle auswirkte, was ursächlich mit der einsetzenden Abwan-

Hanomag R 12 mit Fahrerdach, Baujahr 1957

derung der Landarbeiter zur Industrie und der damit verbundenen Aufgabe vieler landwirtschaftlicher Kleinbetriebe in Zusammenhang gebracht werden musste. Aber gerade in diesem Bereich hatte sich Hanomag mit dem R 12 besonders stark gemacht mit dem Ergebnis, dass 1955 fast jeder zweite Schlepper des Hauses von einem Zweitaktmotor angetrieben wurde.

Leider aber hatte das ventillose Zweitaktverfahren – wie beschrieben – in der Praxis doch so manche Mängel und Schwachstellen und sich nicht wie vorgesehen bewährt, was zu einer Verärgerung vieler Kunden führte. Die geringe Akzeptanz ihrer Zweitaktmodelle durch die Kundschaft traf Hanomag nach den großen in sie gesetzten Erwartungen und Investitionen wie nicht anders zu erwarten war, sehr schwer. Der Misserfolg dieser Modelle war der Hauptgrund dafür, dass Hanomag ihre oben beschriebene Marktführerschaft schnell wieder verlor. War einmal ein derartiger Vertrauensverlust eingetreten, konnte dieser in der Regel nicht so schnell wieder ausgeräumt werden. Deshalb wendeten sich viele Kunden der selbstverständlich wachsamen Konkurrenz zu, für die dieser Misserfolg ein gefundenes Fressen war. Die daraufhin wieder stärker forcierten Viertaktmodelle mit höherer Leistung aber konnten die Absatzeinbrüche nicht auffangen.

	R 12	R 12	C 112	C 115 Greif	R 18	R 24	C 224
Produktionszeitraum	1953-1954	1954-1957	1957-1960	1960-1962	1956-1957	1955-1957	1957-1962
Motorleistung (PS)	12	12	12	14	18	24	24
Hubraum (cm³)	511	511	511	511	1021	1021	1021
Anzahl der Zylinder	1	1	1	1	2	2	2
Anzahl der Gänge	6+2	6+2	6+2	6+2	6+2	6+2	6+2
Eigengewicht (kg)	800	800	820	1075	1300	1440	1550
Höchstgeschwindigkeit	18,8 km/h	19,0 km/h	19,0 km/h	20,0 km/h	18,5 km/h	19,5 km/h	19,9 km/h

Hanomag Tragschlepper C 112

Batterie	❶	⓴	Getriebe-Schaltwelle
Kühlwasser-Einfüllstutzen	❷	㉑	Schalthebel für Getriebe
Thermostat	❸	㉒	Getriebegehäuse
Einspritzdüse	❹	㉓	Arbeitszylinder für Hydraulik
Kraftstoff-Einfüllstutzen	❺	㉔	Kurbelwelle
Kraftstoffbehälter	❻	㉕	Ölwanne mit Ölablaßschraube
Schalthebel für Hydraulik	❼	㉖	Achsschenkel
Lenkung	❽	㉗	Spurstange
Schalthebel für Hydraulik	❾	㉘	Vorderachse (Rohrachse)
Schalthebel für Zapfwelle	❿	㉙	Hydraulik-Pumpe
Schalthebel für Differentialsperre	⓫	㉚	Vordere Zug- und Druckvorrichtung
Tellerrad	⓬	㉛	Scheinwerfer
Antriebsritzel	⓭	㉜	Ventilator
Zapfwelle	⓮	㉝	Luftaufnehmer
Untere Zugvorrichtung	⓯	㉞	Kolben
Bremsbeläge	⓰	㉟	Kühler
Antriebswelle	⓱	㊱	Einspritzpumpe
Ausgleichgetriebe (Differential)	⓲	㊲	Kraftstoffhahn
Vorgelegewelle	⓳		

Hanomag Tragschlepper C 112

Hanomag C 112, Baujahr 1958

Hanomag C 112, Baujahr 1959

Hanomag C 115 Greif, Exportausführung, Baujahr 1960

Hanomag C 115 Greif, Baujahr 1962

Hanomag Combitrac R 18

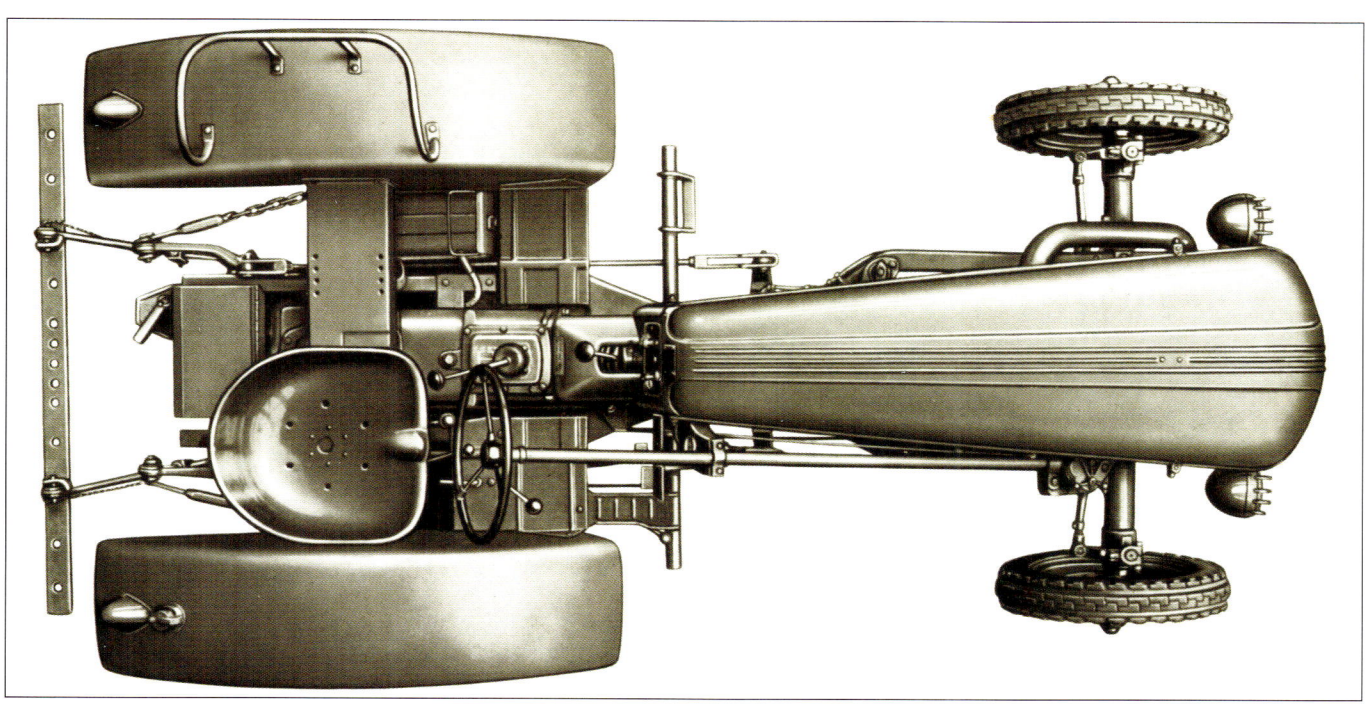

HANOMAG R24 Combitrac

TECHNISCHE EINZELHEITEN

Zweizylinder-Zweitakt-Dieselmotor mit Gebläsespülung

Motorleistung	24 PS
Hubraum	1021,4 cm³
Drehzahlbereich	600—2200 U/min
Kühlsystem	moderne Thermostat-Zweikreiskühlung
Bereifung, vorn	5,50—16 AS Front
Bereifung, hinten	8—36 AS

Serienmäßig Speichenräder, auf Wunsch Scheibenräder

Länge über alles	3100 mm
Breite über alles (schmale Spur)	1600 mm
Radstand	1960 mm
Spurweiten	1250 / 1375 / 1500 mm
Bodenfreiheit	420 mm
Ganganzahl	6 Vorwärtsgänge 2 Rückwärtsgänge
Geschwindigkeiten, vorwärts	1,65 / 3,05 / 4,75 / 6,75 / 12,5 / 19,5 km/st
rückwärts	2,15 und 8,75 km/st
Zapfwellendrehzahl	578 U/min
Eigengewicht	1360 kg

Serienmäßige Ausrüstung: Elektr. Beleuchtung; elektr. Anlasser; obere Zugvorrichtung für Straßenbetrieb; vordere Zug- und Druckvorrichtung; Ackerschiene; Zapfwelle; Sitzbank mit 2 Polstersitzen; Öldruckmesser; Anschluß für Anhängerbeleuchtung, Differentialsperre; Aufstiegbügel; Einzelradfederung, vorn; Vorderradkotflügel; 1 Satz Werkzeuge.

Sonderausrüstung: Fahrerdach mit Windschutzscheibe und Scheibenwischer; elektr. Winkeranlage; elektr. Blinkanlage; verbreiterte Anhängeschiene; elektr. Betriebsstundenzähler; kupplungsunabhängige Zapfwelle; Fernthermometer; Rückscheinwerfer; Riemenscheibe; Zentral-Hydraulik; Seilwinde; Gitterräder; Bereifung 9—36 AS und 10—28 AS.

Alle HANOMAG-Schlepper erhalten eine Kunstharzlackierung, die größte Widerstandsfähigkeit, Kratztestigkeit und spiegelnden Hochglanz besitzt.

Konstruktive Änderungen und Abweichungen vorbehalten.

HANOMAG C 224
Combitrac

STANDARDAUSRÜSTUNG:
Vollständige elektr. Licht- und Anlasseranlage; Anschluß für Anhängerbeleuchtung; Zapfwelle; Differentialsperre; Fußbremse mit Einzelradbremsung; Handbremse; obere Zugvorrichtung; vordere Zug- und Druckvorrichtung; gefederte Vorderachse; Ackerschiene; Parallelogramm-Schwebesitz; Beifahrersitz auf Kotflügel; Aufstiegtritt, Hand- und Fuß-Gas, Oldruckmesser; 1 Satz Werkzeuge.

SONDERAUSRÜSTUNG:
Zentralhydraulik; **Antischlupf**; Frontlader; Mähwerk; Motorzapfwelle mit Doppelkupplung; Fahrerdach mit Windschutzscheibe und Scheibenwischer; Blinkanlage; Vorderradkotflügel; elektr. Betriebsstundenzähler; Fernthermometer; Bereifung hinten 9–36 AS und 10–28 AS.

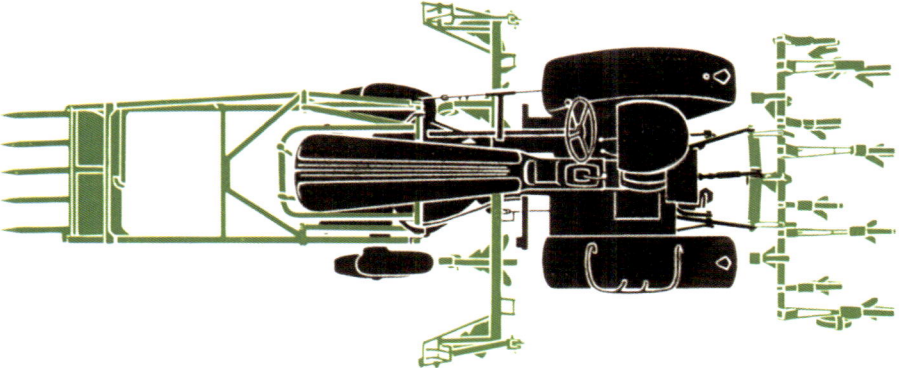

HANOMAG-Combitrac — die ideale Arbeitseinheit von Schlepper und Gerät

Bewährte Technik unter moderner Verkleidung

Die Hanomag-Traktorenmodelle des Jahres 1957 hatten, wie bereits in den vorherigen Kapiteln angeführt, neue dreistellige Typenbezeichnungen erhalten, wobei die erste Ziffer die Anzahl der Zylinder, die beiden letzten Zahlen die Motorleistung in PS ausdrückte und waren nun von ihren Vorgängern durch abgerundete, gefälliger wirkende Kühler- und Motorverkleidungen zu unterscheiden. Die schweren Radschlepper hingegen wurden optisch nicht verändert und behielten ihre kantigen Hauben. Hanomag stellte nun wieder seine leistungsmäßig angehobenen und nach dem Baukastensystem gefertigten Viertaktschleppermodelle in den Vordergrund. Zur gleichen Zeit wurde mit insgesamt 19 für den inländischen Markt angebotenen Typen das umfangreichste Schlepperbauprogramm von Hanomag angeboten. Es erstreckte sich vom 12 PS starken C 112 bis zum schweren 55 PS Radschlepper R 455 S. Doch auch die große Anzahl an unterschiedlichen Typen und Ausführungen konnten nicht über die Tatsache hinwegtäuschen, dass das ab 1. April 1955 unter dem Namen "Hanomag AG" firmierende Unternehmen in eine Krise geraten war, die hauptsächlich durch die in der Praxis nicht bewährten Zweitakt-Dieselmotoren ausgelöst wurde. Der über zwei Jahre gehaltene erste Platz in der Zulassungsstatistik musste im Jahr 1956 an Deutz abgegeben werden, und ein Jahr darauf erfolgte ein ganz gravierender Absturz auf den fünften Rang, weil von Hanomag weit über 50 Prozent weniger Schlepper gegenüber dem Stand vor zwei Jahren zugelassen wurden.

An dieser Stelle seien noch einige Ausführungen zu dem bereits kurz erwähnten "Combitrac-System" von Hanomag gestattet. Dieser in der damaligen Firmenwerbung hervorgehobene Begriff ging von der Tragschlepper-Idee aus und

Hanomag R 217, Baujahr 1957

Hanomag Combitrac R 217/R217E

Bewährte Technik unter moderner Verkleidung **109**

Hanomag R 217, Baujahr 1958

Hanomag R 217 S, Baujahr 1960

wurde zunächst zwar nur für die Zweitakttypen verwendet, später aber auf das gesamte Schlepperprogramm bis zum R 35 übertragen. Diese werbliche Initiative, die herausstellen sollte, dass ein Hanomag-Schlepper über seine Funktion als Zugmaschine hinaus mit seinen vielen, auf den Einsatzzweck abgestimmten Anbaugeräten eine in sich geschlossene Einheit und Mehrzweckmaschine bildete, hatte auch praktische und für den Landwirt durchaus positive Resultate aufzuweisen.

Ausgehend von gewissen Grundeigenschaften des Schleppers selbst, die sich in der Forderung nach hoher Zugkraft, einem geringen Eigengewicht und einer optimalen Gewichtsverteilung auf die beiden Achsen äußerte, musste die Motorkraft über Zapfwelle und Riemenscheibe abgegeben werden können. Zahlreiche Landmaschinenhersteller stimmten ihre Geräteprogramme auf die Eigenschaften der Hanomag-Schlepper ab, so dass mit Hilfe der genormten Anschlüsse für nahezu jede mögliche Arbeit ein perfekt aufeinander abgestimmtes, sozusagen maßgeschneidertes Arbeitsgerät zur Verfügung stand, das weitgehend vollmotorisierte und damit rationelle und kostensparende Arbeitsabläufe gewährleistete.

Der von 1957 bis 1959 gebaute Typ R 217 entstand aus dem Modell R 16, von dem er sich – abgesehen von der verrundeten Motorhaube – unter Verwendung des bisherigen zweizylindrigen D14-Motoraggregats, vor allem durch die um ein PS höhere Motorleistung unterschied. Mit dem von 1959 bis 1962 gebauten Modell R 217 S erfuhr dieser Schlepper, abgesehen von geringen Detailverbesserungen, eine erneute Leistungsanhebung um weitere zwei Pferdestärken. Dieses Modell bildete dann den Abschluss der mit dem R 16 begonnenen Zweizylinderbaureihe.

Analog dazu bezeichnete das 1957er Hanomag-Typenprogramm den bisherigen R 19 als R 217 E, wobei das angehängte "E" in der Typenbezeichnung nicht ganz der neuen Nummernterminologie entsprach. Während aus dem bisherigen R 22 nun der um zwei PS stärkere, mit dem Dreizylindermotor D 21 bestückte R 324 entstand, war bei der Bezeichnungsänderung des stärkeren R 27 in R 324 E das gleiche zu beobachten. Diese Schlepper waren mit einem Fritzmeyer-Wetterschutzverdeck mit klappbarer Panorama-Wind-

Hanomag R 217 S mit Fahrerdach, Baujahr 1960

schutzscheibe ausrüstbar. Die zwischen 1959 und 1962 lieferbare Ausführung R 324 S verfügte infolge durchgeführter Änderungen an der Kurbelwelle des Motors über einen gleichmäßigeren Zündabstand und damit verbesserte Laufruhe als das Vormodell. Ebenso zeichnete sich dieser Typ durch eine eingebaute Motorzapfwelle aus, was konstruktiv bedingt zu einer geringfügigen Verlängerung des Halbrahmens führte.

Ab März 1961 ging mit dem gegenüber dem R 324 S verbesserten Modell R 332 Granit ein weiterer mit Namen bezeichneter Schlepper in Serie. Er besaß einen um fünf PS stärkeren und gleichzeitig höherdrehenden Motor, sowie einen leistungsfähigeren Kraftheber, vergrößerten Kraftstofftank und die ZF-Gemmer-Lenkung. Auch die Motorhaube wies eine stärkere Wölbung auf, wodurch weiterer Raum zur Unterbringung des Motors gewonnen werden konnte. Die neue Hanomag-Pilot-Hydraulik, die am Schluss dieses Kapitels beschrieben werden soll, war ein weiteres fortschrittliches Element, das zum Einbau gelangte. Das in zwei Schaltgruppen aufgeteilte Zehnganggetriebe ermöglichte Fahrgeschwindigkeiten zwischen 1,05 und 20 km/h.

Des weiteren seien noch die Modelle R 35 (neu R 435) und R 35/45 (neu R 435/-45) angeführt. Sie blieben leistungsmäßig unverändert und unterschieden sich von

Hanomag Combitrac R 324/324E

Hanomag R 324 A, Baujahr 1957

	R 217	R 217 S	R 324	R 324 E	R 324 S	R 332 Granit	R 435	R 435/45	R 442 Brillant	R 442/50 Brillant
Produktionszeitraum	1957-1959	1959-1962	1957-1959	1957-1959	1959-1962	1961-1962	1957-1960	1957-1960	1960-1962	1960-1962
Motorleistung (PS)	17	19	24	27	27	32	35	35/45	42	42/50
Hubraum (cm³)	1400	1400	2099	2099	2099	2799	2099	2799	2799	2799
Anzahl der Zylinder	2	2	3	3	3	3	4	4	4	4
Anzahl der Gänge	8+2	10+2	10+2	10+2	10+2	10+2	10+2	10+2	10+2	10+2
Eigengewicht (kg)	1170	1300	1520	1520	1855	2190	2799	2799	2799	2799
Höchstgeschwindigkeit	19,4 km/h	19,8 km/h	19,5 km/h	19,5 km/h	20,0 km/h	20,0 km/h	19,8 km/h	18,0 km/h	19,4 km/h	19,9 km/h

den bisherigen Typen bis auf Detailverbesserungen im Grunde genommen nur durch die aktualisierte abgerundete Motorverkleidung. Der bis 1960 gefertigte R 435 besaß beispielsweise auf Wunsch eine Doppelkupplung für das in zwei Schaltgruppen gegliederte zehngängige Getriebe mit zusätzlichen Rückwärts- und integrierten Kriechgängen mit einem Geschwindigkeitsbereich von 0,85 bis 19,8 km/h. Diese Kupplung ließ das Schalten und Anhalten des Schleppers bei laufender Zapfwelle zu. Auch in diesem Fall musste der Halbrahmen und damit der Radstand um wenige Zentimeter verlängert werden.

Eine technische Besonderheit war der R 435/45, dessen Motor mit einem Roots-Aufladegebläse bestückt war. Das ohne Gebläse 35 PS leistende Triebwerk konnte nach deren Zuschaltung mittels eines Handhebels vom Fahrersitz aus bei unveränderter Drehzahl durch die vergrößerte Kraftstoff-Einspritzmenge eine Mehrleistung von zehn PS mobilisieren. Wie schon der R 35/45 zählte dieser Schlepper im Verhältnis zu seiner hohen Leistung zu den leichtesten Schleppern seiner Klasse. Mit der eingebauten Motorzapfwelle verfügte dieser so genannte "Mähdrusch-Schlepper" über ausreichende Leistungsreserven, um sich besonders unter Zuschaltung des Gebläses gut zum Ziehen schwerer, zapfwellengetriebener Geräte zu eignen. Viele dieser Fahrzeuge wurden bereits ab Werk mit einem Fritzmeyer-Verdeck mit klappbarer Panoramascheibe ausgerüstet.

Das seit Oktober 1960 gebaute Modell R 442 Brillant stellte die technisch weiter optimierte Entwicklungsstufe des R 435 bei einer auf 42 PS angehobenen Motorleistung dar. Die weiteren Unterscheidungsmerkmale waren ein vergrößerter Radstand und Kraftstofftank, die ZF-Gemmer-Lenkung und ein leistungsfähigerer Kraftheber. Darüber hinaus hatte die Frontverkleidung der Motorhaube eine weitere Wölbung erfahren, und der für den Motor zur Verfügung stehende Platz war nun geräumiger geworden. Das in zwei Schaltgruppen unterteilte Zehnganggetriebe war auf Kundenwunsch mit Schnellgang für 25 km/h Höchstgeschwindigkeit zu bekommen. Als Eigengewicht brachte der Brillant – je nach Ausstattung – zwischen 2180 und 2255 kg auf die Waage.

Zur gleichen Zeit wurde mit dem Modell R 442/50 Robust – als drittes Modell unter den mit Namen bezeichneten Schleppern – auch der bisherige mit einem zuschaltbaren Roots-Gebläse bestückte R 435/45 leistungsmäßig auf 42 beziehungsweise 50 PS angehoben. Zu den Sonderausstattungen zählten unter anderen die Riemenscheibe, eine Lenkradschaltung und eine von den Hanomag-Werken entwickelte und patentierte Pilot-Regelhydraulik, welche – unabhängig von der Bodenbeschaffenheit – einen gleichmäßigen Tiefgang des Anbaupfluges, selbst bei Unebenheiten und Hanglagen, gewährleistete. Die Fertigung dieses Schleppermodells endete im September 1962.

Hanomag R 324, Baujahr 1959

Hanomag R 324 SA, Baujahr 1957

Hanomag R 324 S mit Fahrerdach, Baujahr 1961

Hanomag R 324 S mit Frontlader, Baujahr 1962

Hanomag R 228 Exportausführung, Baujahr 1962 (oben), Hanomag R 35 S (Nato-Ausführung), Baujahr 1957 (Mitte) und Hanomag R 435 S (ehemals Nato-Ausführung), Baujahr 1957 (unten)

HANOMAG *Combitrac* R 435

STANDARDAUSRÜSTUNG

Vollständige elektr. Licht- und Anlasseranlage, Anschluß für Anhängerbeleuchtung, Zapfwelle, Differentialsperre, Fußbremse mit Einzelradbremsung, Handbremse, obere Zugvorrichtung, vordere Zug- und Druckvorrichtung, breite Ackerschiene, Parallelogramm-Schwingsitz, gefederte Vorderachse, Hand- und Fuß-Motorregulierung, Oldruckmesser, 1 Satz Werkzeuge, Bereifung vorn 5,50—16 AS Front, hinten 11—28 AS.

SONDERAUSRÜSTUNG

Hydraulischer Kraftheber, **Antischlupf**, Frontlader, komb. Mäh- und Riemenscheibenantrieb, Mähwerk, Kriechgang-Untersetzer, Motorzapfwelle (kupplungsunabhängig), Fahrerdach mit Windschutzscheibe und Scheibenwischer, elektr. Blinkanlage, Vorderradkotflügel, Seilwinde, Polstersitz, Gitterräder hinten, Bereifung vorn 6,00—20 AS Front, hinten 9—42 AS, 11—38 AS oder 13—30 AS.

HANOMAG - Combitrac — die ideale Arbeitseinheit von Schlepper und Gerät

Bewährte Technik unter moderner Verkleidung

Hanomag R 435/45 mit Fahrerdach, Baujahr 1960 (oben), Hanomag R 435, Baujahr 1960 (Mitte) und Hanomag R 435 mit Fahrerdach, Baujahr 1959 (unten)

Hanomag R 440, Exportausführung, Baujahr 1961 (oben), Hanomag R 442 Brillant, Baujahr 1961, war bis 1993 im Einsatz (Mitte) und Hanomag R 440 Exportausführung mit Fahrerdach, Baujahr 1960 (unten)

Hanomag R 332 Granit, Baujahr 1962

Hanomag R 332 Granit mit Fahrerdach, Baujahr 1962

Hanomag Robust 442/50 mit festem Fahrerhaus, Baujahr 1962

Hanomag Perfekt 400, Exportausführung mit Frontlader, Baujahr 1964

Die 60er Jahre – Wirbelkammer-Dieselmotoren und kantige Hauben

Im Jahr 1960 konnten die Hanomag-Werke auf ein 125jähriges Bestehen zurückblicken, und ab Juni des gleichen Jahres wies das geänderte Firmensignet, das aus einer Kombination des bisherigen Hanomag-Firmenzeichens mit dem bogenförmigen Emblem der Rheinstahlwerke bestand, auf die bereits seit 1952 als Tochtergesellschaft des Essener Montan-Konzerns "Rheinstahl" bestehende Zugehörigkeit des Unternehmens hin. Im November lief der 200000ste Schlepper von den Fabrikationsbändern.

Ab 1962 wurden die bisherigen, mit Vorkammer-Dieselmotoren bestückten Schlepper-Typen Granit und Brillant auf neue, nach dem Wirbelkammerverfahren arbeitende Motoren umgestellt, was sich in der höheren Leistung in Verbindung mit niedrigeren Kraftstoffverbräuchen ausdrückte. Diese mit neuen Zylinderköpfen und Kolben konstruierten Motoren arbeiteten nach dem Ricardo-Brennkammer-Verfahren, und die jeweiligen Radschleppertypen erhielten den Zusatz "CR" in der Typenbezeichnung. Im gleichen Jahr fand auch die neunjährige, eher glücklose Epoche der Zweitaktmodelle im Bereich der Radschlepper ihren endgültigen Abschluss.

Mit dem Typ Perfekt 300, dem kleinsten in Blockbauweise konstruierten Modell, fand bei Hanomag der Einstieg in das Schlepper-Programm statt. Der nach dem Wirbelkammerverfahren arbeitende Zweizylinder-Viertaktmotor des Modells D 14 CR konnte mit 25 PS aus 1400 ccm Hubraum bei 2400 U/min im Vergleich zu früheren Modellen eine sehr hohe Leistung zur Verfügung stellen. Gänzlich neu war das mit zwei Schalthebeln ausgebildete Dreiganggetriebe, womit zwischen den Schaltgruppen schnell und langsam gewählt werden konnte. Damit standen dem Fahrer insgesamt zwölf Vorwärts- und vier Rückwärtsgänge – bei Geschwindigkeiten zwischen 1,75 und 26 km/h – zur Verfügung. Der Perfekt 300 besaß eine an Querblattfedern befestigte Teleskopachse und infolge seines langen Kupplungsgehäuses einen mit 2030 mm recht be-

Hanomag Granit 500, Baujahr 1965

Hanomag Granit, Baujahr 1962

Hanomag Granit 500 mit Allwetterverdeck, Baujahr 1962

Hanomag Brillant 600

Hanomag Brillant 600 mit Allwetterverdeck, Baujahr 1965

achtlichen Radstand, welcher nicht nur den Anbau von Zwischenachsgeräten ermöglichte, sondern den Aufstieg des Fahrers wie bisher von hinten und auch von der Seite zuließ. Während die Vorderachse in sieben unterschiedlichen Spurweiten zwischen 1250 und 1950 mm einstellbar war, konnte die Hinterachse sechsfach verändert werden. Der Perfekt 300 trat die Nachfolge des R 217 S an, dem er ein erhebliches Leistungsplus voraus hatte. Von der Konzeption her ersetzte dieses neue Schleppermodell den früheren Zweitakttyp C 224, von dem er sich aber durch den bewährten und zuverlässigen Viertaktmotor unterschied.

Gleichfalls neu vorgestellt wurden die Modelle Granit 500 und Brillant 600, die unter Verwendung der neuen Wirbelkammermotoren die Nachfolge der Typen R 332 Granit und R 442 Brillant antraten. Durch Erhöhung der Nenndrehzahl wurden beim dreizylindrigen Granit zunächst 38, ab 1963 40 PS und beim Vierzylindermodell Brillant 50 PS Leistung aus den Antriebsaggregaten herausgeholt. Beide Schlepper waren in der Halbrahmenbauweise konstruiert und wiesen gleichfalls den schon beim Modell Perfekt 300 beschriebenen längeren Radstand auf. Ab 1964 wurden beide Typen mit neugestalteten Seitenblechen der Motorverkleidung versehen.

Mit dem Erscheinen des technisch optimierten Brillant 600 wurde sowohl das Modell R 442 als auch der mit Auflademotor bestückte R 442/50 ersetzt, weil der neue Vierzylindermotor nun die gleiche Leistung wie der alte Ladermotor erzeugen konnte. Aufgrund seiner sehr vielseitigen Verwendbarkeit als Allzweckschlepper und wegen seiner Leistungsstärke blieb der Brillant 600 bis 1967 im Verkaufsprogramm des Unternehmens.

Auf der 1963 in Paris stattfindenden Internationalen Landwirtschaftsausstellung war auf dem Ausstellungsstand der Rheinstahl-Hanomag AG erstmals ein neuer, als Perfekt 400 bezeichneter, 32 PS starker Tragschlepper zu sehen, der die bestehende Leistungslücke zwischen dem Perfekt 300 und dem Granit 500 abdecken sollte. Das Besondere an diesem, konzeptionell dem Perfekt 300 sehr ähnlichen, Schlepper war der vierzylindrige Viertakt-Wirbelkammer-Dieselmotor des Typs D 301 R 1 von Borgward. Dieses zur Zeit des 1961 erfolgten Konkurses von Borgward noch in der Entwicklung befindliche, ursprünglich als Antrieb für Personenkraftwagen vorgesehene 1,8-Liter-Aggregat, hatte Hanomag im Dezember des gleichen Jahres aus der Borgward-Konkursmasse übernommen und mit Hilfe des für dessen Entwicklung zuständigen Ingenieurs, der gleichzeitig zur Rheinstahl-Hano-

	Perfekt 300	Perfekt 400	Granit 500	Brillant 600	Perfekt 300	Perfekt 400	Granit 500	Granit 500	Granit 500 E
Produktionszeitraum	1962-1964	1963-1964	1962-1966	1962-1967	1964-1968	1964-1968	1966-1967	1968-1970	1967-1970
Motorleistung (PS)	25	32	38/40	50	25/27	32	40	40	48
Hubraum (cm³)	1400	1797	2099	2799	1797	1797	2099	2126/2356	2356
Anzahl der Zylinder	2	4	3	4	4	4	3	3	3
Anzahl der Gänge	6+2	6+2	10+2	10+2	6+2	6+2	9+3	9+3	9+3
Eigengewicht (kg)	1710	1650	2340	2585	1695	1770	2100	2070	2070
Höchstgeschwindigkeit	26,0 km/h	26,0 km/h	23,0 km/h	24,5 km/h	24,6 km/h	26,0 km/h	25,6 km/h	27,2 km/h	25,3 km/h

mag AG überwechselte, überarbeitet und im Hinblick auf den beabsichtigten Einbau sowohl in Schlepper als auch in Hanomag-Lieferwagen zur Serienreife gebracht.

Dieses Tragscheppermodell besaß erstmals eine abnehmbare Blechverkleidung, mit der der Anbauraum zwischen den Achsen optisch abgedeckt wurde. Spätere Modelle erhielten ebenfalls diese Verkleidungsbleche.

Mit dem Übergang zur kantigen, funktionalen Motorhaubenform wartete Hanomag ab Oktober des Jahres 1964 mit einer erneuten, sehr modern wirkenden, Designänderung auf. Es waren die in Blockbauweise konstruierten Modelle Perfekt 301 und 401, die eine neue, niedrigere Haube und einen neugestalteten Fahrerplatz mit geänderter, ergonomisch ausgearbeiteter Sitzposition und neuem Armaturenbrett erhalten hatten und als Zug- und Tragschlepper für Einmann-Bedienung ausgelegt waren. Beide Schlepper besaßen eine hohe Zugkraft und boten die Möglichkeit des Anbaus von Zwischenachsgeräten. Die motortechnischen Daten der beiden Schlepper hingegen blieben unverändert. Die Käufer dieser Traktormodelle konnten auf ein sehr umfangreiches Zubehörprogramm – vom Frontlader bis zu Zusatzgewichten – zurückgreifen. Während beim Perfekt 400 aus dem Vierzylindertriebwerk 32 PS herausgeholt wurden, leistete der kleinere Bruder zunächst 25 PS, und ab dem Jahr 1967 erhöhte man die Leistung auf 27 PS. Die Fertigung beider Typen endete 1968.

Auch der 1966 vorgestellte, in Blockbauweise konstruierte neue Granit 500 erhielt die bereits von den Perfekt-Schleppern her bekannte eckige Haube, die infolge des größeren Motors etwas wuchtiger wirkte. Unter dieser arbeitete anfangs noch der 40 PS starke dreizylindrige D 21CR-Motor, der über 2099 ccm Hubvolumen verfügte. Gegenüber seinem Vorgänger war der Granit 500 mit einem neu entwickelten, in drei Schaltgruppen gegliederten Leichtschaltgetriebe ausgestattet, das die Anpassung an die unterschiedlichen Arbeitsgeschwindigkeiten verbesserte. Damit konnte der Fahrer zwischen insgesamt 18 Vorwärts- und 6 Rückwärtsgängen in den Geschwindigkeitsbereichen von 1,7 bis 25,6 km/h wählen. Außerdem war der Kraftheber des 2100 kg schweren Schleppers verbessert und verstärkt worden, und ab 1967 erfolgte die Ausrüstung mit zwei Zapfwellen unterschiedlicher Drehzahlbereiche, wobei die höhere Drehzahl insbesondere beim Antrieb von Pumpen und Gebläsen von Vorteil war. Zwischen 1968 und 1970 wurde der an der Kühlermaske und an den Seitenblechen den größeren Modellen des Hanomag-Programms optisch angepasste Granit 500 mit dem Dreizylinder-Kurzhubmotor D 131 R gleicher Leistung einer neuen Motoren-Baukastenreihe ausgerüstet, auf welche im folgenden Kapitel noch eingegangen werden soll. Zwischen 1969 und 1970 gelangte der etwas hubraumstärkere D-132-R-1-Motor gleicher Leistung in diesen Schlepper zum Einbau. Ebenso wurde im Laufe des Jahres 1968 die bisher verwendete Pilot-Regelhydraulik durch eine verbesserte Ausführung ersetzt.

Von Juni 1967 bis zum Jahr 1970 füllte der mit dem neuen 48-PS-Kompaktmotor bestückte Granit 500 E die leistungsmäßige Lücke zwischen dem Granit 500 mit 40 PS und dem neuen 58-PS-Modell Brillant 600. Für den Granit 500 E stand ab 1968 ebenfalls die verbesserte Regelhydraulik des Typs Pilot III zur Verfügung und auch dieser Schlepper wurde mit zwei Zapfwellen unterschiedlicher Geschwindigkeiten ausgerüstet. Der Granit 500 E war zugleich das stärkste Tragschleppermodell aus dem Hause Hanomag.

Hanomag Brillant 600 mit Frontlader und Allwetterverdeck, Baujahr 1963

Hanomag Brillant 600, Baujahr 1963 (oben), Hanomag Brillant 600, Baujahr 1965 (Mitte) und Hanomag Brillant 600, Baujahr 1963 (unten)

Hanomag Perfekt 300, 25 PS

Hanomag Perfekt 400 mit Frontlader und Allwetterverdeck, Baujahr 1970 (großes Foto) und Hanomag Perfekt 400 mit Fahrerdach, Baujahr 1968

Hanomag Perfekt 400

Hanomag Granit 500 mit Mähbalken und Allwetterverdeck, noch 1992 im Einsatz (oben links), Hanomag Granit 500, Baujahr 1970 (oben rechts), Hanomag Granit 500, rote Exportausführung, Baujahr 1967 (Mitte rechts) und Hanomag Granit 500 bei der Feldarbeit (unten)

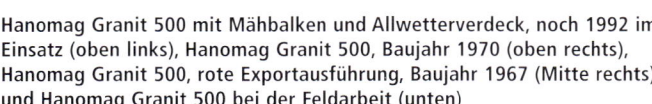

Hanomag Granit 500 E mit Fahrerdach, Baujahr 1969

Hanomag Granit 500 E, Baujahr 1969

Granit, Brillant und Robust – die letzten Hanomag-Schlepperkonstruktionen

Währenddessen hatte Hanomag in den Jahren 1965 bis 1967 alle Anstrengungen auch in finanzieller Hinsicht – man sprach von einem Kostenaufwand von rund 100 Millionen DM für Forschungs- und Entwicklungsarbeiten bis zum Jahr 1969 – unternommen, um sowohl eine neue Motorenbaureihe als auch neue Getriebe auf die Beine zu stellen und zur Serienreife zu bringen. Mit deren Hilfe konnte eine neue, leider aber auch die letzte Schlepperbaureihe von Hanomag realisiert werden. Die Rheinstahl-Hanomag AG erreichte zu jener Zeit mit mehr als 11000 Mitarbeitern den höchsten je in Friedenszeiten erreichten Personalstand. Für die Serienfertigung der neuen Antriebsaggregate war eigens eine neue Fertigungsstraße errichtet worden, die damals zu den modernsten der Welt zählte. Mit dieser Flucht nach vorn hatte das Unternehmen einen letzten Versuch gestartet, den stagnierenden Marktanteil im Traktorenbereich zu einem besseren Ergebnis zu wenden. Es stand um dieses Segment nicht gerade günstig, denn die letztlich neu entwickelten leichteren Modelle hatten bei der Kundschaft zwar Interesse erweckt, was aber leider keine nachhaltigen Absatzsteigerungen hervorrufen konnte.

Getriebe und Motoren, die beiden neuen Baukomponenten, wurden im Herbst 1967 in den neuen Schleppermodellen Brillant 600, Brillant 700 und Robust 900 vorgestellt. Die neuen grünblau lackierten Schleppertypen konnten in technischer Hinsicht den teilweise in Verlust geratenen Anschluss an die Konkurrenz wieder herstellen, in manchen Punkten diese sogar übertreffen. Sie waren in einer sehr ansprechenden und aktu-

Hanomag Brillant 700, Baujahr 1969

Hanomag Brillant 701, Baujahr 1970

ellen nüchtern-kantigen trapezförmigen Bauweise mit tief liegendem Schwerpunkt, langen Radständen und günstigen Gewichtsverteilungen mit weit über die Vorderachse vorgezogenem Motor gestaltet. Auch die Anordnung der Bedienungselemente und Armaturen des Fahrerplatzes hatte man neu durchdacht. Die Traktoren waren mit Lenkradschaltung ausgerüstet. Bei den nach dem Baukastenprinzip entwickelten kurzhubigen, relativ hochdrehenden vier- und sechszylindrigen Motoren war im Hinblick auf die Minimierung der Herstellungs- und Lagerhaltungskosten auf die größtmögliche Austauschbarkeit von Teilen besonderer Wert gelegt worden. Die aus dem Ricardo-Brennverfahren entwickelten Wirbelkammer-Dieselmotoren besaßen ein gutes Kaltstartverhalten und zeichneten sich durch niedrige Kolbengeschwindigkeiten sowie eine ruhige und elastische Arbeitsweise aus.

Mit den Typen Brillant und Robust befanden sich erstmals Allradschlepper bei Hanomag im Angebot, wobei die entsprechenden Planeten-Vorderachsgetriebe nicht der eigenen Fertigung entstammten, sondern von ZF aus Friedrichshafen bezogen wurden, die auch unter Volllast eine gleichmäßige Übertragung der Motorkräfte auf die Achse sicherstellten.

Damit war es gelungen, ein im Gegensatz zu früheren Jahren auf nur wenigen Grundtypen basierendes, sehr übersichtliches und technisch ausgereiftes, den gestiegenen Anforderungen angepasstes Schlepperprogramm auf die Beine zu stellen. Leistungsmäßig wurde eine Bandbreite von 25 PS beim Perfekt 300 bis zum 85 PS starken Robust 900 erreicht. Die bewährten, seit 1963 gebauten Typen der unteren Leistungsklasse bis zum Granit 500 E wurden, bis auf einige technische, dem Fortschritt dienende Verbesserungen, beispielsweise durch die Verwendung von neuen Dreizylinder-Dieselmotoren und teilweise neu abgestufter Getriebe, ohne größere Änderungen weiterhin im Verkaufsprogramm belassen.

Das als Brillant 600 bezeichnete kleinste Modell der 1967 neu vorgestellten Typen verfügte über den 58 PS starken Vierzylindermotor D 142R mit 3142 ccm Hubraum. Drei Getriebegruppen mit insgesamt zwölf Vorwärts- und drei Rückwärtsgangmöglichkeiten standen bis maximal 27,15 km/h Höchstgeschwindigkeit – die letzte Fahrstufe des 12. Gangs war nur unter Verwendung einer Druckluftbremsanlage freigeschaltet – zur Verfügung. In der ersten Ganggruppe waren die Kriechgeschwindigkeiten zwischen 1,64 und 4,33 km/h zusammengefasst. Der Gangwechsel innerhalb der Schaltgruppen wurde mittels einer Bolzenschaltung herbeigeführt. Zunächst auf Wunsch, später

Hanomag Robust 900 im Einsatz

Granit, Brillant und Robust **135**

Hanomag Brillant 700

aber serienmäßig, konnte auch eine Synchronisierung eingebaut werden. Der äußerst geringe Kraftverlust innerhalb des Getriebes war bemerkenswert, so dass fast die gesamte Motorleistung den Antriebsrädern zur Verfügung stand. Der im Vergleich zu den Mitbewerbern technisch recht aufwendig konstruierte Schlepper war mit 2970 kg Eigengewicht entsprechend schwer und teuer. Den Brillant 600 gab es auch in einer werksintern als Brillant 601 A bezeichneten, gleich starken geländegängigen Allradausführung.

Schließlich wurde zwischen 1969 und 1970 das mit 62 PS leistungsstärkere, andererseits aber auch in manchen Teilen, zum Beispiel im Bereich des Differentials und der Achswellen des Getriebes, bewusst vereinfachte Hinterradmodell Brillant 601 L angeboten. Hierdurch konnten sowohl das Gewicht (der 601 L war 170 kg leichter geraten) als auch der Produktionskostenaufwand geringfügig gesenkt werden. Die Allradversion hingegen blieb unverändert; sie wurde ab 1969 ebenfalls auf 62 PS gesteigert.

Bei den beiden Spitzenmodellen Brillant 700 und Robust 900 des neuen Schlepperprogramms, zugstarke Traktoren in erster Linie für landwirtschaftliche Großbetriebe, kamen erstmals bei Hanomag Sechszylindermotoren zum Einbau. Während der Brillant 700 den anfänglich 68 PS leistenden D 161R-Motor mit 4252 ccm Hubraum besaß, war der Robust 900 mit dem D 162R-Motor mit zunächst 85 PS ausgestattet, der über 4712 ccm Rauminhalt verfügte. Die Motorleistung wurde 1969 bei beiden Modellen durch Erhöhung der Einspritzmenge auf 75 beziehungsweise 92 PS angehoben. Beide Ausführungen verfügten über eine ganze Reihe sehr fortschrittlicher, den Fahrkomfort und die Arbeit des Fahrers erleichternde Einbauten und Verbesserungen, angefangen beim ergonomisch gestalteten Komfort-Schwingsitz über blendfreie Armaturen bis hin zur hydraulischen Lenkung, die das Fahren im Gelände oder unter erschwerten Bedingungen leichter machte. Der Brillant 700 und der Robust 900 waren – wie im übrigen die kleineren Modelle ebenfalls –

mit sehr vielen unterschiedlichen Reifengrößen lieferbar. Seit August 1970 wurden die großen Sechszylindertraktoren von Hanomag ausschließlich mit synchronisiertem Getriebe gefertigt.

Der Robust 900 – das stärkste Radschleppermodell, das Hanomag jemals baute – unterschied sich vom kleineren Brillant durch den volumenstärkeren Motor und die höhere Leistung. Auch die in drei Schaltgruppen gegliederten Übersetzungen des G 265/R 270-Getriebes waren baugleich abgestuft und wiesen daher keine Unterschiede auf. Mit den vorhandenen Rückwärtsgangstufen standen insgesamt 15 Gänge – von 1,64 bis 27,15 km/h – zur Verfügung. Die schnellste Gangstufe durfte in Deutschland allerdings nur mit spezieller Ausrüstung wie vorderen Kotflügeln und Druckluftbremsanlage betrieben werden. War diese nicht vorhanden, blieb der Schnellgang gesperrt und die Maximalgeschwindigkeit auf 19,87 km/h begrenzt. Durch Lösen einer Schraube konnte das 120 kg schwere Frontgewicht weiter nach vorn gezogen

Hanomag Brillant 700 Allrad, Baujahr 1970

Hanomag Robust 900, Baujahr 1968

Hanomag Robust 900

und damit die Vorderachse stärker belastet werden, was beispielsweise bei schweren Pflugarbeiten erforderlich sein konnte. War dies immer noch nicht ausreichend, so konnten weitere Gewichte bis maximal 200 kg zwischen Frontgewicht und Kühler angebracht werden, wodurch das Eigengewicht des Robust 900 auf etwa 3,5 t angehoben wurde, was zum Ziehen eines Fünfscharpfluges durchaus ausreiche. Auf Wunsch gab es für dieses größte Modell eine konstruktiv aufwendig, allseits abgeschlossene, winterfeste Fritzmeyer-Kabine mit Heizung, aufklappbarer Frontscheibe und Einstiegstüren.

Den beiden Spitzenmodellen Brillant 700 und Robust 900 wurden leistungsgleiche Allradvarianten zur Seite gestellt. Diese schweren Allradschlepper wurden mit Außenplanetenachsen des Typs APL 3050 von ZF bestückt und waren unter extrem erschwerten Arbeitsbedingungen, ob bei aufgeweichten, schmierigen oder wenig tragenden Böden, bei Holzrückarbeiten im Forst oder im Ödland, im Gebirge oder auf vereisten Straßen, in ihrem Element. Ab 1969 gelangte die bereits im Brillant 600 verwendete, einfachere Achse des Typs AL 1550 (ohne Planetengetriebe) zum Einbau.

Brillant	600 L	600 A	700	700 A	Robust 900	900 A
Produktionszeitraum	1967-1970	1967-1970	1967-1970	1967-1970	1967-1970	1967-1970
Motorleistung (PS)	58/62	62	75	75	85/92	85/92
Hubraum (cm³)	3142	3142	4252	4252	4712	4712
Anzahl der Zylinder	4	4	6	6	6	6
Anzahl der Gäng	12+3	12+3	12+3	12+3	12+3	12+3
Eigengewicht (kg)	2970	3535	3295	3820	3490	4030
Höchstgeschwindigkeit	27,15 km/h	25,5 km/h	27,15 km/h	27,15 km/h	27,15 km/h	27,15 km/h

Trotz fortschrittlicher Technik und ansprechendem Äußeren der Modelle war der Schlepperabsatz bei einem nahezu unveränderten Marktanteil von nur noch sechs Prozent im Jahr 1970 bei einer Stückzahl von 3940 Einheiten, das entsprach dem sechsten Platz in der Zulassungsstatistik, nicht allzu üppig. Leider aber räumten die veränderten Marktbedingungen – nach dem vermeintlich unbegrenzt anhaltendem Wirtschaftswunder waren etwa Mitte der 60er Jahre die ersten Marktsättigungen eingetreten, die zu einem Absatzstau von Produkten führten – Hanomag nur noch geringe Chancen ein. Immerhin gab es die neue Schlepperbaureihe schon seit 1967 und auch die Mitbewerber schliefen selbstverständlich nicht und präsentierten technisch immer weiterentwickeltere, stärkere Schleppertypen, die oftmals schon bei über 100 PS lagen. Daher wären umfangreiche Investitionen für Weiterentwicklungen in absehbarer Zeit aufzubringen gewesen, wobei es bei der unbefriedigenden Marktposition mehr als zweifelhaft gewesen wäre, diese hohen Kosten jemals wieder hereinzuholen. Daher entschloss sich die Rheinstahl-Konzernleitung bereits 1969, als die Fertigung von Traktoren auf unter 7000 Einheiten gesunken war, infolge der unbefriedigenden Ertragslage – man schrieb schon seit längerem rote Zahlen – und nicht mehr vertretbarer hoher Stückkosten, die durch den bestehenden Konkurrenzdruck nicht an die Kundschaft weitergegeben werden konnten, sich ganz aus dem Schleppergeschäft zurückzuziehen. Auch den Kostenersparnismöglichkeiten durch noch stärkere Standardisierungsmaßnahmen waren Grenzen gesetzt.

Im Mai 1970 wurde schließlich die Aufgabe der Traktorenherstellung bekannt gegeben und man kündigte an, auf der in Kürze in Köln stattfindenden DLG-Ausstellung nicht mehr präsent zu sein. Nachdem die Ersatzteilversorgung durch eine Vereinbarung mit dem Marktführer Klöckner-Humboldt-Deutz (KHD), der sich die Händlerorganisation eingliederte, bis zum Jahr 1977 sichergestellt worden war, verließen die letzten, zwecks Beschleunigung des Abverkaufs zu Sonderkonditionen angebotenen Traktoren im März 1971 die Werkshallen. Die nahezu neuwertigen

Hanomag Robust 900 mit Allwetterverdeck, Baujahr 1969, noch 1999 in Belgien im Einsatz (oben) und Hanomag Robust 900 Allrad, Baujahr 1970

Granit, Brillant und Robust **139**

Hanomag Brillant 600 Allrad

Hanomag Robust 900 Allrad, mit Frontlader und Allwetterverdeck, Baujahr 1971 (oben) und Hanomag Robust 900 Allrad mit Allwetterverdeck, Baujahr 1971

Fertigungsanlagen für die Motoren wurden 1972 an Volvo veräußert und im folgenden Jahr nach Schweden transportiert. Damit war die Traktorenherstellung bei der Hanomag endgültig beendet, nachdem das Unternehmen in seiner wechselvollen Geschichte von knapp sechs Jahrzehnten insgesamt mehr als 250000 Exemplare gefertigt hatte.

Nach Übernahme der Hanomag-Lastwagen-Fertigung durch den Daimler-Benz-Konzern war das Werk zu einem reinen Baumaschinenhersteller mit nur noch 2400 Mitarbeitern degradiert worden. Nach einem wechselvollen Auf und Ab in den folgenden Jahren unter verschiedenen neuen Eigner, fanden die auf nur noch wenige hundert Beschäftigte zusammengeschrumpften Reste der einstmals so bedeutenden Hanomag-Werke ab 1989 eine neue Heimat im Rahmen des japanischen Baumaschinenherstellers Komatsu, des weltweit zweigrößten Anbieters dieser Branche. 1996 musste die mehrheitlich von Komatsu übernommene Hanomag ihre Selbständigkeit aufgeben, was in dem neuen Firmennamen "Komatsu-Hanomag AG" seinen Ausdruck fand. Seither werden in Hannover Radlader unterschiedlicher Größe produziert und vertrieben. So kann der einst in vielen Produktionsbereichen so bedeutende Hersteller optimistisch in die Zukunft blicken, dass dessen Name, zumindest im Baumaschinenbereich, erhalten bleibt.

Weitere Literatur für Schlepper-Liebhaber

Fordern Sie kostenlos und völlig unverbindlich unseren neuesten Prospekt an mit Büchern über:

- Traktoren
- Baumaschinen
- Lastwagen
- Omnibusse
- Feuerwehren
- Autos
- Motorräder
- Lokomotiven

Podszun-Verlag GmbH
Postfach 1525
D-59918 Brilon
Telefon 02961 / 53213
Fax 02961 / 9639900

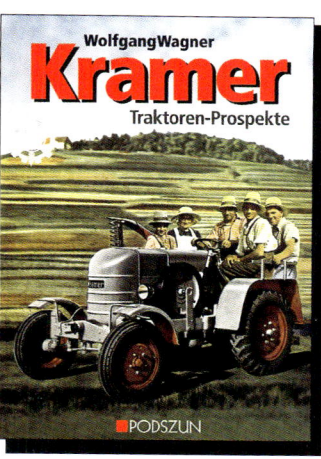

144 Seiten, fester Einband
ISBN 3-86133-261-2
19,90 EUR

144 Seiten, fester Einband
ISBN 3-86133-246-9
19,90 EUR

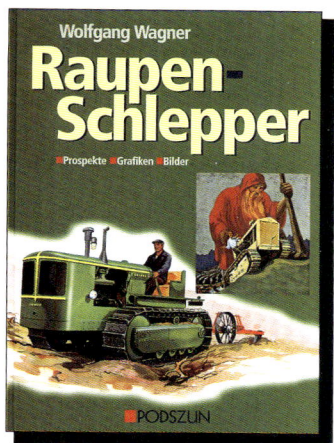

144 Seiten, fester Einband
ISBN 3-86133-278-7
19,90 EUR

144 Seiten, fester Einband
ISBN 3-86133-190-X
24,90 EUR

238 Seiten, fester Einband
ISBN 3-86133-152-7
34,90 EUR

174 Seiten, fester Einband
ISBN 3-86133-272-8
24,90 EUR

erscheint jährlich im Oktober neu

144 Seiten, Broschur
ISBN 3-86133-301-5
14,90 EUR

144 Seiten, fester Einband
ISBN 3-86133-275-2
19,90 EUR

144 Seiten, fester Einband
ISBN 3-86133-239-6
19,90 EUR

2013

CRISC™ Review Manual 2013

CRISC™ Certified in Risk and Information Systems Control™

An ISACA® Certification

ISACA®
Trust in, and value from, information systems

ISACA®

With more than 100,000 constituents in 180 countries, ISACA (*www.isaca.org*) is a leading global provider of knowledge, certifications, community, advocacy and education on information systems (IS) assurance and security, enterprise governance and management of IT, and IT-related risk and compliance. Founded in 1969, the nonprofit, independent ISACA hosts international conferences, publishes the *ISACA® Journal*, and develops international IS auditing and control standards, which help its constituents ensure trust in, and value from, information systems. It also advances and attests IT skills and knowledge through the globally respected Certified Information Systems Auditor® (CISA®), Certified Information Security Manager® (CISM®), Certified in the Governance of Enterprise IT® (CGEIT®) and Certified in Risk and Information Systems Control™ (CRISC™) designations.

ISACA continually updates and expands the practical guidance and product family based on the COBIT® framework. COBIT helps IT professionals and enterprise leaders fulfill their IT governance and management responsibilities, particularly in the areas of assurance, security, risk and control, and deliver value to the business.

Disclaimer

ISACA has designed and created *CRISC™ Review Manual 2013* primarily as an educational resource to assist individuals preparing to take the CRISC certification exam. It was produced independently from the CRISC exam and the CRISC Certification Committee, which has had no responsibility for its content. Copies of past exams are not released to the public and were not made available to ISACA for preparation of this publication. ISACA makes no representations or warranties whatsoever with regard to these or other ISACA publications assuring candidates' passage of the CRISC exam.

Reservation of Rights

© 2012 ISACA. All rights reserved. No part of this publication may be used, copied, reproduced, modified, distributed, displayed, stored in a retrieval system or transmitted in any form by any means (electronic, mechanical, photocopying, recording or otherwise) without the prior written authorization of ISACA.

ISACA

3701 Algonquin Road, Suite 1010
Rolling Meadows, IL 60008 USA
Phone: +1.847.253.1445
Fax: +1.847.253.1443
Email: *info@isaca.org*
Web site: *www.isaca.org*

Participate in the ISACA Knowledge Center: *www.isaca.org/knowledge-center*
Follow ISACA on Twitter: *https://twitter.com/ISACANews*
Join ISACA on LinkedIn: ISACA (Official), *http://linkd.in/ISACAOfficial*
Like ISACA on Facebook: *www.facebook.com/ISACAHQ*

ISBN 978-1-60420-326-4
CRISC™ Review Manual 2013
Printed in the United States of America

CRISC is a trademark/service mark of ISACA. The mark has been applied for or registered in countries throughout the world.

Acknowledgments

This manual is the result of contributions from volunteers across the globe who are actively involved in risk management and information systems control design, implementation, monitoring and maintenance and who generously contributed their time and expertise. This international team exhibited a spirit and selflessness that has become the hallmark of contributors to ISACA manuals. Their participation and insight are truly appreciated.

CRISC Developer
Kevin M. Henry, CISA, CISM, CRISC, CISSP, CBCI, SCF, KM Henry & Affiliates Management, Inc., Canada

CRISC Subject Matter Expert Reviewers
Ronald D. Burns, CRISC, American Express, USA
Alvaro Cayul, CISM, CRISC, Chile
Sandy Fadale, CISM, CGEIT, CRISC, Bell Aliant Regional Communications, Canada
Shawna M. Flanders, CISA, CISM, CRISC, CSSGB, SSBB, PSCU, USA
Robert T. Hanson, CISA, CISM, CRISC, Australia
W. Noel Haskins-Hafer, CISA, CISM, CGEIT, CRISC, CFE, USA
Pokit Lok, CISA, CISSP, CPIM, Hong Kong Productivity Council, Hong Kong

ISACA has begun planning the 2014 edition of the *CRISC™ Review Manual*. Volunteer participation drives the success of the manual. If you are interested in becoming a member of the select group of professionals involved in this global project, we want to hear from you. Please email us at *studymaterials@isaca.org*.

Table of Contents

Starting Page

Introduction
A. Overview vii
B. About This Manual viii
C. About Other CRISC Study Materials xiii

Part I—Risk Management and Information Systems Control Theory and Concepts
Domain 1—Risk Identification, Assessment and Evaluation 1
    A. Chapter Overview 1
    B. Task and Knowledge Statements 4
    C. The Big Picture—Risk Management and Risk Governance 6
    D. Risk Management Frameworks, Standards and Practices 18
    E. Risk Identification, Assessment and Evaluation 21
    F. Risk Scenarios 26
    G. Risk Factors 34
    H. Qualitative and Quantitative Risk Analysis 42
    I. IT Risk Identification and Assessment 52
    J. Suggested Resources for Further Study 70
Domain 2—Risk Response 71
    A. Chapter Overview 71
    B. Task and Knowledge Statements 73
    C. The Risk Response Process 74
    D. Risk Response Process Details 84
    E. Suggested Resources for Further Study 96
Domain 3—Risk Monitoring 97
    A. Chapter Overview 97
    B. Task and Knowledge Statements 99
    C. Essentials of Risk Monitoring 100
    D. Suggested Resources for Further Study 116
Domain 4—Information Systems Control Design and Implementation 117
    A. Chapter Overview 117
    B. Task and Knowledge Statements 120
    C. IS Controls 122
    D. Building Control Design Into the SDLC 131
    E. System Development Life Cycle (SDLC) Phases 134
    F. Managing Project Risk 152
    G. Project Management Tools and Techniques 160
    H. Suggested Resources for Further Study 165
Domain 5—Information Systems Control Monitoring and Maintenance 167
    A. Chapter Overview 167
    B. Task and Knowledge Statements 169
    C. Process Capability Assessment 171
    D. The Control Life Cycle 177
    E. Information Systems Control Monitoring and Maintenance 179
    F. Identify and Assess Information 182
    G. Tools for Monitoring 189
    H. Implementing Control Monitoring Processes 198
    I. Implementing Control Maintenance Processes 204
    J. Suggested Resources for Further Study 208

Table of Contents

Part II—Risk Management and Information Systems Control in Practice
Overview 209
1. Managing the IT Strategy 211
   A. Chapter Overview 211
   B. Related Knowledge Statements 212
   C. Key Terms and Concepts 213
   D. Process Overview 214
   E. Risk Management Considerations 218
   F. Information Systems Control Design, Monitoring and Maintenance 221
   G. The Practitioner's Perspective 224
   H. Suggested Resources for Further Study 226
2. Portfolio, Program and Project Management 227
   A. Chapter Overview 227
   B. Related Knowledge Statements 228
   C. Key Terms and Concepts 229
   D. Process Overview 230
   E. Risk Management Considerations 241
   F. Information Systems Control Design, Monitoring and Maintenance 244
   G. The Practitioner's Perspective 250
   H. Suggested Resources for Further Study 252
3. Change Management 253
   A. Chapter Overview 253
   B. Related Knowledge Statements 254
   C. Key Terms and Concepts 255
   D. Process Overview 256
   E. Risk Management Considerations 259
   F. Information Systems Control Design, Monitoring and Maintenance 261
   G. The Practitioner's Perspective 264
   H. Suggested Resources for Further Study 266
4. Third-party Service Management 267
   A. Chapter Overview 267
   B. Related Knowledge Statements 268
   C. Key Terms and Concepts 269
   D. Process Overview 270
   E. Risk Management Considerations 273
   F. Information Systems Control Design, Monitoring and Maintenance 275
   G. The Practitioner's Perspective 277
   H. Suggested Resources for Further Study 279
5. Continuity Management 281
   A. Chapter Overview 281
   B. Related Knowledge Statements 282
   C. Key Terms and Concepts 283
   D. Process Overview 284
   E. Risk Management Considerations 289
   F. Information Systems Control Design, Monitoring and Maintenance 291
   G. The Practitioner's Perspective 295
   H. Suggested Resources for Further Study 297
6. Information Security Management 299
   A. Chapter Overview 299
   B. Related Knowledge Statements 300
   C. Key Terms and Concepts 301
   D. Process Overview 302
   E. Risk Management Considerations 306
   F. Information Systems Control Design, Monitoring and Maintenance 309
   G. The Practitioner's Perspective 316
   H. Suggested Resources for Further Study 318

Table of Contents

7. Configuration Management 319
    A. Chapter Overview 319
    B. Related Knowledge Statements 320
    C. Key Terms and Concepts 321
    D. Process Overview 323
    E. Risk Management Considerations 325
    F. Information Systems Control Design, Monitoring and Maintenance 327
    G. The Practitioner's Perspective 329
    H. Suggested Resources for Further Study 332
8. Problem Management 333
    A. Chapter Overview 333
    B. Related Knowledge Statements 334
    C. Key Terms and Concepts 335
    D. Process Overview 337
    E. Risk Management Considerations 340
    F. Information Systems Control Design, Monitoring and Maintenance 342
    G. The Practitioner's Perspective 345
    H. Suggested Resources for Further Study 347
9. Knowledge Management 349
    A. Chapter Overview 349
    B. Related Knowledge Statements 350
    C. Key Terms and Concepts 351
    D. Process Overview 352
    E. Risk Management Considerations 355
    F. Information Systems Control Design, Monitoring and Maintenance 357
    G. The Practitioner's Perspective 360
    H. Suggested Resources for Further Study 367
10. IT Operations Management 369
    A. Chapter Overview 369
    B. Related Knowledge Statements 370
    C. Key Terms and Concepts 371
    D. Process Overview 372
    E. Risk Management Considerations 376
    F. Information Systems Control Design, Monitoring and Maintenance 378
    G. The Practitioner's Perspective 381
    H. Suggested Resources for Further Study 382

Study Questions, Answers and Explanations 383
Glossary 395
Suggested Resources for Further Study 407
General CRISC Information 411
List of Exhibits 418
Index 421
Your Evaluation of the CRISC™ Review Manual 425
Prepare for the 2013 CRISC Exam 426

Introduction

A. Overview

Contents

The introduction contains the following topics:

Topic	Starting Page
A. Overview	vii
B. About This Manual	viii
C. About Other CRISC Study Materials	xiii

B. About This Manual

Purpose of This Manual

ISACA is pleased to offer the third edition of the *CRISC™ Review Manual*.

The purpose of the manual is to provide CRISC candidates with information and references to assist in the preparation and study for the Certified in Risk and Information Systems Control (CRISC) exam.

Certification has resulted in a positive impact on many careers, including worldwide recognition for professional experience and enhanced knowledge and skills. The Certified in Risk and Information Systems Control™ certification (CRISC™, pronounced "see-risk") is designed for IT and business professionals who have hands-on experience with risk identification, assessment and evaluation; risk response; risk monitoring; information systems (IS) control control design and implementation; and IS control monitoring and maintenance. We wish you success with the CRISC exam.

> **Note:** The *CRISC Review Manual 2013* ("manual") is intended to assist candidates in preparing for the CRISC exam. The manual is **one** source of preparation for the exam. The manual should **not** be thought of as the only source nor should it be viewed as a comprehensive collection of all the information and experience that is required to pass the exam. No single publication offers such coverage and detail.

Basis for the Content

The content in the manual is based on the current CRISC job practice found at *www.isaca.org/criscjobpractice*.

This manual is the result of contributions from volunteers across the globe who are actively involved in risk management and IS control design, implementation, monitoring and maintenance and who generously contributed their time and expertise.

Organization of This Manual

The *CRISC™ Review Manual 2013* is organized into several parts:
Part I—Risk Management and Information Systems Control Theory and Concepts
Part II—Risk Management and Information Systems Control in Practice
Additional Resources—Study Questions, Glossary, Suggested Resources,
General CRISC Information

Part I

Part I—Risk Management and Information Systems Control Theory and Concepts consists of five chapters, each dedicated to one of the five CRISC domains, as described in the CRISC job practice:
Domain 1—Risk Identification, Assessment and Evaluation
Domain 2—Risk Response
Domain 3—Risk Monitoring
Domain 4—Information Systems Control Design and Implementation
Domain 5—Information Systems Control Monitoring and Maintenance

Exhibit B.1: High-level Relationship Between CRISC Domains

Exhibit B.1 describes the high-level relationship between CRISC domains:

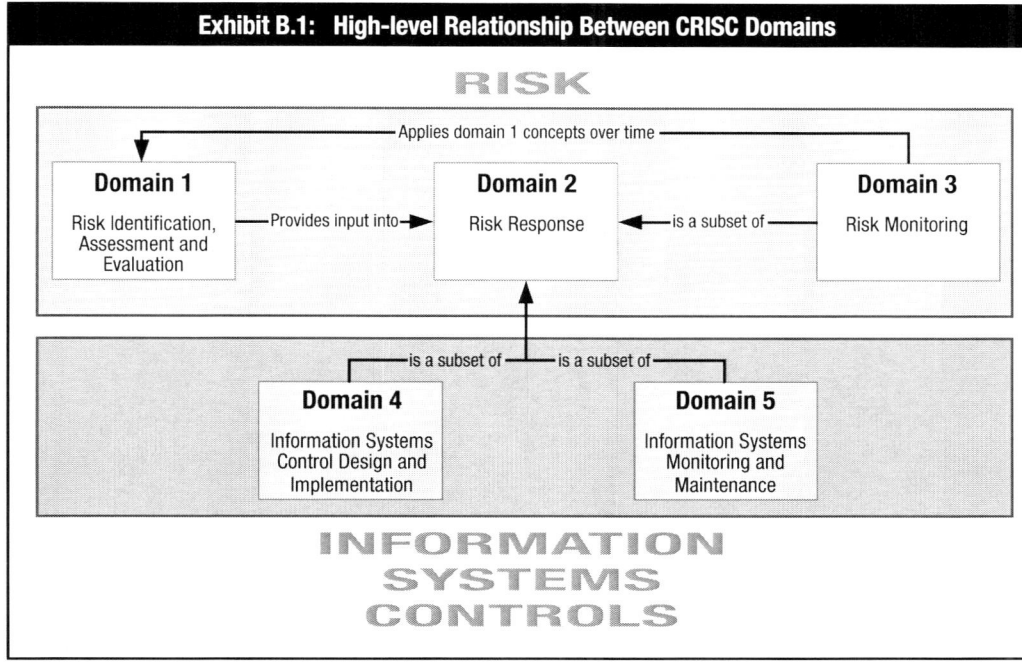

Part I Chapter Structure

Each of the Part I chapters:
- Depicts the tasks performed by individuals who have a management, advisory or assurance role related to risk and IS control
- Describes the knowledge required to perform these tasks
- Serves as a definition of the roles and responsibilities of the professionals performing risk and IS control work

Domain 1 describes how risk management ties into risk governance and introduces essential risk governance concepts, such as risk appetite and tolerance. While not established by the CRISC, these concepts are an essential input into the risk practitioner's activities and help the CRISC candidate better understand the environment to which the risk practitioner contributes.

Note: The knowledge statements from domains 1, 4 and 5 that relate to business or IT process-specific risk, controls, control objectives, activities and metrics are addressed in Part II of this publication.

Introduction
B. About This Manual

Exhibit B.2: Knowledge Statements Addressed in Part II

Exhibit B.2 highlights the process-specific knowledge statements (**in bold**) addressed in Part II:

Exhibit B.2: Process-specific Knowledge Statements Addressed in Part II

Domain 1—Task Statements
1.1 Collect information and review documentation to ensure that risk scenarios are identified and evaluated.
1.2 Identify legal, regulatory and contractual requirements and organizational policies and standards related to information systems to determine their potential impact on the business objectives.
1.3 Identify potential threats and vulnerabilities for business processes, associated data and supporting capabilities to assist in the evaluation of enterprise risk.
1.4 Create and maintain a risk register to ensure that all identified risk factors are accounted for.
1.5 Assemble risk scenarios to estimate the likelihood and impact of significant events to the organization.
1.6 Analyze risk scenarios to determine their impact on business objectives.
1.7 Develop a risk awareness program and conduct training to ensure that stakeholders understand risk and contribute to the risk management process and to promote a risk-aware culture.
1.8 Correlate identified risk scenarios to relevant business processes to assist in identifying risk ownership.
1.9 Validate risk appetite and tolerance with senior leadership and key stakeholders to ensure alignment.

Domain 2—Task Statements
2.1 Identify and evaluate risk response options and provide management with information to enable risk response decisions.
2.2 Review risk responses with the relevant stakeholders for validation of efficiency, effectiveness and economy.
2.3 Apply risk criteria to assist in the development of the risk profile for management approval.
2.4 Assist in the development of risk response action plans to address risk factors identified in the organizational risk profile.
2.5 Assist in the development of business cases supporting the investment plan to ensure that risk responses are aligned with the identified business objectives.

Domain 3—Task Statements
3.1 Collect and validate data that measure key indicators (KRIs) to monitor and communicate their status to relevant stakeholders.
3.2 Monitor and communicate key risk indicators (KRIs) and management activities to assist relevant stakeholders in their decision-making process.
3.3 Facilitate independent risk assessments and risk management process reviews to ensure that they are performed efficiently and effectively.
3.4 Identify and report on risk, including compliance, to initiate corrective action and meet business and regulatory requirements.

Domain 4—Task Statements
4.1 Interview process owners and review process design documentation to gain an understanding of the business process objectives.
4.2 Analyze and document business process objectives and design to identify required information systems controls.
4.3 Design information systems controls in consultation with process owners to ensure alignment with business needs and objectives.
4.4 Facilitate the identification of resources (e.g., people, infrastructure, information, architecture) required to implement and operate information systems controls at an optimal level.
4.5 Monitor the information systems control design and implementation process to ensure that it is implemented effectively and within time, budget and scope.
4.6 Provide progress reports on the implementation of information systems controls to inform stakeholders and to ensure that deviations are promptly addressed.
4.7 Test information systems controls to verify effectiveness and efficiency prior to implementation.
4.8 Implement information systems controls to mitigate risk.
4.9 Facilitate the identification of metrics and key performance indicators (KPIs) to enable the measurement of information systems control performance in meeting business objectives.
4.10 Assess and recommend tools to automate information systems control processes.
4.11 Provide documentation and training to ensure that information systems controls are effectively performed.
4.12 Ensure that all controls are assigned control owners to establish accountability.
4.13 Establish control criteria to enable control life cycle management.

Domain 5—Task Statements
5.1 Plan, supervise and conduct testing to confirm continuous efficiency and effectiveness of information systems controls.
5.2 Collect information and review documentation to identify information systems control deficiencies.
5.3 Review information systems policies, standards and procedures to verify that they address the organization's internal and external requirements.
5.4 Assess and recommend tools and techniques to automate information systems control verification processes.
5.5 Evaluate the current state of information systems processes using a maturity model to identify the gaps between current and targeted process maturity.
5.6 Determine approach to correct information systems control deficiencies and maturity gaps to ensure that deficiencies are appropriately considered and remediated.
5.7 Maintain sufficient, adequate evidence to support conclusions on the existence and operating effectiveness of information systems controls.
5.8 Provide information systems control status reporting to relevant stakeholders to enable informed decision making.

Domain 1—Knowledge Statements—Knowledge of...
1.1 standards, frameworks and leading practices related to risk identification, assessment and evaluation
1.2 techniques for risk identification, classification, assessment and evaluation
1.3 quantitative and qualitative risk evaluation methods
1.4 business goals and objectives
1.5 organizational structures
1.6 risk scenarios related to business processes and initiatives
1.7 business information criteria
1.8 threats and vulnerabilities related to business processes and initiatives
1.9 information systems architecture (e.g., platforms, networks, applications, databases and operating systems)
1.10 information security concepts
1.11 threats and vulnerabilities related to third-party management
1.12 threats and vulnerabilities related to data management
1.13 threats and vulnerabilities related to the system development life cycle
1.14 threats and vulnerabilities related to project and program management
1.15 threats and vulnerabilities related to business continuity and disaster recovery management
1.16 threats and vulnerabilities related to management of IT operations
1.17 the elements of a risk register
1.18 risk scenario development tools and techniques
1.19 risk awareness training tools and techniques
1.20 principles of risk ownership
1.21 current and forthcoming laws, regulations and standards
1.22 threats and vulnerabilities associated with emerging technologies

Domain 2—Knowledge Statements—Knowledge of...
2.1 standards, frameworks and leading practices related to risk response
2.2 risk response options
2.3 cost-benefit analysis and return on investment (ROI)
2.4 risk appetite and tolerance
2.5 organizational risk management policies
2.6 parameters for risk response selection
2.7 project management tools and techniques
2.8 portfolio, investment and value management
2.9 exception management
2.10 residual risk

Domain 3—Knowledge Statements—Knowledge of...
3.1 standards, frameworks and leading practices related to risk monitoring
3.2 principles of risk ownership
3.3 risk and compliance reporting requirements, tools and techniques
3.4 key performance indicators (KPIs) and key risk indicators (KRIs)
3.5 risk assessment methodologies
3.6 data extraction, validation, aggregation and analysis tools and techniques
3.7 various types of reviews of the organization's risk monitoring process (e.g., internal and external audits, peer reviews, regulatory reviews, quality reviews)

Domain 4—Knowledge Statements—Knowledge of...
4.1 standards, frameworks and leading practices related to information systems control design and implementation
4.2 business process review tools and techniques
4.3 testing methodologies and practices related to information systems control design and implementation
4.4 control practices related to business processes and initiatives
4.5 the information systems architecture (e.g., platforms, networks, applications, databases and operating systems)
4.6 controls related to information security
4.7 controls related to third-party management
4.8 controls related to data management
4.9 controls related to the system development life cycle
4.10 controls related to project and program management
4.11 controls related to business continuity and disaster recovery management
4.12 controls related to management of IT operations
4.13 software and hardware certification and accreditation practices
4.14 the concept of control objectives
4.15 governance, risk and compliance (GRC) tools
4.16 tools and techniques to educate and train users

Domain 5—Knowledge Statements—Knowledge of...
5.1 standards, frameworks and leading practices related to information systems control monitoring and maintenance
5.2 enterprise security architecture
5.3 monitoring tools and techniques
5.4 maturity models
5.5 control objectives, activities and metrics related to IT operations and business processes and initiatives
5.6 control objectives, activities and metrics related to incident and problem management
5.7 security testing and assessment tools and techniques
5.8 control objectives, activities and metrics related to information systems architecture (platforms, networks, applications, databases and operating systems)
5.9 control objectives, activities and metrics related to information security
5.10 control objectives, activities and metrics related to third-party management
5.11 control objectives, activities and metrics related to data management
5.12 control objectives, activities and metrics related to the system development life cycle
5.13 control objectives, activities and metrics related to project and program management
5.14 control objectives, activities and metrics related to software and hardware certification and accreditation practices
5.15 control objectives, activities and metrics related to business continuity and disaster recovery management
5.16 applicable laws and regulations

Introduction
B. About This Manual

Part II	*Part II—Risk Management and Information Systems Control in Practice* contains selected process-specific chapters: 1. Managing the IT Strategy 2. Portfolio, Program and Project Management 3. Change Management 4. Third-party Service Management 5. Continuity Management 6. Information Security Management 7. Configuration Management 8. Problem Management 9. Knowledge Management 10. IT Operations Management
Part II Chapter Structure	Each chapter introduces one IT or business process at a high level and address the following, as they relate to that specific process: • Risk factors • Common vulnerabilities • Generic risk scenarios • Key risk indicators (KRIs) • Key control activities • IS control metrics for monitoring This knowledge is necessary to perform the tasks discussed in Part I of the manual and is necessary to pass the CRISC exam. The learning objectives for each of these domains appear at the beginning of each chapter with the corresponding knowledge statements that are tested on the exam. Exam candidates should evaluate their strengths and weaknesses based on their knowledge and experience in each of these processes.
Additional Resources	The manual also contains the following sections: • **Study Questions, Answers and Explanations**—To familiarize candidates with question structure and general content, 25 study questions, sorted by domain with explanations for the correct and incorrect answers, are provided. • **Glossary**—The definitions of key terms and acronyms are provided for a common understanding of key CRISC concepts. • **Suggested Resources for Further Study**—As candidates read through the manual and encounter topics that are new or less familiar to them or ones in which they feel their knowledge and experience are limited, additional references should be sought. Suggested resources for further study are provided in Parts I and II, and a comprehensive list appears at the end of the manual. These references can be used to further acquire and better understand detailed information on the topics addressed in the manual. • **List of Exhibits**—A comprehensive listing of all of the exhibits is provided at the end of the manual.

**Introduction
B. About This Manual**

Written Material Is Not a Substitute for Experience	The CRISC exam is a practice-based exam. Simply reading this manual will not properly prepare candidates for the exam. The study questions included in this manual are designed to provide further clarity to the content presented in the manual and to depict the type of questions typically found on the CRISC exam. The practice questions and answers: • Should *not* be used independently as a source of knowledge • Should *not* be considered a measure of one's ability to answer questions correctly on the exam for any domain • *Are* intended to familiarize candidates with question structure and general content • *May or may not* be similar to questions that will appear on the actual exam
Preparing for the Exam	Good preparation for the CRISC exam can be achieved through an organized plan of study. To assist individuals with the development of a successful study plan, ISACA offers study aids and review courses to exam candidates. See *www.isaca.org/criscprep* to view the ISACA study aids that can help prepare for the exam.
Disclaimer	No representations or warranties are made by ISACA in regard to this or other ISACA publications assuring candidates' passage of the CRISC exam. This publication was produced independently of the CRISC Certification Committee, which has no responsibility for the content of this manual.
Your Feedback Is Requested	The *CRISC Review Manual* will be updated annually to keep pace with rapid changes in the field of IT-related business risk management in information systems controls. As such, your comments and suggestions regarding this manual are welcomed. After reading the manual, please take a moment to complete the online questionnaire. Your observations will be extremely valuable for the preparation of the 2014 edition of the manual. A link to an online feedback questionnaire is posted at *www.isaca.org/studyaidsevaluation*.

C. About Other CRISC Study Materials

Other Study Materials

The CRISC candidate may find it useful to study the *CRISC™ Review Questions, Answers & Explanations Manual 2013* and the *CRISC™ Review Questions, Answers & Explanations Manual 2013 Supplement.*

The *CRISC™ Review Questions, Answers & Explanations Manual 2013* consists of 200 multiple-choice study questions, answers and explanations arranged in the areas of the current CRISC job practice. Many of these items appeared in the previous editions of the *CRISC™ Review Questions, Answers & Explanations Manual* and *CRISC™ Review Questions, Answers & Explanations Manual Supplement.*

The 2013 supplement is a result of ISACA's dedication each year to create 100 new sample questions, answers and explanations for candidates to use in preparation for the CRISC exam. Each year, ISACA develops 100 new review questions, using a strict process of review similar to that performed for the selection of questions for the CRISC exam by the CRISC Certification Committee.

The following resources are available from ISACA in addition to the *CRISC™ Review Manual 2013.*

Resource	Function
The Risk IT Framework	Although knowledge of the Risk IT framework and COBIT is not specifically tested on the CRISC exam, the Risk IT and COBIT processes are reflected in the CRISC job practice knowledge statements. As such, a thorough review of the Risk IT framework and COBIT is recommended for candidates preparing for the CRISC certification.
The Risk IT Practitioner Guide	
COBIT® 5	**Note:** The following three COBIT 5 publications are available from ISACA and can be downloaded at *www.isaca.org/cobit*: • COBIT 5—A Business Framework for the Governance and Management of Enterprise IT (complimentary PDF) • *COBIT® 5: Enabling Processes* (member complimentary PDF) • *COBIT® 5 Implementation* (member complimentary PDF)
CRISC™ Review Questions, Answers & Explanations Manual 2013	Designed to provide CRISC candidates with an understanding of the type and structure of questions and content that will appear on the CRISC exam, the *CRISC™ Review Questions, Answers & Explanations Manual 2013* consists of 200 multiple-choice study questions. To help candidates maximize study efforts, questions are presented in the following two ways: • Sorted by job practice area, allowing CRISC candidates to focus on particular topics • Scrambled as a sample 100-question exam, enabling candidates to effectively determine their strengths and weaknesses and allowing them to simulate an actual exam

Other Study Materials *(cont.)*

Resource	Function
CRISC™ Review Questions, Answers & Explanations Manual 2013 Supplement	The CRISC supplement offers an additional 100 study questions in the same structure as the *CRISC™ Review Questions, Answers & Explanations Manual 2013*.
CRISC™ Exam Self-study Subscription—6 month	The CRISC Exam Self-study subscription offers the 300 questions from the *CRISC Review Questions, Answers & Explanations Manual 2013* and the *CRISC Review Questions, Answers & Explanations Manual 2013 Supplement* in a web-based format. Candidates can take sample exams with randomly selected questions and view the results by job practice, allowing for concentrated study in particular areas.

Part I—Risk Management and Information Systems Control Theory and Concepts
Domain 1—Risk Identification, Assessment and Evaluation
A. Chapter Overview

Part I—Risk Management and Information Systems Control Theory and Concepts

Domain 1—Risk Identification, Assessment and Evaluation

A. CHAPTER OVERVIEW

Introduction

This chapter provides the core practices of risk identification, assessment and evaluation with which the risk practitioner should be familiar.

The chapter also provides information about "The Big Picture" or risk governance to ensure that the risk practitioner can effectively differentiate between governing and managing IT-related business risk.

Learning Objectives

As a result of completing this chapter, the CRISC candidate should be able to:
- Differentiate between risk management and risk governance.
- Identify the roles and responsibilities for risk management.
- Identify relevant standards, frameworks and practices.
- Explain the meaning of key risk management concepts, including "risk appetite" and "risk tolerance."
- Distinguish between threats and vulnerabilities.
- Apply risk identification, classification, quantitative/qualitative assessment and evaluation techniques.
- Describe the key elements of a risk register.
- Discuss risk scenario development tools and techniques.
- Help develop and support risk awareness training tools and techniques.
- Relate security concepts to risk assessment.

Inputs (Tie-back)

The risk governance function is the responsibility of senior management. It shapes the risk culture within which the risk practitioner functions, and provides inputs for many of the tasks performed during risk identification, assessment, response and evaluation. The risk practitioner is expected to provide risk response recommendations in alignment with the risk governance culture of senior management and with regard to their input, particularly regarding risk appetite and risk tolerance levels.

Process Objectives

Risk identification, assessment and evaluation is concerned with correctly determining the risk faced by the enterprise and providing recommendations to senior management on how to effectively maintain risk at an acceptable level, including, but not limited to:
- Identifying risk, including emerging risk and risk associated with people, processes, technology, architecture, applications, information, natural factors, the operating environment, and physical threats
- Assessing the risk levels associated with each threat, including anticipated risk likelihood and impact and evaluating the effectiveness of current and planned controls
- Calculating the risk levels using both quantitative and qualitative metrics and determining the impact of the risk on the ability of the business to meet its goals and objectives

Outputs (Tie-forward)

The output of the risk assessment process helps to identify and recommend appropriate controls for reducing or eliminating risk during the risk response process. To determine the likelihood of a future adverse event, threats to the enterprise must be analyzed in conjunction with the potential vulnerabilities and the controls in place for the system.

This process includes the creation of a comprehensive, prioritized inventory of relevant risk—often in the form of a risk register—as well as risk response recommendations. Other deliverables may include risk awareness programs and training.

Part I—Risk Management and Information Systems Control Theory and Concepts
Domain 1—Risk Identification, Assessment and Evaluation
A. Chapter Overview

Contents

This chapter contains the following sections:

Section	Starting Page
A. Chapter Overview	1
B. Task and Knowledge Statements	4
C. The Big Picture—Risk Management and Risk Governance	6
1. Essentials of Risk Governance	6
2. Essentials of Risk Management	13
D. Risk Management Frameworks, Standards and Practices	18
1. Key Terms and Concepts	18
2. Examples of Frameworks Related to Risk Management and IS Control	19
3. Examples of Standards Related to Risk Management and IS Control	19
4. Examples of Leading Practices Related to Risk Management and IS Control	20
E. Risk Identification, Assessment and Evaluation	21
1. Key Terms and Concepts	21
2. Risk Identification, Assessment and Evaluation Process Objectives	22
3. The Risk Identification, Assessment and Evaluation Process	22
F. Risk Scenarios	26
1. Key Terms and Concepts	26
2. Risk Scenario Development	26
3. Risk Register	32
G. Risk Factors	34
1. Key Terms and Concepts	34
2. External Risk Factors	35
3. Internal Environmental Risk Factors	37
4. Capabilities	41
H. Qualitative and Quantitative Risk Analysis	42
1. Key Terms and Concepts	42
2. Qualitative Risk Analysis	43
3. Quantitative Risk Analysis	46
4. Methods for Discovering High-impact Risk Types	49
5. Risk Associated With Business Continuity and Disaster Recovery Planning	50
I. IT Risk Identification and Assessment	52
1. Threats and Opportunities Inherent in Enterprise Use of IT	52
2. Types of Business Risk and Threats That Can Be Addressed Using IT Resources	60
3. Enterprise Risk Management	61
4. Methods/Frameworks for Describing IT Risk in Business Terms	64
5. Risk Awareness and Communication	66
J. Suggested Resources for Further Study	70

Part I—Risk Management and Information Systems Control Theory and Concepts
Domain 1—Risk Identification, Assessment and Evaluation
A. Chapter Overview

Process-specific controls for the following processes are addressed in Part II of this manual:
1. Managing the IT Strategy
2. Portfolio, Program and Project Management
3. Change Management
4. Third-party Service Management
5. Continuity Management
6. Information Security Management
7. Configuration Management
8. Problem Management
9. Knowledge Management
10. IT Operations Management

B. TASK AND KNOWLEDGE STATEMENTS

Introduction

This section describes the task and knowledge statements for Domain 1, which focuses on identifying, assessing and evaluating risk to enable the execution of the enterprise risk management (ERM) strategy.

Task Statements

The following table describes the task statements for Domain 1 that the CRISC candidate must know how to perform.

No.	Task Statement (TS)
TS1.1	Collect information and review documentation to ensure that risk scenarios are identified and evaluated.
TS1.2	Identify legal, regulatory and contractual requirements and organizational policies and standards related to information systems to determine their potential impact on the business objectives.
TS1.3	Identify potential threats and vulnerabilities for business processes, associated data and supporting capabilities to assist in the evaluation of enterprise risk.
TS1.4	Create and maintain a risk register to ensure that all identified risk factors are accounted for.
TS1.5	Assemble risk scenarios to estimate the likelihood and impact of significant events to the enterprise.
TS1.6	Analyze risk scenarios to determine their impact on business objectives.
TS1.7	Develop a risk awareness program and conduct training to ensure that stakeholders understand risk and contribute to the risk management process and to promote a risk-aware culture.
TS1.8	Correlate identified risk scenarios to relevant business processes to assist in identifying risk ownership.
TS1.9	Validate risk appetite and tolerance with senior leadership and key stakeholders to ensure alignment.

Part I—Risk Management and Information Systems Control Theory and Concepts
Domain 1—Risk Identification, Assessment and Evaluation
B. Task and Knowledge Statements

Knowledge Statements

The following table describes the knowledge statements for Domain 1. The CRISC candidate must have a good understanding of each of the areas delineated by the knowledge statements. These statements are the basis for the exam.

No.	Knowledge Statement (KS) — Knowledge of:
KS1.1	Standards, frameworks and leading practices related to risk identification, assessment and evaluation
KS1.2	Techniques for risk identification, classification, assessment and evaluation
KS1.3	Quantitative and qualitative risk evaluation methods
KS1.4	Business goals and objectives
KS1.5	Organizational structures
KS1.6	Risk scenarios related to business processes and initiatives
KS1.7	Business information criteria
KS1.8	Threats and vulnerabilities related to business processes and initiatives
KS1.9	Information systems architecture (e.g., platforms, networks, applications, databases and operating systems)
KS1.10	Information security concepts
KS1.11	Threats and vulnerabilities related to third-party management
KS1.12	Threats and vulnerabilities related to data management
KS1.13	Threats and vulnerabilities related to the system development life cycle
KS1.14	Threats and vulnerabilities related to project and program management
KS1.15	Threats and vulnerabilities related to business continuity and disaster recovery management
KS1.16	Threats and vulnerabilities related to management of IT operations
KS1.17	The elements of a risk register
KS1.18	Risk scenario development tools and techniques
KS1.19	Risk awareness training tools and techniques
KS1.20	Principles of risk ownership
KS1.21	Current and forthcoming laws, regulations and standards
KS1.22	Threats and vulnerabilities associated with emerging technologies

Note: Knowledge statements 1.8, 1.10-1.16 and 1.22 are process specific and are addressed in Part II of this manual.

C. THE BIG PICTURE—RISK MANAGEMENT AND RISK GOVERNANCE

Introduction

Enterprises continuously plan, operate and deploy business activities and processes to achieve business objectives. The risk practitioner is actively involved in ensuring that the operational risk of each business activity is assessed; monitored; and, if necessary, addressed.

Each business activity carries both risk and opportunity, and the risk practitioner must be aware of the need to balance business needs and productivity with effective controls. Some controls that will be specifically considered are IS controls; however, the risk practitioner must also be familiar with other risk response methods as seen in *Domain Two—Risk Response*.

The risk practitioner must be capable of evaluating risk across the entire enterprise, not just on a systemwide or departmentwide basis. An enterprisewide risk perspective is necessary because a risk in any area of the enterprise may pose a risk to other areas of the enterprise. Having a grasp of the "Big Picture" will allow risk management efforts to be better integrated with business priorities and culture; provide a more efficient, interoperable, measureable and effective risk response; and support governance and compliance requirements more thoroughly.

1. ESSENTIALS OF RISK GOVERNANCE

Introduction

This section contains a brief introduction to risk governance to provide the risk practitioner with a baseline understanding of the holistic environment in which the risk practitioner functions.

Relevance

Risk governance addresses the oversight of the business risk management strategy of the enterprise.

Risk governance is the domain of senior management and the shareholders of the enterprise. They establish the enterprise's risk culture and determine the acceptable levels of risk; set up the management framework; and ensure that the risk management function is operating effectively to identify, manage, monitor and report on current and potential risk facing the enterprise.

1.1 Key Terms and Concepts

Governance

Ensures that stakeholder needs, conditions and options are evaluated to determine balanced, agreed-on enterprise objectives to be achieved; setting direction through prioritization and decision making; and monitoring performance and compliance against agreed-on direction and objectives

> **Note:** Conditions can include the cost of capital, foreign exchange rates, etc. Options can include shifting manufacturing to other locations, subcontracting portions of the enterprise to third-parties, selecting a product mix from many available choices, etc. (from COBIT 5).

Risk Governance

Risk governance is a strategic business function that helps ensure that:
- Risk management activities align with the enterprise's opportunity and loss capacity and leadership's subjective tolerance of it.
- The risk management strategy is aligned with the overall business strategy.

Enterprise decisions consider the full range of (risk) opportunities and consequences.

> **Note:** The enterprise's risk appetite is defined as part of risk governance activities; this risk appetite should be reflected in the policies. A risk-averse enterprise has stricter policies than a risk-aggressive enterprise (from COBIT 5).

Part I—Risk Management and Information Systems Control Theory and Concepts
Domain 1—Risk Identification, Assessment and Evaluation
C. The Big Picture—Risk Management and Risk Governance

Responsibility for Risk Governance	Risk governance is ultimately the responsibility of the board of directors and senior management. They establish the enterprise's risk culture and the acceptable levels of risk; set up the management framework; and ensure that the risk management function is operating effectively to identify, manage, monitor and report on current and potential risk facing the enterprise.
	Note: While risk governance and the decisions made in the execution of risk governance ultimately are not the responsibility of the risk practitioner, the practitioner must nevertheless contribute to and enable sound risk management decisions through the execution of many underlying tasks associated with the risk governance process.
Risk Appetite	The amount of risk, on a broad level, that an entity is willing to accept in pursuit of its mission (or vision)
Risk Culture	Risk culture is the shared values and beliefs that govern the attitudes and behaviors toward risk taking, care and integrity, and determines how openly risk and losses are reported and discussed.
	The risk culture will drive the security culture and reflect the priorities and focus that is placed on designing, implementing and maintaining appropriate security controls.
Risk Tolerance	The acceptable level of variation that management is willing to allow for any particular risk as the enterprise pursues its business objectives
	Example: Standards require projects to be completed within the estimated budgets and time, but overruns of 10 percent of budget or 20 percent of time are tolerated.
	Note: The definitions for risk appetite and risk tolerance are compatible with the Committee of Sponsoring Organizations of the Treadway Commission (COSO) ERM definitions, which are equivalent to the ISO 31000 definition in Guide 73:2009, Risk Management Vocabulary.

1.2 Risk Governance Objectives

Risk Governance Objectives

Effective risk governance helps ensure that risk management practices are embedded in the enterprise, enabling it to secure optimal risk-adjusted return. Risk governance has four main objectives:
1. Establish and maintain a common risk view.
2. Integrate risk management into the enterprise.
3. Make risk-aware business decisions.
4. Ensure that risk management controls are implemented and operating correctly.

Risk Governance Objective	Description
1. Establish and maintain a common risk view.	Effective risk governance establishes the common view of risk for the enterprise. This determines which controls are necessary to mitigate risk and how risk-based controls are integrated into business processes and IS. The risk governance function sets the tone of the business regarding how to determine an acceptable level of risk tolerance. Risk governance is a continuous life cycle that requires regular reporting and ongoing review. The risk governance function must oversee the operations of the risk management team.
2. Integrate risk management into the enterprise.	Integrating risk management into the enterprise enforces a holistic enterprise risk management (ERM) approach across the entire enterprise. It requires the integration of risk management into every department, function, system and geographic location. Understanding that risk in one department or system may pose an unacceptable risk to another department or system requires that all business processes be compliant with a baseline level of risk management. The objective of ERM is to establish the authority to require all business processes to undergo a risk analysis on a periodic basis or when there is a significant change to the internal or external environment.
3. Make risk-aware business decisions.	To make risk-aware business decisions, the risk governance function must consider the full range of opportunities and consequences of each such decision and its impact on the enterprise, its place in society and the environment.
4. Ensure that risk management controls are implemented and operating correctly.	Governance requires oversight and due diligence to ensure that the enterprise is following up on the implementation and monitoring of controls to ensure that the controls are effective to mitigate risk and protect organizational assets.

Foundation for Effective Risk Governance

To effectively govern enterprise and IT risk, there must be an:
- Understanding and consensus with respect to the risk appetite and risk tolerance of the enterprise
- Awareness of risk and the need for effective communication about risk throughout the enterprise
- Understanding of the elements of risk culture

1.3 Risk Appetite and Tolerance

Introduction

"Risk appetite" and "risk tolerance" are concepts that are frequently used, but the potential for misunderstanding is high. Some people use the concepts interchangeably; others see a clear difference.

Major Factors Influencing Risk Appetite	When considering the risk appetite for the enterprise, the following two major factors are important: • The enterprise's objective capacity to absorb loss, e.g., financial loss, reputation damage • The (management) culture or predisposition toward risk taking—cautious or aggressive. (What is the amount of loss that the enterprise wants to accept to pursue a return?) **Note:** Risk appetite and risk tolerance should be applied not only to risk assessments, but also to all risk decision making.

Risk Appetite Variations Between Enterprises	Risk appetite can and will be different among enterprises—there is no absolute norm or standard of what constitutes acceptable and unacceptable risk. Every enterprise has to define its own risk appetite levels and should:
	• Ensure that such definitions/levels are:
	  – In line with the overall risk culture that the enterprise wants to express (i.e., ranging from very risk averse to risk-taking/opportunity-seeking)
	  – Well-defined, understood, communicated and consistent
	  – Reviewed on a regular basis

Exhibit 1.1: Risk Map Indicating Risk Appetite Bands	In practice, risk appetite can be defined in terms of combinations of frequency and magnitude of a risk, using risk maps. Exhibit 1.1 and the following table depict and describe different bands of risk significance, based on frequency and magnitude of risk. The magnitude of risk is defined as "a measure of the potential severity of loss or the potential gain from realized events/scenarios."

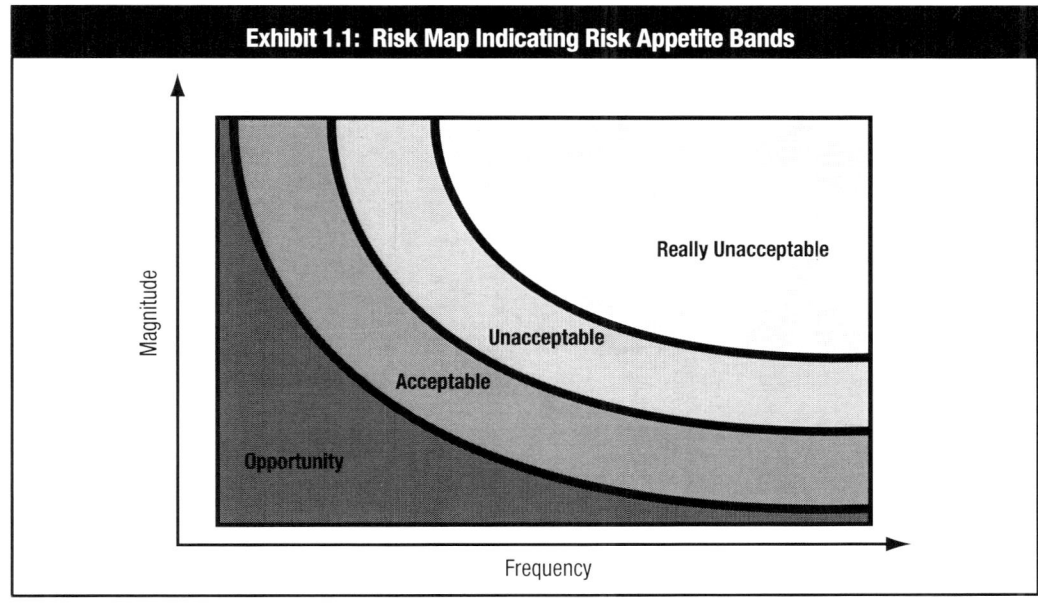

Risk Level	Description
Really unacceptable	Indicates really unacceptable risk. The enterprise estimates that this level of risk is far beyond its normal risk appetite. Any risk found to be in this band is likely to trigger an immediate risk response.
Unacceptable	Indicates elevated risk, i.e., also above acceptable risk appetite. The enterprise may, as a matter of policy, require mitigation or another adequate response to be defined within certain time boundaries.
Acceptable	Indicates a normal, acceptable level of risk, usually with no special action required, except for maintaining the current controls or other responses.
Opportunity	Indicates very low risk, in which cost-saving opportunities may be found by decreasing the degree of control or in which opportunities for assuming more risk may arise.

Note: This risk appetite scheme is provided as an example only. Each enterprise needs to define its own risk appetite levels and review them regularly.

Part I—Risk Management and Information Systems Control Theory and Concepts
Domain 1—Risk Identification, Assessment and Evaluation
C. The Big Picture—Risk Management and Risk Governance

Guidelines for Risk Appetite and Risk Tolerance

The guidelines listed in the following table apply to risk appetite and risk tolerance.

Guidelines for Risk Appetite and Risk Tolerance	
Guideline	**Description**
Risk appetite and risk tolerance must connect.	Risk appetite and risk tolerance go hand in hand. Risk tolerance is defined at the enterprise level and is reflected in policies set by senior management. At lower (tactical) levels of the enterprise, or in some entities of the enterprise, exceptions can be tolerated (or different thresholds defined) as long as the overall exposure does not exceed the risk appetite set at the enterprise level. Any business initiative includes a risk component (both for profit or opportunity and for loss), so management should have the discretion to pursue new opportunities of risk. Enterprises in which policies are inflexible (cast in stone rather than as "lines in the sand"), could lack the agility and innovation to exploit new business opportunities. Conversely, there are situations in which policies are based on specific legal, regulatory or industry requirements and force the organization and management to have no risk tolerance for failure to be compliant with the requirements.
Exceptions to risk tolerance standards must be reviewed and approved.	Risk tolerance is defined at the enterprise level by the board of directors or senior management and clearly communicated to all stakeholders. A process should be in place to review and approve any exceptions to such standards.
Risk appetite and tolerance change over time.	Risk appetite and tolerance change due to: • New technology • New organizational structures • New market conditions • New business strategy • Many other factors Such factors require an enterprise to reassess its risk portfolio at regular intervals and also require the enterprise to reconfirm its risk appetite at regular intervals, triggering risk policy reviews. In this respect, an enterprise also needs to understand that the better risk management it has in place, the more risk can be taken in pursuit of return.
Cost of risk mitigation options can affect risk tolerance.	There may be circumstances in which the cost/business impact of risk mitigation options exceeds an enterprise's capabilities/resources, thus forcing higher tolerance for one or more risk conditions. **Example:** If a regulation states that sensitive data at rest must be encrypted, yet there is no feasible encryption solution or the cost of implementing a solution would have a large negative impact, the enterprise may choose to accept the risk associated with regulatory noncompliance, which is a form of risk trade-off.

1.4 Risk Culture

Relevance

Risk management efforts are directly and indirectly affected by the risk culture.

A risk-aware culture:
- Characteristically offers a setting in which components of risk are discussed openly and acceptable levels of risk are understood and maintained

Risk awareness also implies that all levels within an enterprise are aware of why a response is needed and how to respond to adverse events.

Responsibility for Risk Culture

In a risk-aware culture, the tone is set at the top with the board and business executives who:
- Set direction.
- Communicate risk-aware decision making.
- Reward effective risk management behaviors.

The risk governance function is the body that most significantly affects the risk culture of an enterprise because organizational beliefs are created and shaped at the leadership level.

Exhibit 1.2: Elements of a Risk Culture

Risk culture is a concept that is not easy to describe. Exhibit 1.2 and the following table depict and describe the series of behaviors that are elements of a risk culture.

Source: ISACA, *The Risk IT Framework*, USA, 2009, figure 11

Elements of a Risk Culture	
Behavior toward taking risk	How much risk does the enterprise feel it can absorb, and what specific risk is it willing to take?
Behavior toward policy compliance	To what extent will people embrace and/or comply with policy?
Behavior toward negative outcomes	How does the enterprise deal with negative outcomes, i.e., loss events or missed opportunities? Will it learn from them and try to adjust, or will blame be assigned without treating the root cause?
Symptoms of an Inadequate or Problematic Risk Culture	
Misalignment between real risk appetite and translation into policies	Management's real position toward risk can be reasonably aggressive and risk taking, whereas the policies that are created reflect a much stricter attitude.
Existence of a "blame culture"	This type of culture should, by all means, be avoided; it is the most effective inhibitor of relevant and efficient communication. In a blame culture, business units tend to point the finger at IT when projects are not delivered on time or do not meet expectations. In doing so, they fail to realize how the business unit's involvement up front affects project success. In extreme cases, the business unit may assign blame for a failure to meet the expectations that the unit never clearly communicated. The "blame game" only detracts from effective communication across units, further fueling delays. Executive leadership must identify and quickly control a blame culture if collaboration is to be fostered throughout the enterprise.

Benefits of a Risk-aware Culture

A risk-aware culture offers a setting in which components of risk are discussed openly and acceptable levels of risk are understood and maintained.

Risk awareness also implies that all levels within an enterprise understand why a response is needed and how to respond to adverse IT events.

2. ESSENTIALS OF RISK MANAGEMENT

Introduction

This section clarifies the scope of risk management in comparison to risk governance and provides an overview of the roles and responsibilities related to risk management.

2.1 Key Terms and Concepts

Risk

Reflects the combination of the probability of an event occurring and the impact the event has on the enterprise.

Risk—the potential for events and their consequences—contains both:
- Opportunities for benefit (upside)
- Threats to success (downside)

Part I—Risk Management and Information Systems Control Theory and Concepts
Domain 1—Risk Identification, Assessment and Evaluation
C. The Big Picture—Risk Management and Risk Governance

Management	Often differentiated from governance as the distinction between being "committed" (governance) and "involved" (management), management entails the judicious use of means (resources, people, processes, practices, etc.) to achieve an identified end. Management plans, builds, runs and monitors activities in alignment with the direction set by the governance body to achieve the enterprise objectives. Management is responsible for the execution of prescribed tasks or activities within the direction set by the guiding body or unit. Management is about planning, building, organizing and controlling operational activities to align with the direction set by the governance body. It is a means or instrument by which the governance body achieves a result or objective.
Risk Management	The coordinated activities to direct and control an enterprise with regard to risk Risk management is the identification, assessment and prioritization of risk followed by coordinated and economical application of resources to minimize, monitor and control the probability and/or impact of adverse events or to maximize the realization of opportunities.
Guiding Principles for Effective Risk Management	The following are guiding principles for effective risk management: • Maintain focus on the business mission, goals and objectives. • Integrate IT risk management into enterprise risk management (ERM). • Balance the costs and benefits of managing risk. • Promote fair and open communication. • Establish tone at the top and assign personal accountability. • Promote continuous improvement as part of daily activities. The following table provides further detail.

Guiding Principles for Effective Risk Management	
Principle	**Description**

Maintain focus on the business mission, goals and objectives.	• All risk is treated as a business risk, and the risk management approach must be comprehensive and cross-functional.
	• The focus is on business outcome. Each business function supports the achievement of business objectives; IT-related risk is not only measured according to the impact the risk may have on IT directly but must also be measured according to the impact it can have on the achievement of business objectives or strategy.
	• Every risk analysis considers business and IT-process resilience and contains a dependency analysis of the extent to which the business process depends on IT-related resources, such as:
	  – People
	  – Information
	  – Applications
	  – Infrastructure
	• IT-related business risk is viewed from two angles:
	  – Protection against value destruction
	  – Enablement of value generation

Part I—Risk Management and Information Systems Control Theory and Concepts
Domain 1—Risk Identification, Assessment and Evaluation
C. The Big Picture—Risk Management and Risk Governance

Guiding Principles for Effective Risk Management *(cont.)*

	Guiding Principles for Effective Risk Management *(cont.)*
Principle	**Description**
Integrate IT risk management into enterprise risk management (ERM).	Most enterprises rely so heavily on information systems that a systems failure could have devastating consequences. For that reason, IT risk must be considered as a part of overall business risk. Risk assessments must be built into all aspects of the enterprise's business model. This includes ensuring that: • Business objectives and the amount of risk that the enterprise is prepared to take are clearly defined and documented. • The entity's risk appetite reflects its risk management philosophy and influences the culture and operating style (as stated in the Committee of Sponsoring Organizations of the Treadway Commission [COSO] *Enterprise Risk Management—Integrated Framework*). • Risk issues are integrated for each business unit, i.e., the risk view is consistent and consolidated across the overall enterprise. • Attestation of/sign-off on control environment is provided.
Balance the costs and benefits of managing risk.	• Risk is prioritized and addressed in line with risk appetite and tolerance. • Controls are implemented to prevent or minimize impact and yet must not place an undue burden on business operations. All risk-based controls should be based on a cost-benefit analysis. In other words, controls are not implemented simply for the sake of implementing controls. • Existing controls are leveraged to address multiple risk factors or to address risk more efficiently.
Promote fair and open communication.	• Open, accurate, timely and transparent information related to risk is exchanged and serves as the basis for all risk-related decisions. • Risk issues, principles and risk management methods are integrated across the enterprise and built into all phases of the business and operational life cycles. • Technical findings are translated into relevant and understandable business terms.
Establish tone at the top and assign personal accountability.	• Key personnel, i.e., influencers, business owners and the board of directors, is engaged in risk management. • There is clear assignment and acceptance of risk ownership. • Top management provides direction by means of policies, procedures and the right level of enforcement. • Enterprise leadership actively promotes a risk-aware culture. • Authorized individuals make risk decisions, including business-focused IT risk, e.g., for IT investment decisions, project funding, major IT environment changes, risk assessments, and the monitoring and testing of controls.

Promote continuous improvement as part of daily activities.	• Because of the dynamic nature of risk, risk management is an iterative, perpetual and ongoing process.
	• The enterprise pays attention to consistent risk assessment methods, roles and responsibilities, tools, techniques, and criteria across the enterprise, noting especially:
	  – Identification of key processes and associated risk
	  – Understanding of impacts on achieving business objectives
	  – Identification of triggers that indicate when an update of the framework is required
	• Risk management practices are appropriately prioritized and embedded in enterprise decision-making processes that enable *risk-return*-aware business decisions.
	• Risk management practices are straightforward and easy to use and contain practices to detect, prevent and mitigate threat and potential risk.

Part I—Risk Management and Information Systems Control Theory and Concepts
Domain 1—Risk Identification, Assessment and Evaluation
C. The Big Picture—Risk Management and Risk Governance

Holistic Risk and Opportunity Management

Risk and opportunity go hand in hand. To provide business value to stakeholders, enterprises must engage in various activities and initiatives, all of which carry various degrees of uncertainty and, therefore, risk.

Managing risk and opportunity is a key strategic activity for enterprise success.

Risk management must be integrated across all business processes. Risk must be carefully considered whenever strategic plans are being considered and whenever a change to an enterprise's processes or systems is being implemented. Performing prudent risk management early in each business activity will allow the enterprise to take steps to minimize risk and avoid situations where an unnecessary level of risk would be encountered.

Responsibilities and Accountability for Risk Management

Various members of the enterprise have responsibility or accountability for risk management. In this context:
- *Responsibility* belongs to those who must ensure that the activities are completed successfully.
- *Accountability* applies to those individuals, groups or entities that are ultimately responsible for the subject matter, process or scope.

RACI Chart

RACI charts provide suggested assignment of level of responsibility for process practices to different roles and structures.

Level of Responsibility	Description
R(esponsible)	**Who is getting the task done?** This refers to the roles taking the main operational stake in fulfilling the activity listed and creating the intended outcome
A(ccountable)	**Who accounts for the success of the task?** This assigns the overall accountability for getting the task done (Where does the buck stop?). Note that the role mentioned is the lowest appropriate level of accountability; there are, of course, higher levels that are accountable, too. To enable empowerment of the enterprise, accountability is broken down as far as possible. Accountability does not indicate that the role has no operational activities; it is very likely that the role gets involved in the task. As a principle, accountability cannot be shared.
C(onsulted)	**Who is providing input?** These are key roles that provide input. Note that it is up to the accountable and responsible role(s) to obtain information from other units or external partners, too. However, inputs from the roles listed are to be considered and, if required, appropriate action has to be taken for escalation, including the information of the process owner and/or the steering committee.
I(nformed)	**Who is receiving information?** These are roles who are informed of the achievements and/or deliverables of the task. The role in "accountable," of course, should always receive appropriate information to oversee the task, as does the responsible roles for their area of interest.

Part I—Risk Management and Information Systems Control Theory and Concepts
Domain 1—Risk Identification, Assessment and Evaluation
C. The Big Picture—Risk Management and Risk Governance

RACI Chart for Risk Management

The following RACI chart example provides suggested assignment of level of responsibility for the risk management process. The enterprise roles listed are shaded darker than the IT roles.

AP012 RACI Chart

Key Management Practice	Board	Chief Executive Officer (CEO)	Chief Financial Officer (CFO)	Chief Operating Officer (COO)	Business Executives	Business Process Owners (BPOs)	Strategy Executive Committee	Steering (Programmes/Projects) Committee	Project Management Office (PMO)	Value Management Office (VMO)	Chief Risk Officer (CRO)	Chief Information Security Officer (CISO)	Architecture Board	Enterprise Risk Committee	Head Human Resources (HR)	Compliance	Audit	Chief Information Officer (CIO)	Head Architect	Head Development	Head IT Operations	Head IT Administration	Service Manager	Information Security Manager (ISM)	Business Continuity Manager	Privacy Officer
AP012.01 Collect data.		I				R		R	R		R	R	I			C	C	A	R	R	R	R	R	R	R	R
AP012.02 Analyse risk.		I				R		C			R	C		I		R	R	A	C	C	C	C	C	C	C	C
AP012.03 Maintain a risk profile.		I				R		C			A	C		I		R	R	R	C	C	C	C	C	C	C	C
AP012.04 Articulate risk.		I				R		C			R	C		I		C	C	A	C	C	C	C	C	C	C	C
AP012.05 Define a risk management action portfolio.		I				R		C			A	C		I		C	C	R	C	C	C	C	C	C	C	C
AP012.06 Respond to risk.		I				R		R			R	R		I		C	C	A	R	R	R	R	R	R	R	R

Source: ISACA, *COBIT® 5: Enabling Processes*, USA, 2012

D. RISK MANAGEMENT FRAMEWORKS, STANDARDS AND PRACTICES

Introduction Frameworks, standards and leading practices are tools that enable an efficient and effective risk management process. Their use enables various degrees of homogeneity, scalability, and enables benchmarking across entities.

1. KEY TERMS AND CONCEPTS

Framework A framework is a generally accepted, business-process-oriented structure that establishes a common language and enables repeatable business processes.

> **Note:** This term may be defined differently in different disciplines. This definition suits the purposes of this manual.

Leading Practice A frequent or usual action performed as an application of knowledge

A leading practice is defined as an action that optimally applies knowledge in a particular area.

Practices are issued by a "recognized authority" that is appropriate to the subject matter. Issuing bodies may include professional associations and academic institutions or commercial entities such as software vendors. They are generally based on a combination of research, expert insight and peer review.

> **Note:** Practices usually are derived from and supplement/support standards and frameworks and are the least formal of the three.

Standard A mandatory requirement, code of practice or specification approved by a recognized external standards organization, such as the International Organization for Standardization (ISO)

Standards are usually intended for compliance purposes and to provide assurance to others who interact with a process or outputs of a process (e.g., food and drug quality).

Standards are intended to be implemented in a rigid way and to minimize the number of deviations based on a cost-benefit analysis. Deviations from the standard should only be granted on an "exception" basis and should follow a defined approval process.

Relevance of Risk Management Frameworks, Standards and Practices Frameworks, standards and practices matter to the risk practitioner because they:
- Provide a systematic view of "things to watch" that could result in harm to customers or an enterprise
- Act as a guide to focus efforts of diverse teams
- Save time and costs, such as training costs, operational costs and performance improvement costs
- Help achieve business objectives more quickly and easily
- Provide credibility to engage senior management (e.g., chief financial officer [CFO]), C-suite leadership and functional management teams

Part I—Risk Management and Information Systems Control Theory and Concepts
Domain 1—Risk Identification, Assessment and Evaluation
D. Risk Management Frameworks, Standards and Practices

2. EXAMPLES OF FRAMEWORKS RELATED TO RISK MANAGEMENT AND IS CONTROL

Examples of Risk Management Frameworks

The following table provides examples of frameworks related to risk management.

Issuing Body	Publication
ISACA	The Risk IT Framework
ISACA	Enterprise Value: Governance of IT Investments, The Val IT Framework 2.0
ISACA	COBIT® 5 **Note:** The following three COBIT 5 publications are available from ISACA and can be downloaded at www.isaca.org/cobit: • The COBIT 5 framework (complimentary PDF) • COBIT® 5: Enabling Processes (member complimentary PDF) • COBIT® 5 Implementation (member complimentary PDF)
Committee of Sponsoring Organizations of the Treadway Commission (COSO)	Enterprise Risk Management—Integrated Framework http://coso.org/
US National Institute of Standards and Technology (NIST)	Risk Management Framework, NIST Special Publication 800-39 Managing Information Security Risk http://csrc.nist.gov/publications/PubsSPs.html

Reminder: Frameworks can be applied flexibly within an enterprise.

3. EXAMPLES OF STANDARDS RELATED TO RISK MANAGEMENT AND IS CONTROL

Examples of Risk Management Standards

Standards related to risk management include, but are not limited to, those in the following table.

Issuing Body	Publication
ISACA	IT Audit and Assurance Standards
International Organization for Standardization (ISO)	ISO 31000:2009 **Note:** Unlike other ISO "standards," this was not intended to be used for certification.
ISO/International Electrotechnical Commission (IEC)	ISO/IEC 2700x (for information security management systems [ISMSs])
International Organization for Standardization (ISO)	ISO 22301:2012 (replaces BS 25999-2) Societal security—Business continuity management systems www.iso.org
Payment Card Industry (PCI) Security Standards Council	PCI Data Security Standard (PCI DSS) www.pcisecuritystandards.org

Reminder: Standards—including corporate standards, which are not addressed here—ideally define measurable objectives to enable compliance assessments. Standards are intended to be implemented in a rigid way with variations only as allowed in the standard.

4. EXAMPLES OF LEADING PRACTICES RELATED TO RISK MANAGEMENT AND IS CONTROL

Examples of Risk Management or Control Leading Practices

The following table provides examples leading practices related to risk management or control.

Issuing Body	Publication
ISACA	*The Risk IT Practitioner Guide*
ISO/IEC	ISO/IEC 2700x (for Information Security Management Systems)
NIST	NIST Special Publication (SP) 800-37, Revision 1, *Guide for Applying the Risk Management Framework to Federal Information Systems* NIST Special Publication (SP) 800-30, *Risk Management Guide for Information Technology Systems* (Note: The draft update has been issued.)
Carnegie Mellon University (CMU) Software Engineering Institute (SEI)	Operationally Critical Threat, Asset, and Vulnerability Evaluation℠ (OCTAVE®)
Spanish Ministry for Public Administrations	Methodology for Information Systems Risk Analysis and Management (MAGERIT version 2)

E. RISK IDENTIFICATION, ASSESSMENT AND EVALUATION

Introduction Risk identification, assessment, and evaluation is the first step in the risk management process. It is concerned with the determination of risk levels and providing guidance for the later steps of risk response, monitoring and maintenance.

1. KEY TERMS AND CONCEPTS

Frequency A measure of the rate by which events occur over a certain period of time

IT Risk The business risk associated with the use, ownership, operation, involvement, influence and adoption of IT within an enterprise

Magnitude A measure of the potential severity of loss or the potential gain from realized events/scenarios

Risk Aggregation The process of integrating risk assessments at a corporate level to obtain a complete view of the overall risk for the enterprise

Risk Analysis A process by which frequency and magnitude of risk scenarios are estimated

Risk Assessment A process used to identify and evaluate risk and its potential effects

Scope Note: Risk assessment includes assessing the critical functions necessary for an enterprise to continue business operations, defining the controls in place to reduce exposure and evaluating the cost for such controls. Risk analysis often involves an evaluation of the probabilities of a particular event.

Risk Evaluation The process of comparing the estimated risk against given risk criteria to determine the significance of the risk [ISO/IEC Guide 73:2002]

Risk Identification Risk identification is the process of determining and documenting the risk that an enterprise faces. The identification of risk is based on the recognition of threats, vulnerabilities, assets and controls in the enterprise's operational environment.

Part I—Risk Management and Information Systems Control Theory and Concepts
Domain 1—Risk Identification, Assessment and Evaluation
E. Risk Identification, Assessment and Evaluation

2. RISK IDENTIFICATION, ASSESSMENT AND EVALUATION PROCESS OBJECTIVES

Inputs From Other Domains

Risk management is a never-ending process. Therefore, effective risk identification, assessment and evaluation includes the review of the enterprise's risk response decisions, risk monitoring results, the existing information systems control architecture and the results of information systems control monitoring efforts.

Domain 2—Risk Response provides risk management action plans that affect the enterprise's overall risk profile.
Domain 3—Risk Monitoring provides key risk indicators (KRIs) that enable tracking of specific risk over time.
Domain 4—Information Systems Control Design and Implementation provides the enterprise's current information systems control inventory.
Domain 5—Information Systems Control Monitoring and Maintenance provides timely information on the actual design and operating effectiveness of existing controls.

Process Objective

The purpose of the risk identification, assessment and evaluation process is to document the risk to the enterprise, prioritize the various risk factors and provide recommendations on what risk responses should be implemented.

The risk practitioner needs to ensure that all risk to the enterprise is considered, documented and evaluated. This ensures that the enterprise is aware of all relevant risk and can make risk-aware business decisions.

If the risk practitioner fails to consider both the positive and negative angles of risk during assessment, the credibility of the risk practitioner and the value of the risk response recommendations provided through the risk identification, assessment and evaluation will be incomplete.

Outputs to Other Domains

The primary output of the risk identification, assessment and evaluation process includes a comprehensive, prioritized inventory of relevant risk—often in a form of a risk register—as well as risk response recommendations. These are core inputs for *Domain 2—Risk Response* and *Domain 4—Information Systems Control Design and Implementation.*

3. RISK IDENTIFICATION, ASSESSMENT AND EVALUATION PROCESS

Risk Identification, Assessment and Evaluation Activities

Effective risk identification, assessment and evaluation involves:
- Collecting data on:
  - The enterprise's operating environment (both internal and external)
  - Risk events
- Identifying risk factors (both internal and external)
- Analyzing and estimating risk (through the use of techniques such as risk scenarios)
- Identifying business process resilience level, including the related IT services and supporting assets
- Monitoring threat advisories to be aware of new or evolving threats

Reference: For more information on:
- Risk scenarios, see section F in this chapter
- Risk factors, see section in G in this chapter
- IT risk identification and assessment, see section I in this chapter

Exhibit 1.3: IT Risk in the Risk Hierarchy

IT risk can be categorized in different ways, as depicted in **exhibit 1.3**.

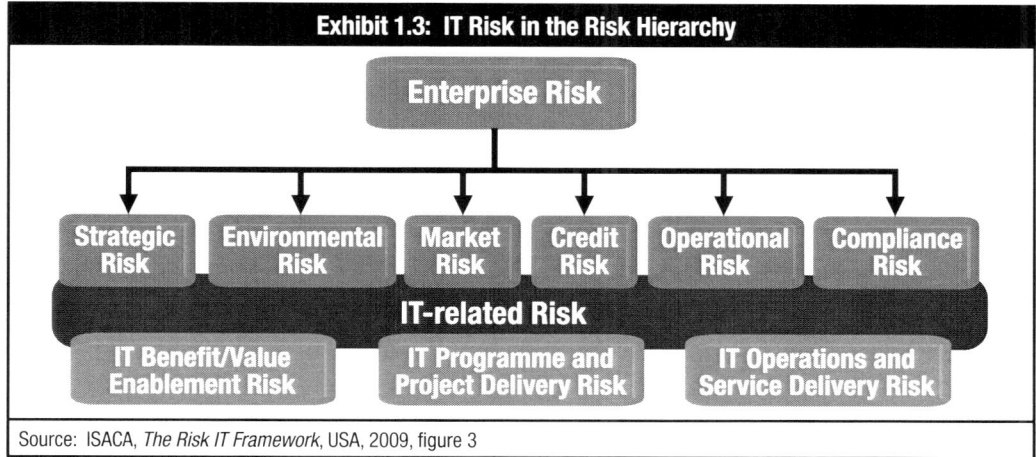

IT Risk Categories

IT risk can be described as follows:
- **IT benefit/value enablement risk**—Associated with (missed) opportunities to use technology to improve efficiency or effectiveness of business processes or as an enabler for new business initiatives
- **IT program and project delivery risk**—Associated with the contribution of IT to new or improved business solutions, usually in the form of projects and programs
- **IT operations and service delivery risk**—Associated with the performance of IT systems and services, which can bring destruction or reduction of value to the enterprise

Exhibit 1.4: IT Risk Categories

Exhibit 1.4 shows that there is an equivalent upside for all risk; for example, successful project delivery brings new business functionality.

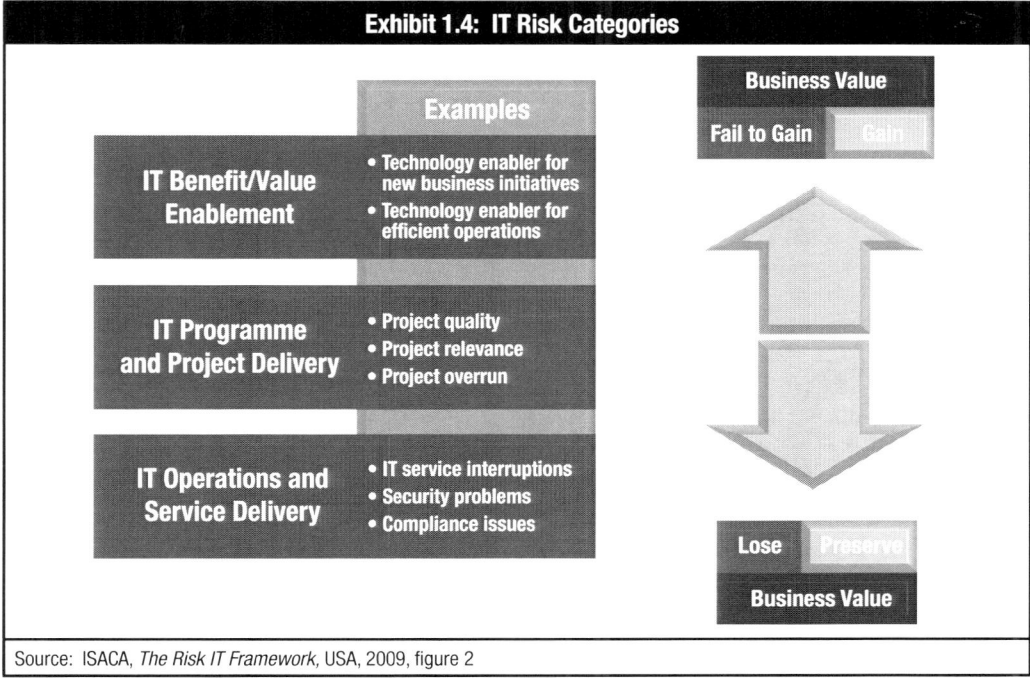

Part I—Risk Management and Information Systems Control Theory and Concepts
Domain 1—Risk Identification, Assessment and Evaluation
E. Risk Identification, Assessment and Evaluation

Exhibit 1.5: Domain 1 High-level Process Phases

Exhibit 1.5 depicts the phases of the risk identification, assessment and evaluation process.

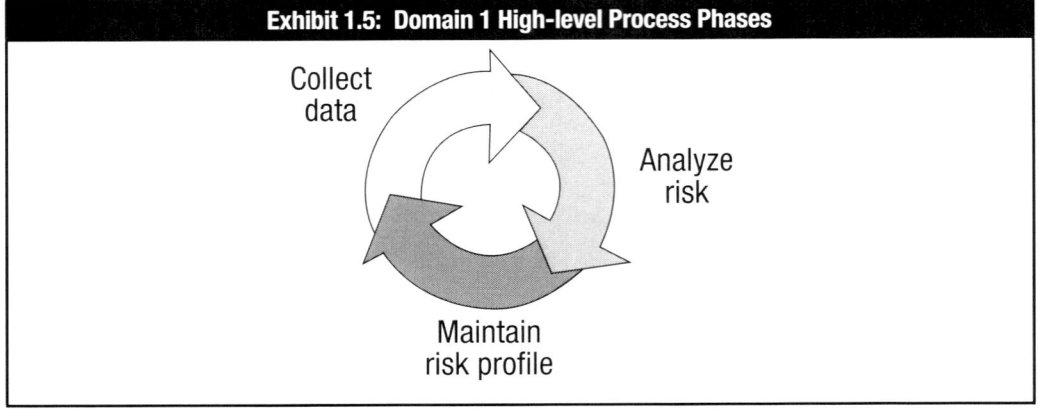

High-level Phases

The following table describes the phases of the risk identification, assessment and evaluation process.

> **Note:** For more information on these phases, see ISACA, The Risk IT Framework, USA, 2009.

Risk Identification, Assessment and Evaluation High-level Phases	
Phase	**Description**
1. Collect data.	Requires collecting data on the business environment, types of events, risk categories, etc., to identify relevant data to enable effective risk identification, analysis and reporting
2. Analyze risk.	Requires analyzing risk to develop information to support risk decisions that take into account the business relevance of risk factors.
3. Maintain a risk profile.	Requires maintaining an up-to-date and complete inventory (in a risk register or similar tool) of known threats and their attributes (e.g., expected likelihood, potential impact, disposition), IT, resources, capabilities and controls as understood in the context of business products, services and processes to effectively monitor risk over time

Collecting Data

Risk assessment will only be effective if the correct data has been gathered prior to conducting the risk analysis.

Some of the methods used to collect risk data include:
- Interviews
- Questionnaires and surveys
- Facilitated workshops
- Observation
- Testing

Describing the Business Impact of IT Risk

As most enterprises are fully dependent on working information systems to support the business processes, the coverage of IT-related business risk (IT risk) is a main task of the risk practitioner.

Meaningful IT risk assessments and risk-based decisions require IT risk to be expressed in clear, business-relevant terms. Effective risk management requires mutual understanding between IT and the business over which risk needs to be managed, the possible ways to manage risk, risk mitigation priorities and supporting rationales.

IT risk is a risk to the business—specifically, the business risk associated with the use, ownership, operation, involvement, influence and adoption of IT within an enterprise. It consists of IT-related events and conditions that could potentially impact the business. It can occur with both uncertain likelihood and magnitude of impact, and it creates challenges in meeting operational and strategic goals and objectives.

F. RISK SCENARIOS

Introduction

One of the challenges for risk management is to identify relevant risk. One of the techniques to overcome this challenge is the development and use of risk scenarios. It is a core approach to bring realism, insight, organizational engagement, improved analysis and structure to the complex matter of enterprise risk.

Once these scenarios are developed, they are used during the risk evaluation, in which likelihood of the risk and its business impacts are estimated.

1. KEY TERMS AND CONCEPTS

Risk Scenario	A description of an event that can lead to a business impact
Purpose of Risk Scenario Analysis	Risk scenario analysis is a technique used to: • Describe risk in a more concrete and tangible manner • Allow for proper risk assessment and analysis

2. RISK SCENARIO DEVELOPMENT

Exhibit 1.6: Risk Scenario Development

Exhibit 1.6 describes both the top-down and bottom-up scenario development approaches and the different categories of risk factors.

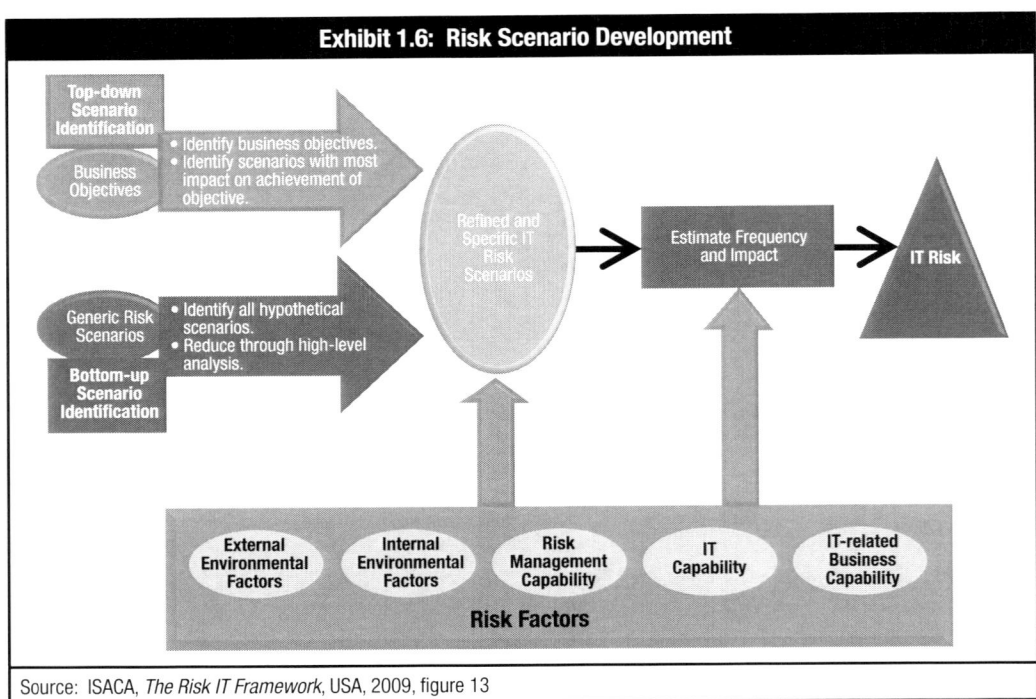

Source: ISACA, *The Risk IT Framework*, USA, 2009, figure 13

Part I—Risk Management and Information Systems Control Theory and Concepts
Domain 1—Risk Identification, Assessment and Evaluation
F. Risk Scenarios

Example	Based on a risk assessment, an enterprise has 20 key applications supported by three distinctly different technology platforms. The number of theoretically possible scenarios approaches 100,000, which is not feasible to maintain. The number of scenarios to be developed and analyzed should be kept to a much smaller, manageable number because every possible combination cannot be retained.
Changes to Risk Factors and the Enterprise Over Time	Because risk factors and the enterprise change over time, scenarios will also change. This requires continuous risk assessment (risk monitoring). Risk assessment should be performed at least on an annual basis, or when important internal or external changes occur.
Approaches to Risk Scenario Development	Risk scenarios can be derived via two different mechanisms: • **Top-down approach**—From the overall business objectives, an analysis of the most relevant and probable risk scenarios impacting the business objectives is performed. If the impact criteria are well aligned with the real value drivers of the enterprise, relevant risk scenarios will be developed. • **Bottom-up approach**—A list of generic scenarios is used to define a set of more concrete and customized scenarios, which are then applied to the individual enterprise situation. **Note:** The approaches are complementary and should be used together. Risk scenarios must be realistic, but thorough. Risk scenarios should be relevant and linked to realistic business risk scenarios. On the other hand, using a wide-ranging set of generic risk scenarios helps to ensure that no risk is overlooked and provides a more comprehensive and complete view of risk. A complex enterprise may consider several hundred different risk scenarios in order to ensure that all significant risk has been considered. An enterprise should also use the lessons learned from previous incidents in risk scenario development.

Risk Scenario Development

Business objectives drive the top-down risk scenario development approach; thus, the approach is unique to each enterprise. The approach is most beneficial in ensuring that the risk scenarios remain relevant and linked to real business risk.

In practice, using a set of generic risk scenarios is suggested since it helps ensure that no risks are overlooked and provides a comprehensive and complete view of risk.

The following table describes the bottom-up risk scenario development process in more detail.

Bottom-Up Risk Scenario Development Process Steps	
Step	**Description**
1	Using a list of generic risk scenarios, define a (manageable) set of concrete risk scenarios for the enterprise. In determining a "manageable" set of scenarios, a business may begin by considering: • Commonly occurring scenarios in its industry or product area • Scenarios representing threat sources that are increasing in number or severity • Scenarios that involve legal and regulatory requirements applicable to the business **Note:** Some less common situations should also be included in the scenarios.
2	Perform a validation against the business objectives of the entity. Do the selected risk scenarios address potential impacts on achievement of business objectives of the entity, in support of the overall enterprise's business objectives?
3	Refine the selected scenarios based on this validation, and detail them to a level in line with the criticality of the entity.
4	Reduce the number of scenarios to a manageable set. "Manageable" does not signify a fixed number, but should be in line with the overall importance (size) and criticality of the unit **Note:** There is no general rule, but if scenarios are reasonably and realistically scoped, the enterprise should expect to develop at least a few dozen scenarios.
5	Keep all risk factors in a register so that they can be reevaluated in the next iteration and included for detailed analysis if they have become relevant at that time.
6	Include in the scenarios an unspecified event—how to address an incident not covered by other scenarios.

Note: Once the set of risk scenarios is defined, it can be used for risk analysis. In risk analysis, likelihood and impact of the scenarios are assessed. Important components of this assessment are the risk factors.

Reference: For more information on risk factors, see section G in this chapter.

F. Risk Scenarios

Exhibit 1.7: Risk Scenario Components

Exhibit 1.7 depicts the components needed to ensure that risk scenarios are complete and usable for risk analysis purposes.

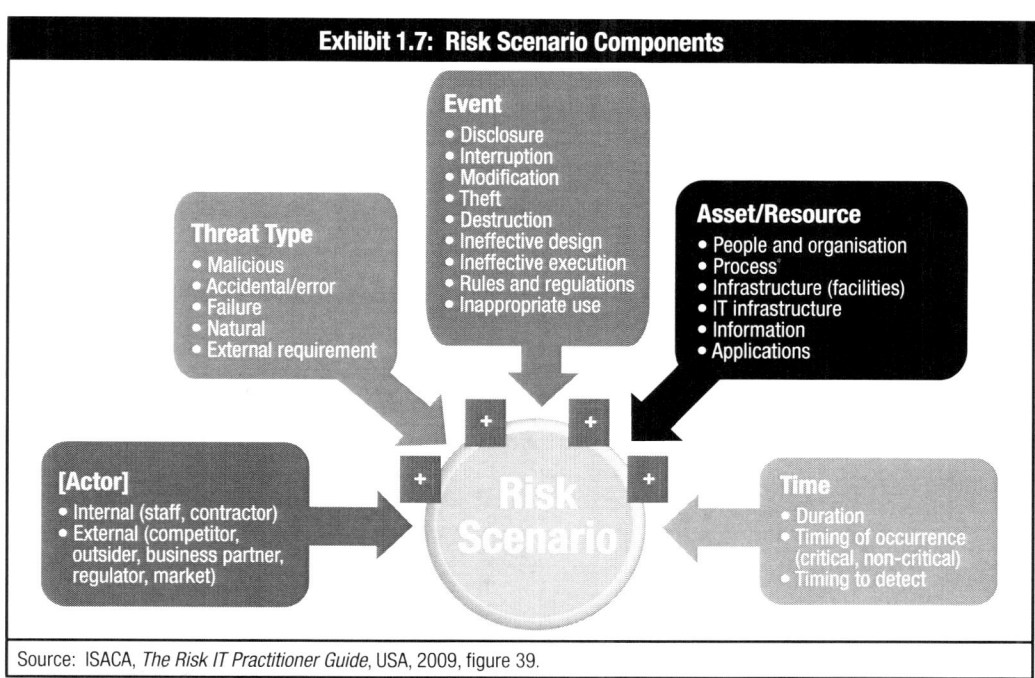

Source: ISACA, *The Risk IT Practitioner Guide*, USA, 2009, figure 39.

Risk Scenario Component Descriptions

The following table describes the different risk scenario components in detail.

Risk Scenario Components	
Component	**Description**
Actor	Actors generate the threat and can be internal or external, human or nonhuman: • Internal actors are within the enterprise, e.g., staff, contractors. • External actors include outsiders, competitors, regulators and the market. **Note:** Not every type of threat requires an actor, e.g., failures or natural causes.
Threat type	The threat type is the nature of the event. Is it malicious? If not, is it accidental or is it a failure of a well-defined process? Is it a natural event (*force majeure*)?
Event	A scenario always has to contain an event. Is it disclosure (of confidential information), interruption (of a system or project), modification, theft, destruction, etc.?

Risk Scenario Component Descriptions *(cont.)*

Risk Scenario Components *(cont.)*			
Component	**Description**		
Asset	An asset is any object of value to the enterprise that can be either tangible or intangible.		
		Tangible Assets	**Intangible Assets**
	Attribute	• Has physical attributes and can be detected with the senses	• Has no physical attributes and cannot be detected with the senses
	Examples	• People • Infrastructure • Finances	• Information • Reputation • Customer trust
	Relevance	• Theft or misuse more easily detected than with intangible assets	• May prove difficult to quantify • Focus on the tangible assets that contain, protect, affect or represent the intangible asset
Timing dimension	The timing dimension is the relevance of the scenario to the: • Time to detect, respond to or recover from the event • Timing: Does the event occur at a critical moment? • Duration of the event (extended outage of a service or data center) • Time lag between the event and the consequence: Is there an immediate consequence, (e.g., network failure, immediate downtime) or a delayed consequence (e.g., wrong IT architecture with accumulated high costs over a time span of several years)?		

> **Note:** The risk scenario structure differentiates between:
> • Loss events (events generating the negative impact)
> • Vulnerabilities or vulnerability events (events contributing to the likelihood or impact of loss events occurring)
> • Threat events (circumstances or events that can trigger loss events)

Managing the Number of Risk Scenarios

One technique of keeping the number of scenarios manageable is to develop:
• A set of generic scenarios throughout the enterprise
• More detailed scenarios in areas in which risk levels are higher

The assumptions made when grouping or generalizing scenarios should be well understood by all and adequately documented.

There should be a sufficient number of risk scenario scales reflecting the complexity of the enterprise and the extent of exposures to which the enterprise is subject.

The enterprise must consider risk that has not yet occurred and consider developing scenarios around unlikely, obscure or nonhistorical events.

Example: The term "insider threat" may not adequately explain whether the scenario addresses the threat of privileged and nonprivileged users.

Part I—Risk Management and Information Systems Control Theory and Concepts
Domain 1—Risk Identification, Assessment and Evaluation
F. Risk Scenarios

Prerequisites for Developing a Manageable Set of Risk Scenarios

Developing a manageable and relevant set of risk scenarios requires:
- Organizational buy-in or support from enterprise entities and business lines, risk management, IT, finance, compliance and other parties
- Expertise and experience to not overlook relevant scenarios and not be drawn into highly unrealistic or irrelevant scenarios
- A thorough understanding of the environment
- The involvement of all stakeholders including:
  - Senior management, which has decision-making authority
  - Business management, which has the best view of business impact
  - IT, which has the understanding of the risks associated with the underlying technology
  - Risk management, which can facilitate the risk management process

Systemic and/or Contagious Risk

Attention should be paid to so-called "systemic" and/or "contagious" risk scenarios:
- **Systemic risk**—Something that happens with an important business system or process that affects a large group of enterprises within an area or industry.

  Example 1: A nationwide air traffic control system goes down for an extended period of time (six hours), which affects air traffic on a large scale and consequently all businesses relying on air traffic.

  Example 2: A strike or work stoppage at a major shipping terminal interrupts the supply chain for many companies.

- **Contagious risk**—Events that happen at several of the enterprise's business partners within a very short time frame

  Example: A financial transaction clearinghouse is prepared for an emergency by having sophisticated disaster recovery measures in place. However, when a catastrophe occurs, it finds that no transactions are sent by its providers and, consequently, with no supply chain, is temporarily out of business.

Obscure Risk Identification

The enterprise must consider risk that has not yet occurred and consider developing scenarios around unlikely, obscure or nonhistorical events.

Developing such scenarios requires two considerations: visibility and recognition.

The enterprise must:
- Be in a position that it can observe anything going wrong
- Have the capability to recognize an observed event as something that is going wrong

Risk Scenarios: Future Circumstances

Scenario analysis should cover threats and vulnerabilities of current and possible future circumstances. Future risk could be related to emerging technologies, new regulations, demographic changes and new business initiatives.

3. RISK REGISTER

Introduction

Building a scenario requires determination of the value of an asset or a business process at risk and the potential threats and vulnerabilities that could lead to a loss event. The risk scenario should then be assessed for relevance and realism and, if found to be relevant, entered into the risk register.

Creation of a Risk Register

A risk register (or risk log) is a listing of all risks identified for the enterprise. The risk register records:
- All known risk
- Priorities of risk
- Likelihood of risk
- Potential risk impact
- Status of the risk mitigation plans
- Contingency plans
- Ownership of risk

Part I—Risk Management and Information Systems Control Theory and Concepts
Domain 1—Risk Identification, Assessment and Evaluation
F. Risk Scenarios

Exhibit 1.8: The following table in **Exhibit 1.8** is an excerpt of a single risk register entry.
Risk Register Entry

Exhibit 1.8: Risk Register Entry

Part I—Summary Data

Risk statement						
Risk owner						
Date of last risk assessment						
Due date for update of risk assessment						
Risk category	☐ Strategic (IT benefit/value enablement)		☐ Project Delivery (IT programme and project delivery)		☐ Operational (IT operations and service delivery)	
Risk classification (copied from risk analysis results)	☐ Low		☐ Medium		☐ High	☐ Very high
Risk response	☐ Accept		☐ Transfer		☐ Mitigate	☐ Avoid

Part II—Risk Description

Title		
High-level scenario (from list of sample high-level scenarios)		
Detailed scenario description—scenario components	Actor	
	Threat type	
	Event	
	Asset/resource	
	Timing	
Other scenario information		

Part III—Risk Analysis Results

Frequency of scenario (number of times per year)	0	1	2	3	4	5
	N≤0.01 ☐	0.01<N≤0.1 ☐	0.1<N≤1 ☐	1<N≤10 ☐	10<N≤100 ☐	100<N ☐
Comments on frequency						
Impact of scenario on business	0	1	2	3	4	5
1. Productivity	Revenue loss over one year					
Impact rating	I≤0.1% ☐	0.1%<I≤1% ☐	1%<I≤3% ☐	3%<I≤5% ☐	5%<I≤10% ☐	10%<I ☐
Detailed description of impact						

Part III—Risk Analysis Results *(cont.)*

2. Cost of response	Expenses associated with managing the loss event (US $)					
Impact rating	I≤$10k ☐	$10K<I≤$100K ☐	$100K<I≤$1M ☐	$1M<I≤$10M ☐	$10M<I≤$100M ☐	$100M<I ☐
Detailed description of impact						
3. Competitive advantage	Drop in customer satisfaction ratings					
Impact rating	I≤0.5 ☐	.05≤I≤1 ☐	1<I≤1.5 ☐	1.5<I≤2 ☐	2<I≤2.5 ☐	2.5<I ☐
Detailed description of impact						
4. Legal	Regulatory compliance—Fines (US $)					
Impact rating	None ☐	<$1M ☐	<$10M ☐	<$100M ☐	<$1B ☐	>$1B ☐
Detailed description of impact						
Overall Impact rating (average of four impact ratings)						
Overall rating of risk, obtained by combining frequency and impact ratings on risk map	☐ Low		☐ Medium		☐ High	☐ Very high

Part IV—Risk Response

Risk response for this risk	☐ Accept	☐ Transfer	☐ Mitigate	☐ Avoid
Justification				

Detailed description of response (not in case of 'accept')	Response Action	Completed	Action Plan
	1.	☐	☐
	2.	☐	☐
	3.	☐	☐
	4.	☐	☐
	5.	☐	☐
	6.	☐	☐
Overall status of risk action plan			
Major issues with risk action plan			
Overall status of completed responses			
Major issues with completed responses			

Part V—Risk Indicators

Key risk indicators for this risk	1.
	2.
	3.
	4.
	5.
	6.

Source: ISACA, *The Risk IT Practitioner Guide*, USA, 2009, figure 36, page 48

G. RISK FACTORS

Introduction The importance of risk factors lies in the influence they have on risk. They are heavy influencers of the likelihood and impact of risk scenarios and should be taken into account during every risk analysis, when likelihood and impact are assessed.

1. KEY TERMS AND CONCEPTS

Risk Factors A condition that can influence the frequency and/or magnitude and, ultimately, the business impact of IT-related events/scenarios

Threats and Risk Factors When considering risk, the risk practitioner must be aware of the threat agents that pose a threat to the assets of the enterprise. Threats may be internal or external, intentional or accidental, skilled or amateur, motivated or curious, and natural, man-made, physical, or related to equipment or utility failure.

Thorough risk analysis will consider all types of threats and the risk that those threats pose to the enterprise. Obviously, a skilled, highly-motivated agent working for another government or enterprise (often referred to as an Advanced Persistent Threat [APT]) will be a more serious threat than a low-skilled, curious individual casually wandering through the Internet looking for targets of opportunity.

Exhibit 1.9: Risk Factors in Detail Exhibit 1.9 depicts different risk factors that are discussed in more detail in the subsequent topics in this section.

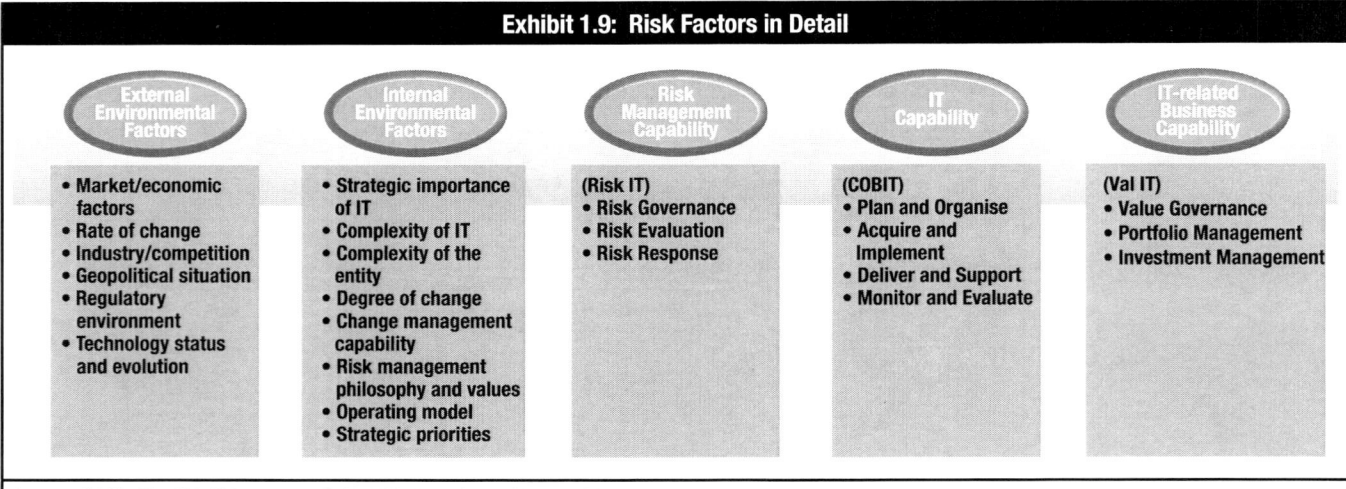

Source: ISACA, *The Risk IT Practitioner Guide*, USA, 2009, figure 36, page 38

2. EXTERNAL RISK FACTORS

Introduction

Shifts in global events, regulation and the economy including financial, supply chain, regulatory and competitive market changes are key external factors.

Importance of Understanding the External Business Environment

The external context is the external environment in which the enterprise seeks to achieve its objectives. Understanding the external context is important to ensure that external stakeholders and their objectives and concerns are considered when developing risk criteria.

The external context is based on the enterprisewide context, but with specific details of legal and regulatory requirements, stakeholder perceptions, and other aspects of risk specific to the scope of the risk management process.

The external context can include:
- The cultural, political, legal, regulatory, financial, technological, economic, natural and competitive environment, whether international, national, regional or local
- Key drivers and trends that impact the objectives of the enterprise
- Perceptions and values of external stakeholders

External Risk Factors

External risk factors are those circumstances that can increase the likelihood or impact of an event and that are not always directly controllable by the enterprise. Because external risk lies outside the enterprise's control, the enterprise is limited in the direct actions that can be taken to handle such risk. Nonetheless, the risk can still be managed by developing strategies to prevent exposures, avoid risk and deal with and contain an incident effectively once the risk materializes.

Example: Building dikes to prevent flooding, moving to an area not subject to flooding, and procuring insurance all can be used to contend with natural disasters such as floods

External risk factors are listed in the following table.

External Risk Factors	
Factor	**Description**
Market/economy	This includes the industry sector, in which the enterprise operates, i.e., operating in the financial sector requires different risk requirements and compliance or risk management capabilities than operating in a manufacturing environment.
	Other economic factors also can be included, e.g., nationalization, mergers and acquisitions, and consolidations.
	Example: Demographic trends—the aging of the population reduces demand for youth-oriented products and may increase reliance on health care or benefits programs..
Rate of change in the market in which the enterprise operates	Are business models changing fundamentally?
	Is the product or service at the end of an important life-cycle moment?
Competition	This is the competitive environment in which the enterprise operates.
	Example: Actions of competitors—the introduction of a new product that makes one of the enterprise's product lines obsolete

Part I—Risk Management and Information Systems Control Theory and Concepts
Domain 1—Risk Identification, Assessment and Evaluation
G. Risk Factors

External Risk Factors *(cont.)*

External Risk Factors *(cont.)*	
Factor	**Description**
Geopolitical situation	Is the geographic location subject to frequent natural disasters? Does the local political and overall economic context represent an additional risk? **Example:** Acts of nature—a sustained drought causing a dramatic drop in the output of agricultural products.
Regulatory environment	Is the enterprise subject to new or stricter regulations? Are there any other compliance requirements beyond regulation (e.g., industry-specific, contractual)? **Example:** Government regulations—concern of enterprises (such as chemical companies that produce hazardous substances) that government will change environmental laws so that it becomes difficult for them to produce their products in a cost-effective manner.
Technology innovation and evolution	Is the enterprise using state-of-the art technology, and more important, how fast are relevant technologies evolving? Is the enterprise capable of applying new technologies at the same pace or faster than its competitors?

Note: The types of risk that impact enterprises vary depending on the home country location, industry, level of globalization and many other factors. For example, financial services enterprises tend to be concerned about credit and market risk. Many enterprises are concerned with reputation and legal risk. However, one area of risk that impacts all enterprises is operational risk.

Example: Today, financial services enterprises, particularly banks, are addressing operational risk due to the capital adequacy accord known as Basel II. Basel II defines operational risk as the risk of losses resulting from inadequate or failed internal processes, people and systems or from external events. Although designed for banking, this definition holds true for practically all industries. The adoption of Basel II by financial services enterprises is partly dependent on the posture of the local regulatory agency (usually central bank) in mandating or promoting this risk-based standard.

Impact of Governance Codes and Local Laws

Governance codes as well as regional and local laws can be generic or industry-specific, and they create additional demands on enterprises—normally in response to heightened expectations from society or as a result of corporate scandals that revealed a need to tighten up existing regulations. An assortment of local laws also adds to the compliance framework within which enterprises must operate. Some professions, such as law, medicine and accounting, provide various codes of conduct and specific regulations that must be adhered to by their practicing members.

The presence of diverse laws and regulations places a challenge on enterprises to either be compliant with an increasingly complex variety of conditions and restrictions or to face major sanctions and/or penalties.

Part I—Risk Management and Information Systems Control Theory and Concepts
Domain 1—Risk Identification, Assessment and Evaluation
G. Risk Factors

3. INTERNAL ENVIRONMENTAL RISK FACTORS

Introduction

The internal environment is shaped by an enterprise's history and culture and affects all components of enterprise risk management (ERM).

These things influence how strategies and objectives are established; business activities are structured; and risk is identified, assessed and acted on. They also influence the design and functioning of control activities, information and communication systems, and monitoring activities.

The internal environment of the enterprise is comprised of many elements, including:
- The enterprise's ethical values
- The competence and development of personnel
- Management's philosophy for managing risk
- How the enterprise assigns authority and responsibility

A board of directors is a critical part of the internal environment and significantly influences other internal environment elements. Although all elements are important, the extent to which each is addressed will vary with the enterprise. Nevertheless, the enterprise should have an internal environment that provides an appropriate foundation for ERM.

Internal Risk Factors

Various internal risk factors are listed in the following table.

> **Note:** Additional internal risk factors are also discussed in this section.

Internal Risk Factors	
Factor	**Description**
Strategic importance of IT in the enterprise	Is IT a strategic differentiator, functional enabler or supporting function?
Complexity of IT	Is IT highly complex (e.g., complex architecture, recent mergers), or is it simple, standardized and streamlined?
Complexity of the enterprise	This includes geographic spread and value chain coverage (e.g., in a manufacturing environment). Does the enterprise manufacture and distribute parts, and/or is it also doing assembly activities?
Enterprise change	This is the degree of internal change the enterprise is experiencing, such as reorganizations and new leadership.
Change management capability	To what extent is the enterprise capable of organizational change?
Risk management philosophy	This includes the risk management philosophy of the enterprise (risk averse or risk taking) and, linked with that, the values of the enterprise.
Operating model	This is the degree to which the enterprise operates independently or is connected to its clients/suppliers—the degree of centralization/decentralization.
Enterprise priorities	These are the strategic priorities of the enterprise.
Technology innovation and evolution	Is the enterprise keeping abreast with technology evolution, and is it capable of supporting evolving technologies over time?

Importance of Integrity and Ethics of Enterprise Risk Management (ERM)	An enterprise's strategy and objectives, and the way they are implemented, are based on preferences, value judgments and management styles. Management's integrity and commitment to ethical values influence these preferences and judgments, which are translated into standards of behavior. Because an enterprise's good reputation is so valuable, the standards of behavior must go beyond mere compliance with the law. Management values must balance the concerns of the enterprise, employees, suppliers, customers, competitors and the public. Managers of well-run enterprises increasingly have accepted the view that good ethics pays, and that ethical behavior is good business. Management integrity is a prerequisite for ethical behavior in all aspects of an enterprise's activities. The effectiveness of ERM cannot rise above the integrity and ethical values of the people who create, administer and monitor enterprise activities. An enterprise that operates with a high degree of ethics may have a lower incidence of risk related to fraud or misappropriation. Integrity and ethical values are essential elements of an enterprise's internal environment and affect the design, administration and monitoring of other ERM components.

Role of Enterprise Management in Determining Enterprise Culture	Top management—starting with the chief executive officer (CEO)—plays a key role in determining the corporate culture (tone at the top). As the dominant personality in an enterprise, the CEO often sets the ethical tone. Certain organizational factors also can influence the likelihood of fraudulent and questionable financial reporting practices. Those same factors are also likely to influence ethical behavior. Individuals may engage in dishonest, illegal or unethical acts simply because the enterprise gives them strong incentives or temptations to do so. Undue emphasis on results, particularly in the short term, can foster an inappropriate internal environment. Focusing solely on short-term results can hurt the enterprise, even in the short term. Concentration on the bottom line—sales or profit, at any cost—often evokes unsought actions and reactions. High-pressure sales tactics, ruthlessness in negotiations or implicit offers of kickbacks, for instance, may evoke reactions that can have immediate (and lasting) effects.
	
	Ethical values must be:
	• Communicated
	• Accompanied by explicit guidance regarding what is right and wrong
	• Established in formal codes of corporate conduct, which address a variety of behavioral issues such as:
	  – Integrity and ethics
	  – Conflicts of interest
	  – Illegal or otherwise improper payments
	  – Anticompetitive arrangements

Management Determination of Competency Levels	Competence reflects the knowledge and skills needed to perform assigned tasks. Management decides how much to invest in making sure that tasks are executed properly using skilled resources, equipment and defined processes. This requires weighing the enterprise's strategy and objectives against plans for their implementation and achievement. A trade-off often exists between competence and cost. The risk of failure is higher with untrained staff, poorly maintained or old equipment, or undefined procedures.

Part I—Risk Management and Information Systems Control Theory and Concepts
Domain 1—Risk Identification, Assessment and Evaluation
G. Risk Factors

Board of Directors' Role in the Internal Environment	An enterprise's board of directors is a critical part of the internal environment and significantly influences its elements. The board's role in risk governance through independent oversight of management, scrutiny of activities, and appropriateness of the enterprise's risk appetite strategy all play a role. Other factors include the: • Degree to which difficult questions are raised and pursued with management regarding strategy, plans and performance • Interaction between the board or audit committee and internal and external auditors An active and involved board of directors, board of trustees or comparable body should possess an appropriate degree of management, and technical and other expertise, coupled with the mindset necessary to perform its oversight responsibilities. This is critical to an effective ERM environment. Because the board must be prepared to question and scrutinize management's activities, present alternative views, and act in the face of wrongdoing, the board must also include outside directors. There must be a sufficient number of independent outside directors—not only to provide sound advice, counsel and direction, but also to serve as a necessary check and balance on management.
Impact of the Enterprise Organizational Structure	An enterprise's organizational structure provides the framework to plan, execute, control and monitor its activities. A relevant organizational structure includes defining key areas of authority and responsibility and establishing appropriate lines of reporting. **Example:** An enterprise that must report on its compliance with regulations will have staff that gather the necessary data and other staff that compile the reports and then submit them to senior management for approval. The reports given to senior management are often not as detailed as the reports first generated and used by local managers to evaluate the performance of their departments. For normal reporting, a scheduled reporting process is used to gather data and generate and submit the reports. In the case of a serious incident, another process may be used to escalate the issue to management and initiate recovery activities. An enterprise develops an organizational structure suited to its needs. Some are centralized; others are decentralized. Some have direct reporting relationships, while others are more of a matrix organization. Some entities are organized by industry or product line, geographic location, business function or a particular distribution or marketing network. Whatever the structure, an enterprise should be organized to enable effective ERM and to carry out its activities to achieve its objectives.

Assignment of Authority and Responsibility	Assignment of authority and responsibility:
	• Involves the degree to which individuals and teams are authorized (and limited by their authority) and encouraged to use initiative to address issues and solve problems
	• Includes establishing reporting relationships and authorization protocols and also policies that describe:
	  – Appropriate business practices
	  – The knowledge and experience of key personnel
	  – The resources provided for carrying out duties
	
	Some entities push authority downward to bring decision making closer to frontline personnel—to the individuals who are closest to everyday business transactions. An enterprise may take this route to become more market-driven or quality-focused. This may involve empowerment to sell products at discount prices; negotiate long-term supply contracts, licenses or patents; or enter alliances or joint ventures. Effective delegation aligns authority with accountability to encourage individual initiative within limits.

Impact of Delegation	A critical challenge is to delegate only to the extent required to achieve objectives. This means ensuring that: • Decision making is based on sound practices for risk identification and assessment, including sizing risk and weighing potential losses vs. gains in determining which risk to accept and how it is to be managed. • All personnel must understand the enterprise's objectives. It is essential that individuals know how their actions are related to one another and contribute to achievement of the objectives. Increased delegation is sometimes the result of streamlining or flattening the organizational structure. Purposeful structural change to encourage creativity, initiative and faster response times can enhance competitiveness and customer satisfaction. Increased delegation may: • Carry an implicit requirement for a higher level of employee competence and greater accountability • Require effective procedures for management to monitor results so that decisions can be reviewed, overruled or accepted as necessary Along with better, market-driven decisions, delegation may increase the number of undesirable or unanticipated decisions. The internal environment is greatly influenced by the extent to which individuals recognize that they will be held accountable. This holds true all the way to the chief executive, who, with board oversight, has ultimate responsibility for all activities and protection of assets within an enterprise.
Impact of Human Resources (HR) Practices	HR practices pertaining to hiring, orientation, training, evaluating, counseling, promoting, compensating and taking remedial actions send messages to employees regarding expected levels of integrity, ethical behavior and competence. • **Hiring standards** with emphasis on educational background, prior work experience, past accomplishments and evidence of integrity and ethical behavior demonstrate an enterprise's commitment to competent and trustworthy people. • **Transfers and promotions** driven by periodic performance appraisals demonstrate the enterprise's commitment to advancement of qualified employees. • **Competitive compensation programs** that include bonus incentives serve to motivate and reinforce outstanding performance—although reward systems should be structured and controls should be put in place to avoid undue temptation to misrepresent reported results. • **Performance improvement processes and disciplinary actions** send a message that underperformance and violations of expected behavior are not tolerated. • **Mentoring and support** helps provide employees with the skills and behaviors necessary to tackle new challenges as issues and risk throughout the enterprise change and become more complex—driven in part by rapidly changing technology and increasing competition. • **Education and training** helps personnel keep pace and deal effectively with the evolving environment. Hiring competent people and providing one-time training are not enough. The education process must be ongoing.

4. CAPABILITIES

Risk Management Capability

Risk management capability is an indication of how well the enterprise is executing the core risk management processes. The better executed or more mature the processes, the more capable the risk management program. This factor is correlated with the capability of the enterprise to recognize and detect risk and adverse events and should not be neglected.

Risk management capability is a very significant element in the likelihood and impact of risk events in an enterprise because it is responsible for:
- Management's risk decisions (or lack thereof)
- The presence, absence and/or effectiveness of controls that exist within an enterprise

Risk management capability is also an important component of the overall risk profile of the enterprise.

IT Capability

In the context of risk management, IT capabilities may be associated with the maturity level of IT processes and IT controls. Mature and well-controlled IT processes are equivalent to high IT capabilities, which can have a positive influence on reducing the:
- **Likelihood of events**—e.g., having good software development processes in place to deliver high-quality and stable software or having good security measures in place to reduce the number of security-related incidents
- **Business impact when events happen**—e.g., having good disaster recovery planning (DRP) in place to restore IT functionality when disaster strikes

The IT sourcing model is often seen as a separate risk factor. There is no doubt that the sourcing model, e.g., keeping IT in-house or outsourcing parts or complete IT departments, has an important impact on risk and how to measure it. The COBIT process model contains several processes dealing with the selection and management of sourcing models.

IT-related Business Capability

The degree to which business management is capable of managing the direction and performance of IT is an important risk factor.

Mature IT value management processes are associated with a high capability of the business to manage IT-related affairs. The enterprise will generate more value from IT and will miss fewer opportunities *if*:
- The business is capable of making the right IT investments.
- Correct IT partners are selected.
- Programs are well selected and managed.

> **Note:** IT-related business capability is especially a risk factor in the risk associated with IT benefit/value enablement.

H. QUALITATIVE AND QUANTITATIVE RISK ANALYSIS

Introduction

After collecting data, risk assessment is the first process in the risk management process. Enterprises use risk assessment to determine the:
- Extent of the potential threat
- Risk associated with business processes, operations and IT systems throughout their development life cycle and use

The entire risk management process should be managed at multiple levels in the enterprise, including the operational, project and strategic levels, and should form part of the risk management practice.

IT risk should be treated in the same manner as any other business risk and should be analyzed and assessed using similar approaches. Accepted business risk assessment practices and methods should be applied consistently across different IT systems.

> **Note:** Risk management frameworks and standards such as the Committee of Sponsoring Organizations of the Treadway Commission (COSO) provide appropriate methods for risk assessment.

1. KEY TERMS AND CONCEPTS

Impact

The magnitude of harm that could be caused by a threat's exploitation of a vulnerability

The level of impact is governed by the potential mission impacts and, in turn, produces a relative value for the IT assets and resources affected (e.g., the criticality and sensitivity of the IT system components and data).

Likelihood

In nontechnical terms, "likelihood" is usually a synonym for "probability;" but in statistical usage, a clear technical distinction is made. It would be improper to switch "likelihood" and "probability" in the two following sentences:
- "If I were to flip a fair coin 100 times, what is the *probability* of it landing heads-up every time?"
- "Given that I have flipped a coin 100 times and it has landed heads-up 100 times, what is the *likelihood* that the coin is fair?"

To determine the likelihood of a future adverse event, enterprises must analyze threats to a system, potential vulnerabilities and the controls in place.

Qualitative Risk Analysis

Defines risk using a scale or comparative values (i.e., defining risk factors in terms of high/medium/low or on a numeric scale from 1 to 10). It is based on judgment, intuition and experience rather than on financial values.

Quantitative Risk Analysis

The use of numerical and statistical techniques to calculate likelihood and impact of risk. It uses financial data, percentages and ratios to provide an approximate measure of the magnitude of impact in financial terms.

Residual Risk	The remaining risk after management has implemented a risk response Residual risk can be used by management to determine: • Which areas require more controls (Is the residual risk within an acceptable level of risk?) • Whether the benefits of such controls outweigh the control costs
US National Institute of Standards and Technology (NIST) Recommended Risk Assessment Methodology	The following table outlines the NIST SP800-30 risk assessment methodology for IT-related risk at a very high level.

NIST Risk Assessment Methodology	
Step	Description
1	System characterization
2	Threat identification
3	Vulnerability identification
4	Control analysis
5	Likelihood determination
6	Impact analysis
7	Risk determination
8	Control recommendations
9	Results documentation

2. QUALITATIVE RISK ANALYSIS

Overview of Qualitative Risk Analysis	Qualitative analysis defines risk using a scale or comparative values (i.e., defining risk factors in terms of high/medium/low or on a numeric scale from 1 to 10). It is based on judgment, intuition and experience rather than on numbers or financial values.
Benefits of Qualitative Risk Analysis	Benefits of qualitative risk analysis include: • The cost of conducting a qualitative analysis is generally significantly lower than the cost of calculating quantitative analysis. • Qualitative risk analysis usually results in a better understanding of business unit dependencies and interactions. • Qualitative risk analysis is more consensus-based and often reflects the input of business units more accurately than quantitative analysis. • It is often better for evaluating "soft" (intangible) risk, such as morale or reputation.
Challenges of Qualitative Risk Analysis	When applying subjective information to analyzing risk, the following challenges must be considered: • Subjectivity or bias in data collected • Overemphasis on minor events • Does not provide good data for cost-benefit analysis • Ranking levels may not be meaningful to data providers

Typical Qualitative Risk Analysis Methods

The following table describes typical qualitative methods used in risk assessment.

Typical Qualitative Risk Analysis Methods	
Method	**Description**
Risk Control Self-assessment (RCSA)	RCSA: • Is based on the evaluation of risk levels by line managers who are expected to be familiar with threats and incidents within their areas of responsibility • Involves all levels of the enterprise in risk management and creates a risk awareness culture and increased risk ownership • May be subjective and provide inaccurate results depending on the culture and openness of the enterprise • Is typically a bottom-up process by business managers, but may be a top-down process by senior stakeholders, which provides a good blend—a granular view from the bottom up and an enterprise view from the top down **Note:** It is vital for executive management to be open and active to assure RCSA participants that they will not suffer from speaking candidly.

Scorecards	Scorecards:
	• Consist of generic questionnaires containing weighted, risk-based questions with multiple-choice responses
	• Create qualitative assessments that can be:
	  – Translated into quantitative measures, such as a ranking of risk factors
	  – Used to adjust capital reserve levels
	• Include rewards for internal control improvements
	
	Note: Problems can arise due to the subjective nature of scorecards and manipulation of the process to artificially lower capital charges.

Key risk indicators (KRIs)	KRIs: • Are used to alert the enterprise to critical changes in risk, especially early warning alerts to changes in the control environment • Can be improved beyond after-the-fact loss indicators to predict KRI challenges • Cannot be expected to capture all potential losses

Part I—Risk Management and Information Systems Control Theory and Concepts
Domain 1—Risk Identification, Assessment and Evaluation
H. Qualitative and Quantitative Risk Analysis

Typical Qualitative Risk Analysis Methods *(cont.)*

Method	Description
Likelihood-impact matrix	The matrix offers a good way to categorize risk events qualitatively in terms of their probability of occurrence (likelihood) and their consequences (impact) Example of a Risk Level Matrix From NIST SP800-30: **Typical Qualitative Risk Analysis Methods** **Description** *See matrix below*
Attribute analysis	Attribute analysis: • Is a creative problem-solving technique that can be employed productively when exploring possible qualitative risk impacts • Enables the development of creative solutions for problems by the application of unconventional perspectives Its main value is that it subjects something familiar to examination in different ways in a structured way.
Delphi forecasting	• This was developed as a forecasting tool involving subject matter experts (SMEs) who anonymously provide their responses to a questionnaire, and these responses are tabulated and portrayed graphically. • Participants in the exercise may be asked to forecast such things as the cost, schedule and resource impacts associated with the occurrence of a risk event. • First-round responses may be highly divergent; however, after several rounds of responses, the responses may show that a consensus has emerged on the expected impact of the risk event. • With the statistical summary of responses, the exercise is repeated. • The objective is to see whether SMEs can achieve a consensus on an issue after they have had a chance to reflect on feedback provided by the responses of their colleagues. **Note:** In a typical Delphi exercise, each round of questionnaire distribution leads to increased conformity in the SMEs' views on an issue. After several rounds, when it becomes clear that additional consensus will not occur, the exercise is ended.

Impact

Threat Likelihood	Low (10)	Medium (50)	High (100)
High (1.0)	Low $10 \times 1.0 = 10$	Medium $50 \times 1.0 = 50$	High $100 \times 1.0 = 100$
Medium (0.5)	Low $10 \times 0.5 = 5$	Medium $50 \times 0.5 = 25$	High $100 \times 0.5 = 50$
Low (0.1)	Low $10 \times 0.1 = 1$	Medium $50 \times 0.1 = 5$	High $100 \times 1.0 = 10$

Risk scale: High (>50 to 100); Medium (>10 to 50); Low (1 to 10)

Typical Qualitative Risk Analysis Methods (cont.)

Typical Qualitative Risk Analysis Methods (cont.)

Method	Description
Failure Modes and Effects Analysis (FMEA)	**FMEA:** • Evaluates the impact of risk from an immediate, near-term and long-term perspective. This methodology will assess risk from both local business unit and enterprisewide perspectives. • Acknowledges that a failure in one area may result in unforeseen consequences in other areas or departments • Results in the prioritization of risk—giving each risk a risk priority number (RPN)

3. QUANTITATIVE RISK ANALYSIS

Overview of Quantitative Analysis

Quantitative analysis is the use of numerical and statistical techniques to calculate likelihood and impact of risk. It uses financial data, percentages and ratios to provide an approximate measure of the magnitude of impact in financial terms.

This measure can be used in the cost/benefit analysis of the recommended controls. Although a computation is used to arrive at various risk aspects, the approach is still subjective to some extent. The values used to calculate likelihood, impact and asset value are subject to speculation and may be quite difficult to quantify. The results of a quantitative risk analysis must, therefore, allow for some margin of error.

Benefits of Quantitative Analysis

Since quantitative analysis is data driven, it:
• Allows for:
  – Data to be classified and counted
  – Statistical models to be constructed to explain what is being observed
  – Findings to be generalized to a larger population and for direct comparison to be made between two different sets of data or observations
• Produces statistically reliable results
• Allows discovery of phenomena which are likely to be genuine versus those which are merely chance occurrences

Challenges of Quantitative Analysis

When measuring, modeling and managing operational risk, the following challenges must be considered:
• It is not always easy to collect data on each and every process.
• Data may not be in the desired format or may not meet the needs of quantitative analysis.
• Reliable historical data are not always available for analysis to allow for quantification of process failures or the risk induced by these failures.
• Past data do not necessarily help predict future events (extremely rare occurrence or black swan phenomenon).
• It is difficult to apply statistical models for events that happen infrequently.
• The cost of quantitative analysis is generally significantly higher than the cost of qualitative analysis.

Part I—Risk Management and Information Systems Control Theory and Concepts
Domain 1—Risk Identification, Assessment and Evaluation
H. Qualitative and Quantitative Risk Analysis

Typical Quantitative Risk Analysis Methods

The following table describes typical quantitative methods used in risk assessment.

Typical Quantitative Risk Analysis Methods	
Method	**Description**
Internal loss data	**Used by:** Financial services enterprises—These are key to any financial services enterprise's efforts to improve operational risk management. **Issue:** The biggest issue most of these enterprises face is the lack of reliable and consistent operational risk data. The following table outlines internal loss data considerations. `<table>`<table><tr><th>Loss Type</th><th>Description</th></tr><tr><td>Internal loss data quality</td><td>• Many banks started accumulating internal loss data to prepare for Basel II, which requires: – A minimum of three years' of data to start – Five years' of data on an ongoing basis as part of the advanced measurement approach (AMA) • The quality of internal loss data: – Is a factor and must be available across all business lines and geographic locations – Should include near-loss data</td></tr><tr><td>Economic losses</td><td>• It is critical to capture all economic losses, not just major or material losses, with a large impact on the bottom line. **Rationale:** • These data predict expected losses (ELs), even though they typically represent less than 25 percent of all losses. • Classification is difficult because: – Many loss events result from a variety and combination of factors. – The same loss event could fall into credit, market and operational risk buckets.</td></tr><tr><td>Operational risk losses</td><td>• There is an issue as to the enterprise's acceptance of risk. • Many enterprises are hesitant to capture operational risk losses as a negative reflection on their performance. Many enterprises view market and credit risk as an acceptable cost of doing business.</td></tr></table> **Alternative method:** Another method of validating internal loss data is to compare it with peer enterprises via externally available data and then scale the data to reflect the enterprise's environment.

CRISC Review Manual 2013

Typical Quantitative Risk Analysis Methods (cont.)

Typical Quantitative Risk Analysis Methods (cont.)	
Method	**Description**
External data	**Used by:** Financial services enterprises External data are needed because there is typically a lack of internal data—especially around unexpected losses, which represent the majority of losses in most banks. **Issue:** The use of external data stems from their sources, which include data providers or bank consortia. **Relevance:** External data must be mapped, scaled and adapted to each bank's business, legal, regulatory, technical, control and cultural environment.
Business process modeling (BPM) and simulation	Although it may appear that most operational risk is preventable with the implementation of procedures and controls, it is not an easy task to identify and control all risk. The BPM and simulation discipline: • Is an effective method of identifying and quantifying the operational risk in enterprise business processes • Improves business process efficiency and effectiveness The process simulation model aids the enterprise in: • Developing insights into the operations of the business • Leveraging assets and reducing costs • Testing process changes before implementation (change management) • Experimenting with process improvements to reduce cycle times and manage operational risk • Conducting stress tests and scenario analysis
Statistical process control (SPC)	The typical SPC is the cumulative sum (cusum) control chart: • It is very effective in detecting small process shifts, which need to be stable and should operate with minor variability. • All cusum statistics incorporate all the information known about the process. • The plain-vanilla cusum simply plots the cumulative sums of the deviations of the observed values and a target value. SPC is most often used by enterprises with manufacturing environments; however, it can be very effective in general BPM.

4. METHODS FOR DISCOVERING HIGH-IMPACT RISK TYPES

Introduction

This section contains information on methods for discovering high-impact risk types.

The uncovering of uncommon, but potentially high-impact, risk needs to be recognized as an important area of focus within risk assessment. This is especially true of process-based activity and in process improvement (as in business process improvement).

Relevance

Without consideration of low-probability, high-impact events, there may be serious gaps in the assessment of risk and, consequently, the completeness of coverage of risk mitigation strategies.

Methods for Uncovering Less Obvious Risk Factors

The following table describes other methods for uncovering less obvious risk factors.

These methods may uncover less obvious risk factors by identifying all the dependencies and inter-relationships that support the business objectives or a business process, and thereby show how a minor or easily overlooked process may play a significant role in contributing to risk.

Methods for Uncovering Less Obvious Risk Factors	
Method	**Description**
Cause-and-effect analysis	A predictive or diagnostic analytical tool used: • To explore the root causes or factors that contribute to positive or negative effects or outcomes • For identifying potential risk **Note:** A typical form is the Ishikawa diagram, also known as the fishbone diagram.

Fault tree analysis	A technique that:
	• Provides a systematic description of the combination of possible occurrences in a system, which can result in an undesirable outcome (top-level event)
	• Combines hardware failures and human failures
	
	A fault tree is constructed by:
	• Relating the sequences of events that, individually or in combination, could lead to the top-level event
	• Deducing the preconditions for the:
	  – Top-level event
	  – Next levels of events, until the basic causes are identified (elements of a "perfect storm" [unlikely simultaneous occurrence of multiple events that cause an extraordinary incident])
	
	Note: The most serious outcome is selected as the top-level event.

Sensitivity analysis	A quantitative risk analysis technique that: • Helps to determine which risk factors potentially have the most impact • Examines the extent to which the uncertainty of each element affects the object under consideration when all other uncertain elements are held at their baseline values **Note:** The typical display of results is in the form of a tornado diagram.

5. RISK ASSOCIATED WITH BUSINESS CONTINUITY AND DISASTER RECOVERY PLANNING

Introduction

Business continuity planning (BCP) and disaster recovery planning (DRP) complement the risk management process by ensuring preparedness at times when rare, but high-impact, events occur and allow the organization to be prepared to handle unforeseen events.

Both risk management and business continuity disciplines can (and should) leverage each other to ensure that the assets of the enterprise are protected.

Business continuity is concerned with ensuring the continuity of business operations in the event of an adverse event occurring that could otherwise cause the critical business functions (CBFs) of the business to be disrupted for a time period.

Disaster recovery planning develops the plans necessary to recover IT operations and infrastructure in the event that the IT functions are interrupted.

Business impact analysis (BIA) is a key business continuity planning (BCP) activity that focuses on determining the impact of an event over time—how the impact level increases over time and at what point in time the business is most likely no longer able to recover from the incident.

Relevance

Since BCP and DRP are the last line of defense in the event of a catastrophic impact on the enterprise, it is imperative that these plans be robust enough to work effectively in a crisis.

The risk for many enterprises is that the business continuity (BC) and disaster recovery (DR) plans are not adequate to maintain or recover business operations in a crisis.

The risk practitioner should ensure that BC and DR plans are kept up to date, are approved by management and are tested on a regular basis.

Business Impact Analysis (BIA)

A process to determine the impact of losing the support of any resource

A BIA is a discovery process meant to:
- Reveal the importance of a process and the potential impact that any disruption to that process would have on the enterprise.
- Establish the escalation of loss over time.
- Answer questions about actual procedures, shortcuts, workarounds and the types of failure that may occur. It involves determining:
  - What the process does
  - Who performs the process
  - What the output is
  - The value of the process output to the enterprise
  - How the impact of the loss would escalate over time and at which point failure of the process might threaten the viability of the enterprise.

> **Note:** The use of qualitative and quantitative assessment for risk discovery and assessment has already been mentioned in the previous section. The same qualitative and quantitative techniques can be used in the BIA.

Part I—Risk Management and Information Systems Control Theory and Concepts
Domain 1—Risk Identification, Assessment and Evaluation
H. Qualitative and Quantitative Risk Analysis

Discovery Exercises for Proper BIA Execution

Proper execution of the BIA process entails a series of discovery exercises, including:
- Asking questions that focus on identifying trigger events and the current sequence of the enterprise's critical processes
- Interviewing key personnel
- Reviewing existing documentation
- Collecting data by observing business processes and personnel performing actual processes (job shadowing)
- Looking for existing workarounds and alternate procedures

> **Note:** Using surveys can raise issues of accuracy and consistency.

The BIA process will:
- Verify critical success factors (CSFs)
- Identify the components necessary for the business to operate:
  - Facilities/premises
  - People
  - Equipment
  - Information/data
  - Supply chain
- Identify vital materials and records necessary for recovery, including:
  - Data backups
  - The vendor list
  - Inventory records
  - The customer list
  - Employee records with contact data
  - Bills of materials
  - Procedure documents
  - The business continuity (BC) plan
  - Banking information
  - Copies of all contracts and legal documents

The data gathered during the BIA can be used later to guide the formulation of the risk response strategy.

I. IT RISK IDENTIFICATION AND ASSESSMENT

Introduction

IT plays a critical role in the operation of nearly every modern business process and enterprise. It is a key part of strategic decisions, operational continuity and projects that bring about organizational change. Therefore, there are many risk factors that pertain specifically to enterprise IT systems.

The strategic importance of IT to the modern enterprise is seen through:
- High investment
- The pervasiveness of IT
- Reliance on IT's continuing operation
- The impact caused when IT does not perform as expected
- IT's critical role in realizing efficiencies
- The ways in which IT enables business to take strategic action

These possibilities carry with them significant risk—risk that is inherent from the use of IT in itself. Only by proactively recognizing and addressing this risk can business interests be safeguarded.

1. THREATS AND OPPORTUNITIES INHERENT IN ENTERPRISE USE OF IT

Inherent Risk and Rewards of IT Use

The use of IT obviously carries a risk, just as it has potential rewards. Ignoring or focusing only on the most obvious risk is dangerous because IT has become a utility that underpins practically every business activity.

Risk is typically not easily measured, reported or monitored. This is compounded by the fact the word "risk" is often incorrectly applied to both eventualities and their likelihood with statements such as: "This risk is low-risk."

Monitoring and reporting on risk is made even more difficult because there is no shared language between those estimating the risk and those making risk response decisions. Clarity in defining the business impact (both positive and negative) of IT-related risk is, therefore, critical for understanding where there are threats and vulnerabilities and where there are opportunities.

Relevance: Being able to clearly differentiate between the terms "risk," "threat" and "vulnerability" is crucial:
- **Risk**—A derived value that refers to the likelihood (or frequency) and magnitude of loss that exists from a combination of asset(s), threat(s) and control conditions
- **Threat**—An action or actor that/who may act in a manner that can result in loss or harm
- **Vulnerability**—A weakness in design, implementation, operation or internal control

Part I—Risk Management and Information Systems Control Theory and Concepts
Domain 1—Risk Identification, Assessment and Evaluation
I. IT Risk Identification and Assessment

Measuring the Adverse Impact of IT Risk

The adverse impact of a risk event can be described in terms of loss or degradation of any, or a combination of any, of the following three basic IT risk goals:
- Integrity
- Availability
- Confidentiality

The following table provides a brief description of each business information requirement and the consequence (or impact) of the goal's not being met.

Business Information Requirements and Related Impacts	
Requirement	**Description/Impact of Unmet Goals**
Integrity	Relates to the accuracy and completeness of information and its validity in accordance with business values and expectations. System and data integrity refers to the requirement that information be processed correctly and protected from improper modification. **Impact:** Loss of integrity—If unauthorized changes are made to the data or information system by either intentional or accidental acts and the loss of system or data integrity is not corrected: • Continued use of the contaminated system or corrupted data may result in inaccuracy, fraud or erroneous decisions. • Violation of integrity may be the first step in a successful attack against system availability or confidentiality.
Availability	Relates to the information being accessible, when required by the business process, and also concerns the safeguarding of necessary resources and associated capabilities. **Impact:** Loss of availability—If a mission-critical IT system is unavailable to its end users, the enterprise's objectives may be affected. Loss of system functionality and operational effectiveness, for example, may result in loss of productive time, thus impeding the end users' performance of their functions in supporting the enterprise's objectives. Loss of data may result in incorrect responses or decisions being made by management.
Confidentiality	Relates to the protection of information from unauthorized disclosure **Note:** Data must be protected from improper disclosure depending on the sensitivity of the data and applicable legal requirements. **Impact:** Loss of confidentiality—Unauthorized disclosure of confidential or sensitive information can range from jeopardizing national security to the disclosure of data covered under the local privacy law. Unauthorized, unanticipated or unintentional disclosure of such information can result in loss of public confidence loss of competitive advantage, embarrassment or legal action against the enterprise.

Measuring the Business Impact of IT-related Risk

Some tangible impacts of IT risk can be measured quantitatively as in:
- Lost revenue
- The cost of repairing the system
- The level of effort required to correct problems caused by a successful threat action

Other impacts are difficult to measure in specific units, but can be qualified or described in terms of high, medium and low impacts. Such impacts include:
- Loss of public confidence
- Loss of credibility
- Damage to an enterprise's interest
- Impact on morale in the enterprise

Business-related IT Risk Types

The business-related IT risk types listed in the following table can be used as a guide.

Business-related IT Risk Types	
Type	**Description**
Investment or expense risk	The risk that the IT investment fails to provide value for money or is otherwise excessive or wasteful This includes consideration of the overall IT investment portfolio.
Access or security risk	The risk that confidential or otherwise sensitive information may be divulged or made available to those without appropriate authority An aspect of this risk is noncompliance with local, national and international laws related to privacy and protection of personal information.
Integrity risk	The risk that data cannot be relied on because it is unauthorized, incomplete or inaccurate
Relevance risk	The risk associated with not getting the right information to the right people (or process or systems) at the right time to allow the right action to be taken
Availability risk	The risk of loss of service or the risk that data are not available when needed
Infrastructure risk	The risk that an enterprise does not have an IT infrastructure and systems that can effectively support the current and future needs of the business in an efficient, cost-effective and well-controlled fashion (includes hardware, networks, software, people and processes)
Project ownership risk	The risk of IT projects failing to meet objectives through lack of accountability and commitment

Threats and Vulnerabilities Associated with SDLC

Most enterprises follow some form of a system development life cycle (SDLC) for IT project management. This structure is intended to ensure that the IT project is properly managed and will meet business requirements.

An important function for the risk practitioner during any IT project—or business process reengineering project—is to ensure that the system is designed and built with adequate levels of protection so that the end product will be built to mitigate any risk that the new system may pose to the enterprise.

This requires that risk assessment be built into each phase of the SDLC—starting with the functional requirements definition phase and continuing on to the implementation and operation of the system.

Functional Requirements Definition Phase	The key task related to risk assessment during the functional requirements definition phase is to ensure that the risk associated with the system is identified. The level of risk associated with the system will depend on the criticality of the system (how important the system is to support business operations) as well as the sensitivity (privacy) needs and criticality needs of the data being processed on the system. Once the risk requirements are known, they must be mitigated through the design of appropriate controls. Some of the controls will be to build redundancy and access controls (to prevent improper modification or disclosure) into the systems design.
Risk Associated With Outsourcing	Many enterprises are choosing an outsourcing strategy to provide IT, data management, web hosting and other supporting services. This presents a new risk to the enterprise since the enterprise now becomes reliant on another enterprise for its supporting infrastructure. There is also risk related to privacy laws that may affect where data are stored and the need to protect sensitive data in transit, storage and processing from inadvertent or unauthorized disclosure. One risk that the enterprise must be aware of is that in many jurisdictions the responsibility for protection of data remains with the original enterprise, not the firm providing the outsourcing services, so the original enterprise must ensure that the security requirements and the right to audit are included in contracts.
IT Project-related Risk Factors	IT projects—the implementation of new technology, business process reengineering, or upgrades and modification to existing systems—are subject to risk of failure due to factors such as: • Unrealistic delivery schedules or budget • Lack of skilled resources • Unclear or changing business requirements • Challenges with technology • Poor project management • Resistance from users
IT Project-related Risk	The risk associated with an IT project that fails to deliver the expected results on time or on schedule may have a serious impact on business operations. Such impact may be related to: • Loss of opportunity or market share • Inability to meet customer or regulatory demand • Lost revenue • Other tangible or intangible consequences Risk associated with changing an existing process must be identified. The ISACA IT Audit and Assurance Standards, Guidelines, and Tools and Techniques identify several risk areas to consider when planning a BPR project. The risk factors can be broken down into the three broad areas, as listed in the following table.

IT Project-related Risk *(cont.)*

IT Project-related Risk		
Risk Area	**Description**	
Design risk	A good design can improve profitability while satisfying customers. Conversely, a design failure would spell doom to any BPR project. It would be reckless to undertake new projects without dedicated resources capable of committing the time and attention necessary to develop a quality solution. Often, this type of detailed planning may consume more money and time than is available from key personnel. Recognition should be given to the types of risk that may occur in the BPR design listed in the following table.	
	Risk Type	**Description**
	Sponsorship risk	• C-level management is not supportive of the effort. • There is insufficient commitment from the top or inappropriate project leadership. • Poor communication is a major problem.
	Scope risk	• The BPR project must be related to the vision and the specifications of the strategic plan. • Serious problems will arise if the scope is improperly defined. • It is a design failure if politically sensitive processes and existing jobs are excluded from the scope of change.
	Skill risk	• Absence of radical, out-of-the-box thinking will create a failure by dismissing new ideas that should have been explored. • "Thinking big" is the most effective way to achieve the highest return on investment (ROI). • Participants without broad skills will experience serious difficulty because the project vision is beyond their ability to define an effective action plan.
	Political risk	• Sabotage or passive resistance is always possible from people fearing a loss of power or being resistant to change. • Uncontrolled rumors lead to fear and subversion of the concept. • People will resist change unless the benefits are well understood and accepted.

Part I—Risk Management and Information Systems Control Theory and Concepts
Domain 1—Risk Identification, Assessment and Evaluation
I. IT Risk Identification and Assessment

IT Project-related Risk *(cont.)*

\multicolumn{2}{c}{IT Project-related Risk *(cont.)*}
Risk Area	**Description**
Implementation risk	The implementation risk factors represent another source of potential failures that could occur during the BPR project. The most common implementation risk factors include those listed in the following table.

Risk Type	Description
Leadership risk	• Leadership failures include disputes over ownership and project scope. • Management changes during the BPR project may signal wavering needs that may cause the loss of momentum. • Strong sponsors will provide money, time and resources while serving as project champions with their political support.
Technical risk	• Technical complexity may exceed the initial project scope. • The required capability may be beyond that of prepackaged software. • Custom functions and design may exceed IT's creative capability or available time. • Delays in implementation could signal that the complexity of scope was underestimated. • If the key issues are not fully identified, disputes will arise about the definitions of deliverables, which leads to scope changes during implementation.
Transition risk	• The loss of key personnel may create a loss of focus during implementation.
Personnel risk	• Personnel may feel burned out because of workload or their perception that the project is not worth the effort. • Reward and recognition are necessary during transition to prevent the project from losing momentum.
Scope risk	• Improperly defined project scope will produce excessive costs with schedule overruns (variance from schedule). • Poor planning may neglect the human resources (HR) requirements, which will lead team members to feel that the magnitude of effort is overwhelming. • The reaction will cause a narrowing of the scope during implementation, which usually leads to a failure of the original BPR objectives.

IT Project-related Risk *(cont.)*

	IT Project-related Risk *(cont.)*	
Risk Area	**Description**	
Operation or rollout risk	It is still possible for the BPR project to fail after careful planning. Common failures during production implementation include negative attitudes and technical flaws. These problems manifest in the form of management risk, technical risk and cultural risk, as described in the following table.	
	Risk Type	**Description**
	Management risk	• Strong, respected leadership is required to resolve power struggles over ownership. • Communication problems must be cured to prevent resistance and sabotage. • Executive sponsors need to provide sufficient training to prevent an unsuccessful implementation.
	Technical risk	• Insufficient support is the most obvious cause of failure in a rollout. • Inadequate testing leads to operational problems caused by software problems. • Data integrity problems represent a root problem capable of escalating into user dissatisfaction. • Data may not be migrated correctly from legacy systems to the new IT platform. • Perceptions of a flawed system will undermine everyone's confidence.
	Cultural risk	• Resistance in the enterprise is a result of failing to achieve user buy-in. • Resistance will increase to erode the benefits. • Effective training is often successful in solving user problems. • Dysfunctional behavior will increase unless the new benefits are well understood and achieved.

Part I—Risk Management and Information Systems Control Theory and Concepts
Domain 1—Risk Identification, Assessment and Evaluation
I. IT Risk Identification and Assessment

Risk Components

In assessing risk, each of the risk components listed in the following table can be considered to determine the levels of component risk within each category

Risk Components	
Component	**Description**
Inherent risk	The risk level/exposure without taking controls or other management actions into account **Example:** • The inherent risk associated with operating system (OS) security is ordinarily high because changes to, or even disclosure of, data or programs through OS security weaknesses could result in system failure, security breach or regulatory penalties. Inherent risk for most areas of IT is ordinarily high because the potential effect of errors ordinarily spans several business systems and many users. In assessing the inherent risk, there should be consideration for pervasive and detailed IT controls, as outlined in the following table.

	Control Type	**Investigation Area**
	Pervasive IT controls	• Integrity of IT management and IT management experience and knowledge • Pressures on IT management that may predispose it to conceal or misstate information (e.g., large business-critical project overruns, hacker activity) • Nature of the enterprise's business and systems (e.g., plans for electronic commerce [e-commerce], complexity of the systems, lack of integrated systems) • Factors affecting the enterprise's industry as a whole (e.g., changes in technology, IS staff availability)
	Specific IT controls	• Complexity of the systems involved • Level of manual intervention required • Susceptibility to loss or misappropriation of the assets controlled by the system (e.g., inventory, payroll) • Likelihood of activity peaks at certain times in the period of investigation • Poor change control procedures • Integrity, experience and skills of the management and staff involved in applying the IT controls

CRISC Review Manual 2013
ISACA. All Rights Reserved.

Risk Components (cont.)

Component	Description
Residual risk	The risk that remains after management has implemented a risk response
Control risk	The risk of a failure of the internal control systems to prevent, detect or correct an incident in a timely manner. **Example:** • The control risk associated with manual reviews of computer logs can be high because activities requiring investigation are often easily missed due to the volume of logged information. The control risk associated with computerized data validation procedures is ordinarily low because the processes are applied consistently.
Detection risk	The risk that the prescribed controls, substantive testing procedures, or monitoring will not detect an error that could be material, individually or in combination with other errors. **Example:** An intrusion detection system (IDS), an antivirus system or firewall is unable to detect or notice an adverse condition and trigger an adequate response (sometimes called a false negative—an indication that everything is fine when there actually is a problem).

Information Systems Architecture

The calculation of risk for IT systems is directly affected by the type of architecture that the enterprise is using. Some types of architecture are much more robust or secure than others. The risk associated with a ring topology or a centralized system are different and may be less than the risk associated with a bus, star or tree topology. The risk practitioner must consider many IT factors when determining IT risk including the age of equipment, maintenance schedules, location and users.

Risk management also considers the risk at all levels of the IT infrastructure. The risk at the application, database, network, operating system, utility and hardware levels all must be considered.

2. TYPES OF BUSINESS RISK AND THREATS THAT CAN BE ADDRESSED USING IT RESOURCES

Introduction

Whether for compliance, effectiveness or efficiency, IT enablement of business has dramatically increased in recent years. As complexities of business evolve, the integral role of IT is extended to that of assisting business in handling risk that is an inevitable part of business strategies, processes and operations.

Therefore, included as part of business enablement, IT has the task of assisting in the management of business risk.

Part I—Risk Management and Information Systems Control Theory and Concepts
Domain 1—Risk Identification, Assessment and Evaluation
I. IT Risk Identification and Assessment

Key Role of IT in the Enterprise Control Environment	When IT is used in the execution of business strategies and operations, it is inevitably drawn into the arena of business risk. A majority of large enterprises have implemented stringent internal controls and empowered internal auditors to conduct more intensive auditing of internal business processes and supporting IT processes and systems. IT is intimately associated with a range of business activities that are sources of risk and, as such, has a key part to play in the enterprise's control environment. IT risk managers, teaming with personnel managing enterprise risk from other perspectives, can ensure that IT risk is given the right priority, and that opportunities for IT systems and services to assist in managing risk factors of different types are leveraged. Just as IT can be applied to yield results that were not previously possible in many fields, IT can also prove itself in the field of risk management. IT can facilitate the wiring up, locking down and constant surveillance of the business; specifically in the domain of risk management information systems (RMISs), IT is relied on for advanced risk analytics and reporting.

3. ENTERPRISE RISK MANAGEMENT

Enterprise Risk Management (ERM) Model Control Objectives for Risk	An ERM model must address the enterprise's objectives with control objectives for risk in the following categories of business: • **Planning**—High-level planning, resource allocation and budgeting • **Operational**—Day-to-day activities • **Financial reporting**—Presentation of financial results • **Compliance**—Adherence to statutory requirements of all jurisdictions within which the enterprise does business The internal controls in each area ensure that: • The business is being run in accordance with the overall plan. • Financial statements and management reporting present an accurate view of the operations. • All activities (including reporting) that are covered by statutory regulations are being carried out within the constraints of those regulations.
Segregation of Duties (SoD) as a Key Component of a Strong Internal Control Environment	Separation of duties (also called segregation of duties) is a key component to maintaining a strong internal control environment because it reduces the risk of errors and fraudulent transactions. When duties for a business process or transaction are segregated so that it requires the involvement of more than one person to accomplish a task, it becomes more difficult for fraudulent activity to occur because it would require collusion among several employees. There are a wide variety of automated (i.e., IT-based) compliance solutions that address the issue of SoD, as seen in the next section. Prior to these tools being available, enterprises typically addressed SoD through a combination of controls, such as: • Defining transaction authorizations • Assigning custody of assets • Granting access to data • Reviewing or approving authorization forms • Creating user authorization tables • Creating manual SoD tables

Automated Tools to Assist in SoD	Automated tools are typically used to address SoD and also to provide the enterprise with reporting functionality on SoD violations (i.e., detective controls) and to implement preventive controls.

In general, automated control systems contain the following three elements:
- **Access controls**—Restrict access to the underlying business systems and data to ensure that only authorized individuals have access, and that each user is only granted the minimal level of access require to perform their job function
- **Process controls**—Restrict the activities performed by authorized users. This employs techniques such as dual control (requiring two people to take action simultaneously to perform a task) or mutual exclusivity (if one person has executed one task, they are prohibited from executing subsequent or supporting tasks).
- **Continuous monitoring**—Employs automation to detect system transactions, setup or data changes that contravene corporate policy. These systems may be used to block certain activities from unauthorized users or to limit the ability of a user to make a change without higher-level approval.

Example: Each of these elements may be subject to access control protection to ensure that only authorized individuals can view or change the access rules. Similarly, process controls ensure that only correct actions are permitted and monitoring controls will track any invalid operations after the fact. |
| **Difficulties in Maintaining Manual Controls** | Automated compliance solutions aim to provide enterprises with timely and efficient internal controls that do not disrupt their normal business process.

Many systems can now be updated automatically—such as antivirus signatures, software patches and access controls.

As enterprises grow and the reliance on IT for both internal and external users increases, it becomes impossible to manage access rights and privileges in a manual fashion. Resources, systems, employees and business partners are added to the infrastructure and employee job functions are changed to mirror the ongoing changes within an enterprise. This causes manual access privileges to become quickly outdated.

Without automating identity management and user access controls, there is the potential of impacting the ability of employees and customers to access enterprise systems and data in a timely manner. Manual authorizations are often time consuming and require significant administrator time. It may not even be possible to maintain a manual system and the time taken for system administration tasks is time taken away from other important tasks. |

Part I—Risk Management and Information Systems Control Theory and Concepts
Domain 1—Risk Identification, Assessment and Evaluation
I. IT Risk Identification and Assessment

Contribution Areas for IT in Managing Business Risk

The following table lists the two specific contribution areas for IT in managing business risk.

IT Contributions to Manage Business Risk	
Contribution Area	**Description**
Locked-down operating	IT can be used to build in business process controls. Historically, for business processes with low levels of automation, there was an ad hoc option—at the discretion of the operators and done on an ad hoc basis. However, automation with IT requires precision. As a consequence of automation, the routine aspects of business are increasingly locked down to a repeatable and predictable pattern. Where variation and variance is to be avoided as much as possible, predictable behavior translates as low risk. Automation with IT is the preferred route toward a six-sigma (one defect per million cycles) or zero-defect goal. Applications enforce business rules, such as: • Mandatory fields are required before a record can be saved. • Lookup fields can be used to ensure that valid codes are entered. • Approvals above a certain value can be routed via work flow for management approval. • Automated teller machines (ATMs) will not discharge money without a valid account and personal identification number (PIN) combination. This is an essential part of controlling normal business operations. It also allows HR to be channeled to doing other things, as long as the IT systems reliably perform the handling and also the checking and balancing. In the creative realm, in which predictability rapidly leads to commoditization and loss of competitiveness, IT tools are available to knowledge workers as enablers. Even here, prescribed forms are typically common, suggesting or encouraging through IT rather than explicit enforcement. **Example:** It is much easier to create a letter according to an enterprise's template than to start one from scratch. Also, in using the template, it is far more likely to achieve a compliant standard result. Constant surveillance over IT operations can maintain a watchful eye on the enterprise information systems and maintain records needed for the provision of evidence in litigation or with which to prosecute.

IT Contributions to Manage Business Risk (cont.)

Contribution Areas for IT in Managing Business Risk (cont.)

Contribution Area	Description
Decision support, risk analytics and reporting	Advanced risk/return decision making requires advanced IT support. It is not feasible to manually calculate the risk/opportunity related to today's credit portfolio. Data volumes are huge, and the sophisticated models require precise calibration and consistent fine-tuning. Quantitative analysis will inevitably turn to IT for the large-scale analysis of risk factors because the use of IT is most feasible when a large number of inputs and mathematical complexity are involved. The objective of all management information systems is to enable faster and better decision making. In the case of risk management information, the decision making relates to known and potential risk. The goal of risk management for information systems is to achieve compatible and efficient IT monitoring and reporting processes for capturing; analyzing; and, ultimately, reporting risk factors of all types across the entire enterprise. The consumers of output from the reporting systems are both internal—across all layers of management and across all lines of the business—and external. Automating risk information management can assist in the embedding of required practices into the enterprise by making "business as usual" risk management activities efficient rather than onerous.

Threat and Vulnerabilities Related to IT Operations Management

A system must be deployed in a secure manner that has adequate controls to mitigate risk and allow the system to be implemented without causing an undue level of risk to the enterprise as a whole, to other systems or networks or to individuals or departments.

However, once installed, the system must continue to operate in a secure manner. This requires the use of operational controls to ensure that the risk management controls built into the system continue to operate correctly. An enterprise that has poor IT management or change control procedures may be exposed to the risk that the controls will not work properly or that they may be bypassed by future development projects.

4. METHODS/FRAMEWORKS FOR DESCRIBING IT RISK IN BUSINESS TERMS

Describing the Business Impact of IT Risk

Meaningful IT risk assessments and risk-based decisions require IT risk to be expressed in unambiguous and clear, business-relevant terms. Effective risk management requires mutual understanding between IT and the business over which risk needs to be managed and why.

All stakeholders must have the ability to understand and express how adverse events may affect business objectives. This means that:
- IT personnel should understand how IT-related failures or events can impact enterprise objectives and cause direct or indirect loss to the enterprise.
- Business personnel should understand how IT-related failures or events can affect key services and processes.

> **IMPORTANT:** The link between IT risk and the ultimate business impact needs to be established to understand the effects of adverse events.

Exhibit 1.10: Expressing IT Risk in Business Terms

Several techniques and options exist that can help the enterprise describe IT risk in business terms. **Exhibit 1.10** and the following table depict and describe some available methods.

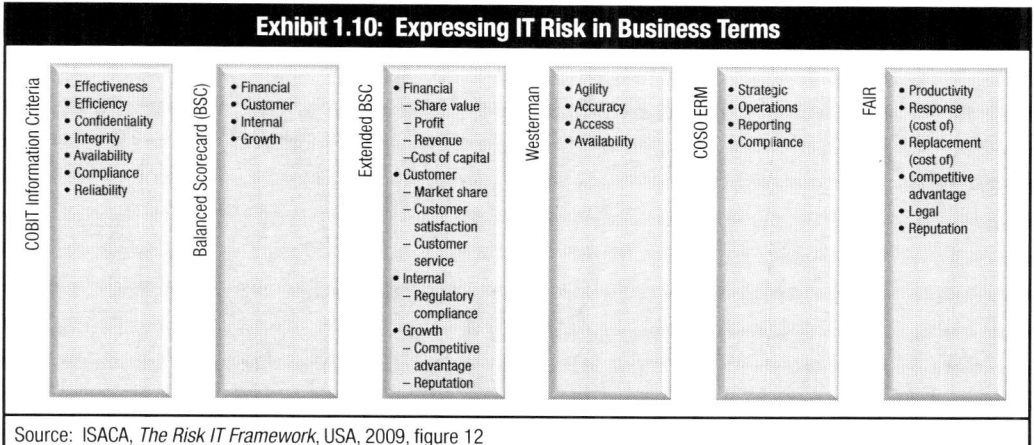

Source: ISACA, *The Risk IT Framework*, USA, 2009, figure 12

Techniques for Expressing IT Risk in Business Terms	
Technique	**Description**
COBIT Information Criteria (business requirements for information)	Allow for the expression of business aspects related to the use of IT They express a condition to which information (in the widest sense), as provided through IT, must conform for it to be beneficial to the enterprise. The business impact of any IT-related event lies in the consequence of not achieving the information criteria. By describing impact in these terms, this remains a sort of intermediate technique that does not fully describe business impact, e.g., impact on customers or in financial terms.
COBIT Business Goals and Balanced Scorecard (BSC)	Based on the "business goals" concept introduced in COBIT Business risk lies in any combination of those business goals not being achieved. The COBIT business goals are structured to align with the four classic BSC perspectives: financial, customer, internal and growth.
Extended BSC Criteria	A variant of the approach described in COBIT business goals and balanced scorecard The extended BSC criteria go one step further, linking the BSC dimensions to a limited set of more tangible criteria.
Westerman 4As	Based on the 4A framework, which defines IT risk as the potential for an unplanned event involving IT to threaten any of the following four interrelated enterprise objectives: • **Agility**—Possess the capability to change with managed cost and speed • **Accuracy**—Provide correct, timely and complete information that meets the requirements of management, staff, customers, suppliers and regulators • **Access**—Ensure appropriate access to data and systems so that the right people have the access that they need and the wrong people do not have access • **Availability**—Keep the systems (and their business processes) running and recover from interruptions

Part I—Risk Management and Information Systems Control Theory and Concepts
Domain 1—Risk Identification, Assessment and Evaluation
I. IT Risk Identification and Assessment

Exhibit 1.10: Expressing IT Risk in Business Terms *(cont.)*

Techniques for Expressing IT Risk in Business Terms *(cont.)*	
Technique	**Description**
The Committee of Sponsoring Organizations of the Treadway Commission (COSO) *Enterprise Risk Management—Integrated Framework*	Lists the following criteria that pertain to: • **Strategic**—High-level goals, aligned with and supporting the enterprise objectives. Strategic objectives reflect management's choice as to how the enterprise will seek to create value for its stakeholders. • **Operations**—The effectiveness and efficiency of the enterprise's operations, including performance and profitability goals, and safeguarding resources against loss • **Reporting**—The reliability of reporting, including internal and external reporting and may involve financial and nonfinancial information • **Compliance**—Adherence to relevant laws and regulations
Factor Analysis of Information Risk (FAIR)	While security-oriented in origin, impact criteria (productivity, cost of response, cost of replacement, competitive advantage, legal and reputational) apply to all IT-related risk factors.

Note: The challenge of describing IT risk in business terms requires, among other things:
• Identifying various risk factors
• Developing risk scenarios to make risk more specific and allow for proper risk analysis and assessment

5. RISK AWARENESS AND COMMUNICATION

Defining Risk Awareness

Risk awareness acknowledges that risk is an integral part of the business. This does not imply that all risk is to be avoided or eliminated, but rather that:
• Risk is well understood and known.
• IT risk issues are identifiable.
• The enterprise recognizes and uses the means to manage risk.

Awareness education and training can serve to mitigate some of the biggest organizational risk and achieve the most cost-effective improvement in risk and security. This can generally be achieved by educating an organization's staff in required procedures and policy compliance, as well as ensuring that staff can identify and understand the risk that threatens the organization. It is critical that the training effectively communicate the risk and its potential impact in order for staff to understand the justification for what many see as inconvenient extra steps that risk mitigation and security controls often require.

The CRISC must also understand the organization's structure and culture, as well as the types of communication that are most effective, in order to develop awareness and training programs that will be effective in the environment. Periodically changing risk awareness messages and the means of delivery will help maintain a higher level of risk awareness. Procedural controls can be complex, and it is essential to provide training as needed to ensure that staff understand the procedures and can correctly perform the required steps.

Awareness of information security policies, standards and procedures by all personnel is essential to achieving effective risk management. However, employees cannot be expected to comply with policies or standards that they are not aware of, or follow procedures they do not understand. The risk manager must devise a standardized approach, such as short computer- or paper-based quizzes to gauge awareness levels. Periodic use of a standardized testing approach provides metrics for awareness trends and training effectiveness. Further training needs can be determined by a skills assessment or employing a testing approach. Indicators for additional training requirements can come from various sources such as tracking help desk activity, operational errors, security events and audits

Importance of Risk Communication	Risk communication is a critical part of the risk management process. People are naturally uncomfortable talking about risk and tend to put off admitting that risk is involved—let alone communicating about issues, incidents or crises. If risk is to be managed and mitigated, it must first be discussed and effectively communicated at an appropriate level to the various stakeholders and personnel throughout the enterprise.
Benefits of Effective Risk Communication	The benefits of open communication on risk include: • Assistance in executive management's understanding of the actual exposure to risk, enabling the definition of appropriate and informed risk responses • Awareness among all internal stakeholders of the importance of integrating risk management into their daily duties • Transparency to external stakeholders regarding the actual level of risk and risk management processes in use

Consequences of Poor Risk Communication	The consequences of poor communication of risk include:
	• A false sense of confidence at all levels of the enterprise and a higher risk of a breach or incident that could have been prevented. Risk ignorance is an unacceptable risk management strategy.
	• Lack of direction or strategic planning to mandate risk management efforts.
	• Unbalanced communication to the external world on risk, especially in cases of high, but managed, risk, which may lead to an incorrect perception on actual risk by third parties such as:
	   − Clients
	   − Investors
	   − Regulators
	• The perception that the enterprise is trying to cover up known risk from stakeholders

Exhibit 1.11: IT Risk Communication Components	**Exhibit 1.11** and the following table depict and describe the broad array of information flows and the major types of IT risk information that should be communicated. 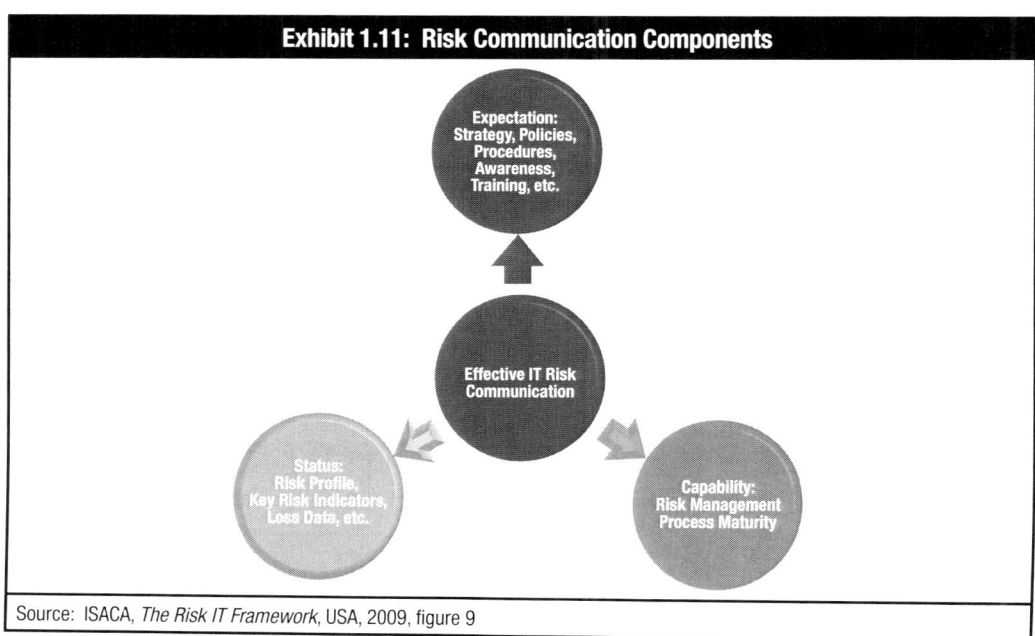 Source: ISACA, *The Risk IT Framework*, USA, 2009, figure 9

Part I—Risk Management and Information Systems Control Theory and Concepts
Domain 1—Risk Identification, Assessment and Evaluation
I. IT Risk Identification and Assessment

Exhibit 1.11: IT Risk Communication Components (cont.)

Description—Risk Components to Be Communicated	
Expectations from risk management	The risk components that must be communicated throughout the enterprise include risk strategy, policies, procedures, awareness training, continuous reinforcement of principles, etc. This is essential communication regarding the enterprise's overall strategy toward IT risk and: • Drives all subsequent efforts on risk management • Sets the overall expectations about the risk management program
Current risk management capability	This information: • Allows for monitoring of the state of the "risk management engine" in the enterprise • Is a key indicator for good risk management • Has predictive value for how well the enterprise is managing risk and reducing exposure
Status	This includes the actual status with regard to IT risk, including information such as: • The risk profile of the enterprise, i.e., the overall portfolio of (identified) risk to which the enterprise is exposed • Key risk indicators (KRIs) to support management reporting on risk • Event/loss data • The root cause of loss events • Options to mitigate risk (including cost and benefits)

Effective Communication

The following table lists the required elements for effective communication.

Effective Communication	
Communication Element	Description
Clear	Risk information must be known and understood by all stakeholders.
Concise	Information or communication should not inundate the recipients. All ground rules of good communication apply to communication on risk. This includes the avoidance of jargon and technical terms regarding risk that may not be understood by the intended audience.
Useful	Any communication on risk must be relevant to the audience. Technical information that is too detailed and/or is sent to inappropriate parties will hinder, rather than enable, a clear view of risk.
Timely	The timing of when to communicate about risk is important for the effectiveness of the communication. Communication at the wrong time may be too late to be effective, whereas communication that is too early may be ignored as being irrelevant. **Examples:** • A risk that is not addressed when an IT system is being set up may result in an unacceptable level of risk associated with system operations and expensive rework to implement controls postimplementation. • Failure to anticipate risk during project planning may result in project failure; the business consequence would be delayed business initiatives. Communication is timely when it allows action to be taken at the appropriate moments to identify and treat the risk. It serves no useful purpose to communicate about a project delay a week before the deadline.

Part I—Risk Management and Information Systems Control Theory and Concepts
Domain 1—Risk Identification, Assessment and Evaluation
I. IT Risk Identification and Assessment

Effective Communication *(cont.)*	\multicolumn{2}{l|}{**Effective Communication** *(cont.)*}

	Communication Element	Description
Effective Communication *(cont.)*	Aimed at the correct target audience	Information must: • Be communicated at the right level of detail • Be adapted for the audience • Enable informed decisions In this process, aggregation of data must not hide root causes of risk. **Example:** A security officer needs technical IT data on intrusions and viruses to deploy solutions. An IT steering committee may not need this level of detail, but it does need aggregated information to decide on policy changes or additional budgets to treat the same risk.
	Available on a need-to-know basis	Information related to risk should be known and communicated to all parties with a genuine need. A risk register with all documented risk is not public information and should be properly protected against internal and external parties with no need of access. Communication does not always need to be formal, through written reports or messages. Timely face-to-face meetings between stakeholders are an important means of communication for information related to business or IT-related risk.
Reference:	\multicolumn{2}{l|}{For more information on: • Risk scenarios, see section F in this chapter • Risk factors, see section G in this chapter}

J. SUGGESTED RESOURCES FOR FURTHER STUDY

Suggested Resources for Further Study

In addition to the resources cited throughout this manual, the following resources are suggested for further study of risk management; risk governance; and related frameworks, standards and leading practices:
- ISACA:
  - COBIT 5, 2012
    Note: The COBIT 5 framework is available at no charge from ISACA and can be downloaded at www.isaca.org/cobit
  - *Enterprise Value: Governance of IT Investments, The Val IT Framework 2.0*, 2008
  - *Implementing and Continually Improving IT Governance*, 2010
  - *The Risk IT Framework*, 2009
  - *The Risk IT Practitioner Guide*, 2009
- Committee of Sponsoring Organizations of the Treadway Commission (COSO), *Enterprise Risk Management—Integrated Framework*, USA, 2004
- International Organization for Standardization (ISO):
  - ISO/IEC 27001:2005, *Information technology—Security techniques—Information security management systems—Requirements*, Switzerland, 2005
  - ISO/IEC 27002:2005, *Information technology—Security techniques—Code of practice for information security management*, Switzerland, 2005
  - ISO/IEC 27004:2009, *Information technology—Security techniques—Information security management—Measurement*, Switzerland, 2009
  - ISO/IEC 27005:2011, *Information technology—Security techniques—Information security risk management*, Switzerland, 2011
- Jones, J.; *An Introduction to Factor Analysis of Information Risk (FAIR)*, Risk Management Insight LLC, USA, November 2006
- Kovacich, Gerald L.; Edward Halibozek; *The Manager's Handbook for Corporate Security: Establishing and Managing a Successful Assets Protection Program*, Butterworth-Heinemann, USA, 2003
- National Institute of Technology and Standards (NIST), *Risk Management Guide for Information Technology Systems*, Special Publication (SP) 800-30, www.csrc.nist.gov
- Peltier, Thomas A.; *Information Security Risk Analysis, 3rd Edition*, Auerbach Publications, USA, 2010
- Westerman, George; Richard Hunter; *IT Risk—Turning Business Threats Into Competitive Advantage*, Harvard Business School Press, USA, 2007

Domain 2—Risk Response

A. CHAPTER OVERVIEW

Introduction

This chapter provides an introduction to the principles of risk response.

The focus of this chapter is to evaluate the various risk response recommendations provided to the analyst and to select the best mitigation strategy—considering all of the factors that will influence the risk response decision.

The risk practitioner may assist in the investigation and evaluation of the risk response options—cost, time, skill, effectiveness, etc.—so that the enterprise can put in place a risk response strategy that is ideal for the enterprise. The best response will be based on several factors such as risk levels, urgency and impact, the cost of the risk response strategy chosen, and the enterprise's risk appetite and risk response capability.

Learning Objectives

As a result of completing this chapter, the CRISC candidate should be able to:
- Define various parameters for risk response selection.
- List the different risk response options.
- Recognize risk responses that may be most suitable for a high-level risk scenario.
- Express how exception management relates to risk management.
- Explain how residual risk relates to inherent risk, risk appetite and risk tolerance.
- Discuss the need for performing a cost-benefit analysis when determining a risk response.
- Describe the attributes of a business case to support project management.
- Identify standards, frameworks and leading practices related to risk response.

Inputs (Tie-back)

Risk response activities rely directly on the input from the previous chapter—risk identification, assessment and evaluation—where the risk that the enterprise is facing is identified and the risk response strategies are recommended for the enterprise's consideration.

Relevance

This domain is an important function of the risk practitioner through the determination of the cost-benefit analysis and recommendation of the:
- Prioritization of risk responses, including IS control implementation efforts
- Appropriate risk response
- Necessary level of control

Outputs (Tie-forward)

The risk response decision process results in the selection of the best risk response strategy for a specific risk, for a group of risk and for the risk profile of the enterprise as a whole.

IT Risk mitigation is addressed in more detail in Domain 4—Information Systems Control Design and Implementation. Concepts related to IS control design and implementation can, in most cases, be transferred to non-IS controls for a wider applicability.

Contents

This chapter contains the following sections:

Section	Starting Page
A. Chapter Overview	71
B. Task and Knowledge Statements	73
C. The Risk Response Process	74
1. Overview of the Risk Response Process	74
2. Risk Response Options	76
3. Risk Response Selection and Prioritization	78
4. Risk Response Implementation and Reporting	82
D. Risk Response Process Details	84
Phase 1—Articulate Risk	84
Phase 2—Manage Risk	88
Phase 3—React to Risk Events	93
E. Suggested Resources for Further Study	96

B. TASK AND KNOWLEDGE STATEMENTS

Introduction

This section describes the task and knowledge statements for Domain 2, which focuses on the development and implementation of risk responses to ensure that risk issues, opportunities, and events are addressed in a cost-effective manner and in line with business objectives.

Task Statements

The following table describes the task statements for Domain 2 that the CRISC candidate must know how to perform.

No.	Task Statement (TS)
TS2.1	Identify and evaluate risk response options and provide management with information to enable risk response decisions.
TS2.2	Review risk responses with the relevant stakeholders for validation of efficiency, effectiveness and economy.
TS2.3	Apply risk criteria to assist in the development of the risk profile for management approval.
TS2.4	Assist in the development of risk response action plans to address risk factors identified in the organizational risk profile.
TS2.5	Assist in the development of business cases supporting the investment plan to ensure risk responses are aligned with the identified business objectives.

Knowledge Statements

The following table describes the knowledge statements for Domain 2. The CRISC candidate must have a good understanding of each of the areas delineated by the knowledge statements. These statements are the basis for the exam.

No.	Knowledge Statement (KS) Knowledge of:
KS2.1	Standards, frameworks and leading practices related to risk response
KS2.2	Risk response options
KS2.3	Cost/benefit analysis and return on investment (ROI)
KS2.4	Risk appetite and tolerance
KS2.5	Organizational risk management policies
KS2.6	Parameters for risk response selection
KS2.7	Project management tools and techniques
KS2.8	Portfolio, investment and value management
KS2.9	Exception management
KS2.10	Residual risk

C. THE RISK RESPONSE PROCESS

Introduction

This section provides knowledge on developing and selecting an appropriate response to a given risk based on cost/benefit analysis, business objectives and risk culture.

Importance of Defining a Risk Response

The purpose of defining a risk response is to ensure that the residual risk is within the limits of the risk tolerance of the enterprise.

Risk response is based on selecting the correct, prioritized response to risk, based on the level of risk, the enterprise's risk tolerance and the cost-benefit advantages of the selected risk response option.

The risk response process integrates with the other risk management processes (identification, assessment, evaluation and monitoring) to ensure that management is provided accurate reports on:
- The level of risk faced by the enterprise
- The types of incidents that have occurred
- Any change to the enterprise's risk profile based on changes in the (internal and external) risk environment

> **Note:** Risk should always be reported based on the risk to the business, the ability of the business to meet its objectives and the risk to IT systems.

1. OVERVIEW OF THE RISK RESPONSE PROCESS

Description of the Risk Response Process

The risk response process is triggered when a risk exceeds the enterprise's risk tolerance level.

The prioritization of the risk responses and development of the risk response plan is influenced by several parameters:
- Cost of the response to reduce risk to within tolerance levels
- Importance of the risk
- Capability to implement the response
- Effectiveness of the response
- Efficiency of the response

Since not all risk can be addressed at the same time and remediation may require considerable investment in time and resources, a risk prioritization strategy is used to create a risk response plan and implementation schedule. Risk with a greater likelihood and impact on the enterprise will—in most cases—be prioritized above other risk that is considered less likely or less damaging.

Exception Management as a Part of Risk Management

Under certain circumstances, an organization may decide to grant an exception to a normal control requirement. The exception may be granted due to extraordinary business requirements, extraordinary cost, or unavailability of a feasible countermeasure or solution. Exceptions should only be granted through a formal exception management process that requires the review, tracking and approval of the exception request by a senior manager or other appropriate authority. Exceptions must be tracked and reviewed on a periodic basis to ensure that the exception is still required and is being applied correctly.

An exception management process will work with the risk management effort to ensure that all exceptions are considered when determining levels of risk and effectiveness of controls.

Part I—Risk Management and Information Systems Control Theory and Concepts
Domain 2—Risk Response
C. The Risk Response Process

Exhibit 2.1: Risk Response Process

Exhibit 2.1 illustrates how the risk response process is driven by input from the risk assessment process.

Exhibit 2.1: Risk Response Process

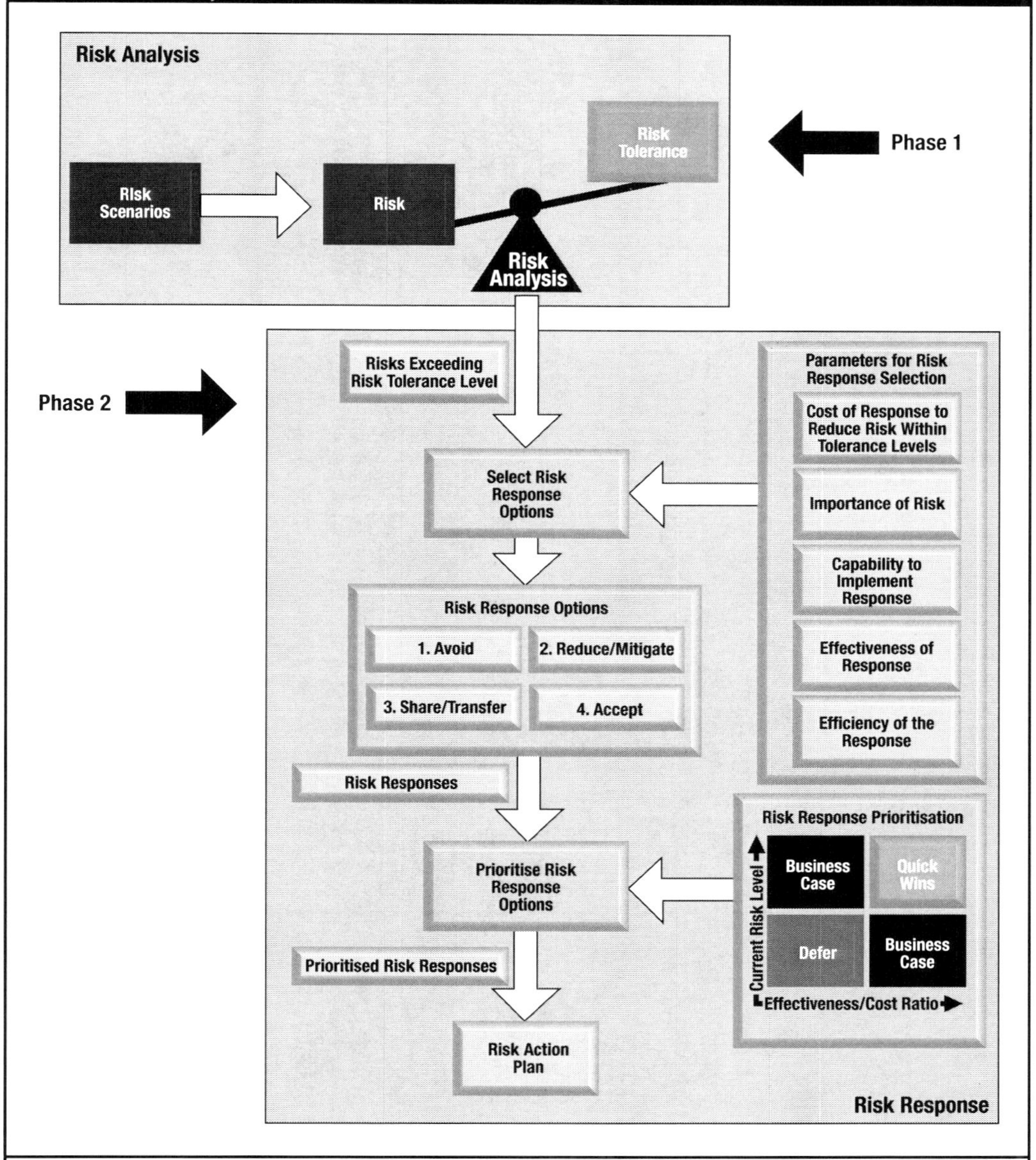

Source: ISACA, *The Risk IT Practitioner Guide*, USA, 2009, figure 46

High-level Risk Response Process

The following table describes the risk response process at a high level.

High-level Risk Response Process Phases		
Phase	**Description**	**Notes**
1	Review results of the risk analysis. Determine if the risk level exceeds the risk tolerance level: • If **yes**, go to phase 2. • If **no**, no action is required.	Risk analysis identifies and prioritizes risk levels according to risk scenarios and makes assessments based on the likelihood and magnitude of the risk, taking into account potential business impact. The risk assessment should also provide recommendations for risk response.
2	Select risk response options.	In instances where the risk analysis shows that risk is not within the defined risk tolerance levels, weigh projected risk vs. the potential cost of implementing and maintaining controls and select the most appropriate response.
3	Prioritize the risk response option.	Select a risk response implementation strategy according to the priorities for addressing risk, and develop the risk action plan.
4	Implement the risk action plan.	Implement the selected risk response according to the risk action plan. This is covered in more detail in domain 4, with a focus on IS control design and implementation.

2. RISK RESPONSE OPTIONS

Introduction

This section describes the four key risk response options utilized to ensure that the enterprise follows the best possible risk response strategy and manages its risk environment to gain the positive benefits of risk while minimizing the negative effects.

Risk Response Options

Enterprises must choose between the following risk response options:
• Risk avoidance (avoid)
• Risk mitigation (reduce/mitigate)
• Risk sharing (share/transfer)
• Risk acceptance (accept)

Part I—Risk Management and Information Systems Control Theory and Concepts
Domain 2—Risk Response
C. The Risk Response Process

Risk Avoidance (Avoid)

Risk avoidance means that activities or conditions that give rise to risk are discontinued.

Risk avoidance applies when the level of risk, even after the selection of controls, would be greater than the risk tolerance level of the enterprise. This is the case when:
- There is no other cost-effective response that can succeed in reducing the likelihood and magnitude below the defined thresholds for risk tolerance.
- The risk cannot be shared or transferred.
- The risk is deemed unacceptable by management.

Examples:
- Not engaging in electronic commerce (e-commerce) to avoid the risk associated with that line of business
- Not engaging in a very large project when the business case shows a significant risk of loss or failure
- Not operating in some countries or regions due to security concerns

Risk Mitigation (Reduce/Mitigate)

Risk mitigation means that actions are taken to reduce:
- The likelihood and/or
- The impact of risk.

Risk mitigation can utilize various forms of control. A complete risk mitigation portfolio will consist of all types of controls carefully integrated together. The main control types to be considered are:
- Managerial (e.g., policies)
- Technical (e.g., tools such as firewalls and intrusion detection systems [IDSs])
- Operational (e.g., procedures, separation of duties)
- Preparedness activities

Examples:
- Strengthening overall risk management practices, such as implementing sufficiently mature risk management processes
- Deploying new technical, management or operational controls that reduce either the likelihood or the impact of an adverse event
- Installing a new access control system
- Implementing policies or operational procedures
- Developing an effective incident response and business continuity plan

Risk Sharing (Share/Transfer)

Risk sharing means that risk impact is reduced by transferring or otherwise sharing a portion of the risk with an external enterprise or another internal entity.

> **IMPORTANT:** In both a physical and legal sense, these techniques **do not relieve an enterprise of a risk**, but can involve the skills of another party in managing the risk and can reduce the financial consequence if an adverse event occurs.

Examples:
- Taking out insurance coverage for disasters or incidents
- Outsourcing unique business processes
- Sharing project risk with other organizations through fixed price arrangements or shared investment arrangements

Risk Acceptance (Accept)	Risk acceptance means that no action is taken relative to a particular risk; loss is accepted when/if it occurs.

> **Note:** This is different from being ignorant of risk. Accepting risk assumes that the risk is known; that is, an informed decision has been made by management to accept it as such.

If an enterprise adopts a risk acceptance stance, it should carefully consider who can accept the risk. Risk should be accepted only by senior business management in collaboration with senior management and the board.

Examples:
- Choosing not to implement costly controls to comply with regulatory requirements and paying the penalty for noncompliance, as applicable
- Selecting a product from a start-up as a software supplier, which is not necessarily a "going concern" or well established yet, because the potential opportunity is promising
- Opting to conduct prosperous business in a politically volatile country due to the potential for higher profit margins

3. RISK RESPONSE SELECTION AND PRIORITIZATION

Introduction	This section contains information on the selection of an appropriate response and the prioritization of risk responses.
Risk Response Considerations	Consider the goals and objectives of an enterprise when selecting any of the risk mitigation options. It may not be practical to address all identified risk, so priority should be given to the threat and vulnerability pairs that have the highest potential to cause significant impact or harm to business objectives. Also, in safeguarding an enterprise's objectives and assets, the option used to mitigate the risk, and the methods used to implement controls, may vary because of each enterprise's unique environment and objectives. The "best in class" approach is to use appropriate technologies from among the various vendor solutions, along with the appropriate risk mitigation options and nontechnical, administrative measures.

Part I—Risk Management and Information Systems Control Theory and Concepts
Domain 2—Risk Response
C. The Risk Response Process

Exhibit 2.2: Risk Response Options and Parameters

Exhibit 2.2 illustrates different high-level risk response options and the parameters that influence the selection of these options.

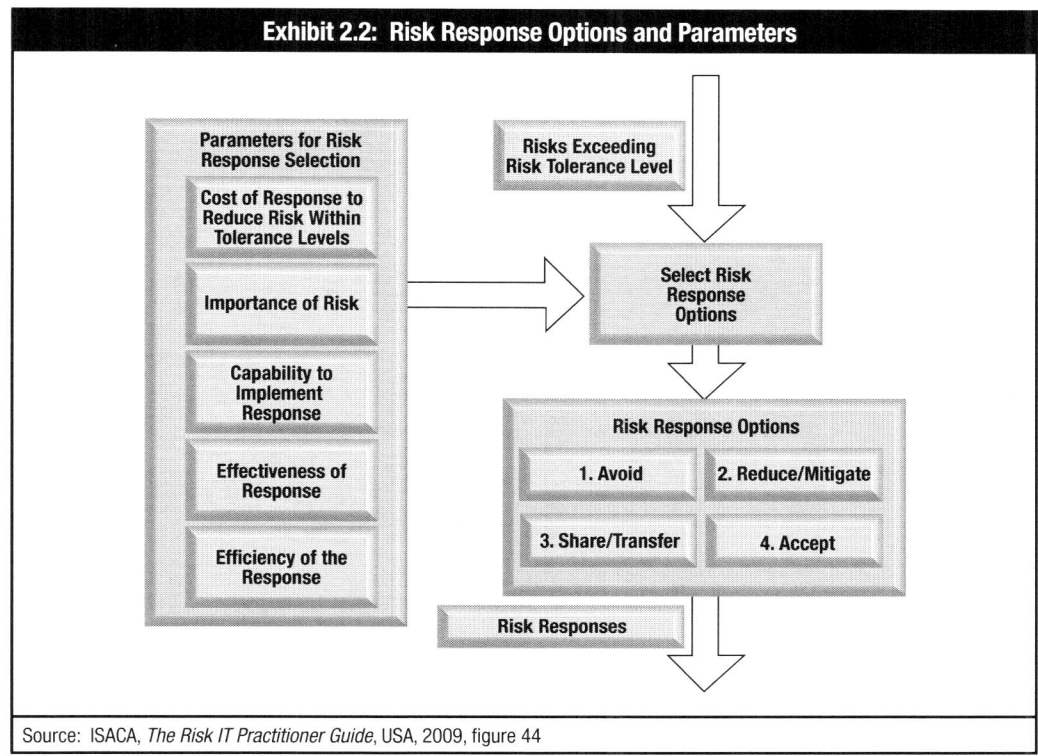
Source: ISACA, *The Risk IT Practitioner Guide*, USA, 2009, figure 44

Parameters for Risk Response Selection

The following table lists the parameters to be considered when selecting a specific risk response option, as illustrated in **exhibit 2.3**.

Risk Response Selection Parameters		
Parameter	**Description**	
Cost of response	The cost of the response to reduce risk to within tolerance levels	
	In the case of …	**The cost of response is the cost …**
	Risk transfer,	Of the insurance premium.
	Risk mitigation,	To implement and maintain control measures. **Examples:** Capital expense, salaries and consulting, licensing, maintenance, training
Importance of risk	The importance of the risk is reflected by: • The combination of likelihood and magnitude (impact) levels (both quantitative and qualitative impact measures) • Its position on the risk map compared to other risk	
Capability to implement response	The enterprise's capability to implement the response	
	When the risk management process is …	**The appropriate risk response may be…**
	Mature,	More sophisticated.
	Immature,	Very basic.
Effectiveness of response	The extent to which the response will reduce the likelihood and/or the impact of the risk	
Efficiency of response	The relative benefits promised compared to those listed in the "Risk Response Prioritization Option Descriptions" table (following) **Example:** One type of risk response control may effectively address several risk factors while another may not.	

Need for Risk Response Prioritization

It is likely that the aggregated required effort for the mitigation responses, i.e., the collection of controls that need to be implemented or strengthened, will exceed available resources.

In this case, prioritization of the risk response is required.

Exhibit 2.3: Risk Response Prioritization Options

Exhibit 2.3 and the following table depict and describe the prioritization of risk responses based on the outcomes that they offer by placing the probable outcomes in a quadrant.

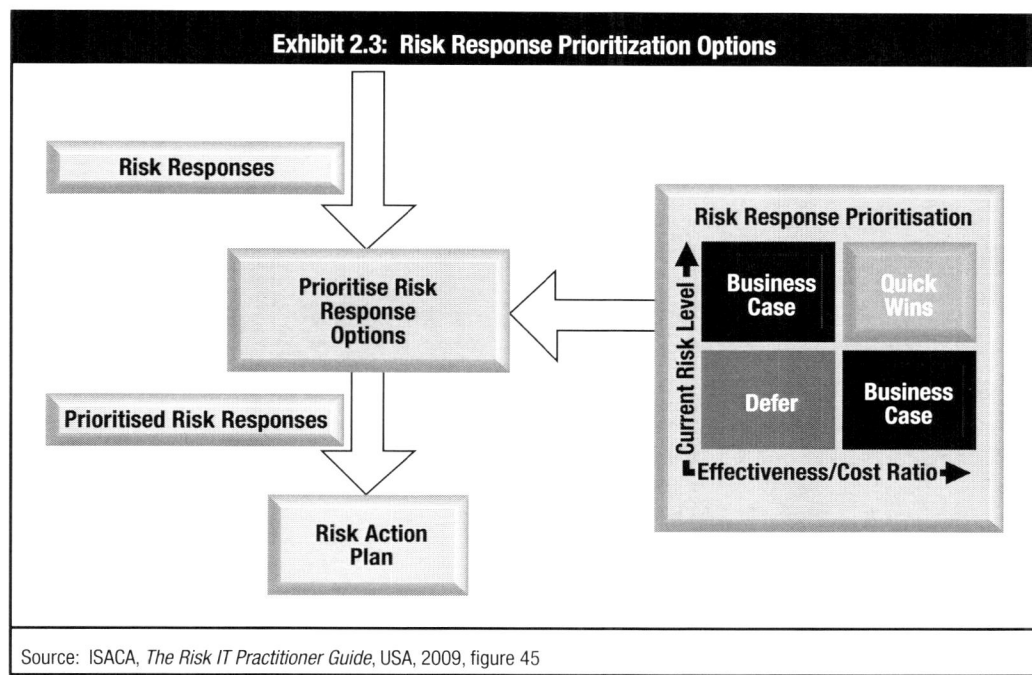

Source: ISACA, *The Risk IT Practitioner Guide*, USA, 2009, figure 45

Risk Response Prioritization Options	
Option	**Description**
Quick win	Very effective and efficient response that addresses medium to high risk
Business case to be made	Requires careful analysis and management decisions on investments: • More expensive or difficult risk responses to medium to high risk • Efficient, effective responses to low to medium risk
Deferral	Costly risk response to a low risk

Considerations of Various Risk Response Selection Alternatives	**Example 1**: A risk has been identified: The equipment being used in manufacturing is reaching the end of its life and is likely to cause production failures or extraordinary repair costs.
	• The response alternatives consist of:
	  – Replacing the equipment with new updated equipment
	  – Major refurbishing of existing equipment
	  – Deciding to discontinue production of the item if the equipment fails.
	  – Deciding to outsource manufacturing of the product to another company
	  – Accepting the risk and deciding what course of action to take when or of the equipment fails
	
	The selection of the best alternative is based on the cost, impact, availability of solutions, market environment, risk appetite and strategic plan of the organization.

IT-related Risk Response Prioritization Examples	**Example 1:** A risk has been identified that the enterprise's IT system and application landscape is so complex that, within a few years, extending capacity will become difficult and maintaining software will become very expensive.
	• The response alternatives consist of:
	  – Major rearchitecture and redesign of the existing system
	  – Purchase of a new, integrated system
	
	This is categorized as a "business case to be made" because of the project cost.
	
	Example 2: A risk of noncompliance with regulations is identified because a number of relatively simple procedures are missing.
	• The response consists of creating the missing procedures and implementing them.
	• This is categorized as a "quick win" because the allocation of existing resources or a minor resource investment provides measurable (and potentially immediate) benefits.
	
	Note: This topic is dealt with in more detail in Domain Four.
	
	> **Note:** Risk response should be selected with careful consideration of the impact of any controls on the business. Controls should not be so restrictive that they impede the ability of the business to be productive or profitable. Controls should also be reviewed to ensure that they are legal, implemented fairly, and accepted by management and employees.

4. RISK RESPONSE IMPLEMENTATION AND REPORTING

Introduction	This section contains information on risk prioritization, development of a risk response plan and reporting of risk response status to relevant stakeholders.
Risk Response Plan	Once the risk response strategy has been decided, the risk practitioner will create a risk response plan. This plan will outline the steps, timelines, budgets, people and tools needed to implement the risk response strategy.

Part I—Risk Management and Information Systems Control Theory and Concepts
Domain 2—Risk Response
C. The Risk Response Process

Risk Response Selection and Prioritization Guidelines	Risk response prioritization considerations should take into account factors such as: • Stakeholder interests • Acceptance of change • Balance of technical and nontechnical solutions • Cost • Impact on productivity • Ownership of controls • Ability to audit and monitor risk • Regulations • Changing market conditions • Resource optimization
Risk Response Tracking	As part of risk response, the ongoing status of risk mitigation processes must be tracked. This tracking is often done using a risk register. This is important to ensure that the risk response strategy remains active and that proposed controls are implemented according to schedule. When an enterprise is aware of a risk, but does not have a justifiable risk response strategy, or is not following its strategy, the liability of the enterprise to adverse publicity or even civil or criminal penalties increases.
Risk Response Integration	The risk practitioner should always look for opportunities to achieve greater efficiency by integrating risk response options to address more than one risk. The use of techniques that are versatile and enterprisewide, rather than individual solutions, provides better justification for risk response strategies and related costs. **Example:** Deploying an access control system that supports more than one system
Risk Response Implementation	The implementation of IS controls should consider the following: • Controls are tested prior to implementation whenever possible. • People are trained in the use of the tools. • A control owner is clearly identified and responsible for the control. • The control is measureable. • The control is documented. • The control is monitored to ensure that it remains effective over time (see Domain 5—Information Systems Control Monitoring and Maintenance). **Note:** The implementation of IS controls is addressed in detail in Domain 4—Information Systems Control Design and Implementation.

D. RISK RESPONSE PROCESS DETAILS

Introduction — This section provides information on the end-to-end risk response process.

Risk Response Process — The overview below outlines the phases, tasks and steps in the risk response process.

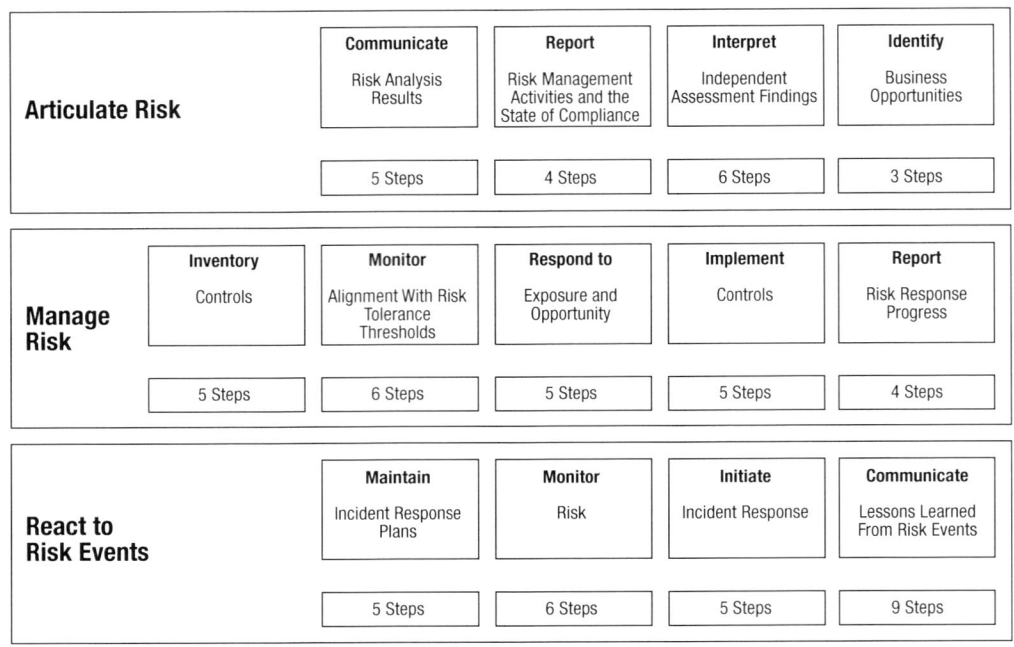

PHASE 1—ARTICULATE RISK

Introduction — Phase 1 of the risk response process requires articulating (documenting and reporting) risk to ensure that information on the true state of exposures and opportunities is made available:
- In a timely manner
- To the right people to enable the appropriate response

Tasks Associated With Phase 1 — The following table lists the tasks to articulate risk.

Tasks to Articulate Risk	
Task	Name/Description
1	Communicate risk analysis results.
2	Report risk management activities and the state of compliance.
3	Interpret independent risk assessment findings.
4	Identify business opportunities.

Part I—Risk Management and Information Systems Control Theory and Concepts
Domain 2—Risk Response
D. Risk Response Process Details

Steps Associated With Task 1

The following table describes the steps to communicate risk analysis results.

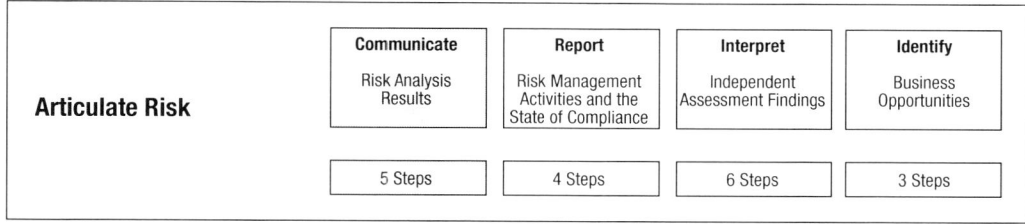

	Steps to Communicate Risk Analysis Results
Step	**Action**
1	Report the results of risk analysis in terms and formats useful to support business and risk management decisions.
2	Coordinate additional risk analysis activity as required by decision makers. **Example:** Report rejection and scope adjustment
3	Clearly communicate the risk-return context, including, wherever possible, probabilities of loss and/or gain, ranges, and confidence levels that enable management to balance risk-return ratios.
4	Identify the: • Negative impacts of events/scenarios that drive response decisions • Positive impacts of events/scenarios that represent opportunities that management should channel back into the strategy- and objective-setting process.
5	Provide decision makers with an understanding of: • Worst-case and most probable scenarios • Due diligence exposures • Significant reputation, legal or regulatory considerations including: – Key components of risk (e.g., likelihood, magnitude, impact) and key risk factors and their estimated effects – Estimated probable loss magnitude or probable future gain – Estimated high-end loss/gain potential and the most probable loss/gain scenario(s) (e.g., a probable loss likelihood of between three and five times per year and a probable loss magnitude of between US $50,000 and US $100,000 with 90 percent confidence) – Additional relevant information to support the conclusions and recommendations of the analysis

Steps Associated With Task 2

The following overview and table describe the steps to report risk management activities and the state of compliance.

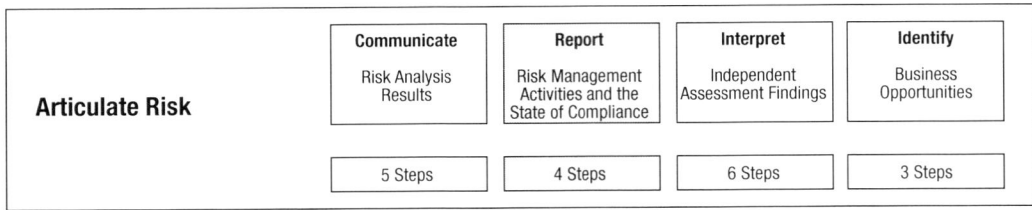

	Steps to Report Risk Management Activities and the State of Compliance
Step	**Action**
1	Meet the risk reporting needs of various stakeholders. **Example:** Board, risk committee, risk control functions, business unit management
2	Apply the principles of relevance, efficiency, timeliness and accuracy to ensure strategic and efficient reporting on risk issues and status.
3	When reporting, include the following: • Control effectiveness and performance • Issues and gaps • Remediation status • Events and incidents • Impacts of events and incidents on the risk profile • Performance of risk management processes
4	Provide inputs to integrated enterprise reporting.

Steps Associated With Task 3

The following overview and table describe the steps to interpret independent risk assessment findings.

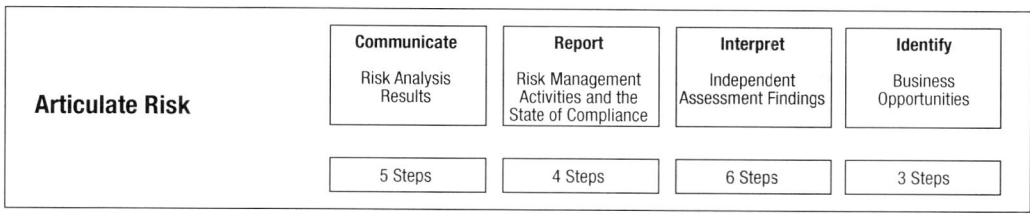

Steps to Interpret Independent Risk Assessment Findings	
Step	**Action**
1	Review the results and specific findings of objective third parties, internal audit, quality assurance, self-assessment activities, etc.
2	Map the results/findings to the risk profile and the risk and control baseline.
3	Consider the established risk tolerance.
4	Take gaps and exposures to the business for its decision on disposition or the need for risk analysis.
5	Help the business understand how corrective action plans will affect the overall risk profile.
6	Identify opportunities for integration with: • Other remediation efforts • Ongoing risk management activities

Steps Associated With Task 4

The following overview and table describe the steps to identify business opportunities.

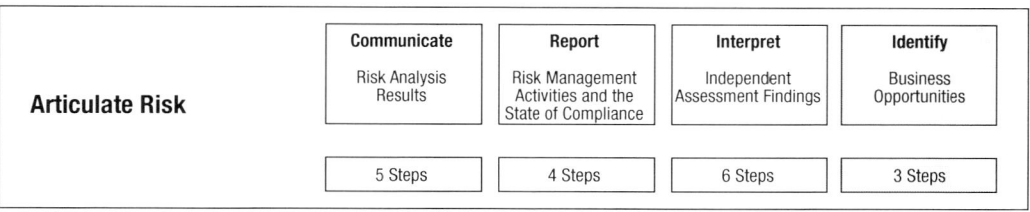

Steps to Identify Business Opportunities	
Step	**Action**
1	On a recurrent basis, consider the relative levels of risk to risk management capacity for specific business processes, business units, products, etc.
2	For areas with relatively high-risk capacity capability (i.e., indicating an ability to take on more risk), identify opportunities that could: • Enable the area to accept greater risk • Enhance growth and return
3	Look for opportunities in which resources can be leveraged to: • Create competitive advantage (e.g., use existing information in new ways, better leverage human and business resources)Reduce enterprise coordination costs • Exploit scale and scope economies in certain key strategic resources common to several lines of business • Coordinate activities among business units or in the value chain

PHASE 2—MANAGE RISK

Introduction

Phase 2 of the risk response process requires managing risk to ensure that measures for seizing strategic opportunities and reducing risk to an acceptable level are managed as a portfolio.

Tasks Associated With Phase 2

The following table lists the tasks to manage risk.

> **Note:** It must be understood that the tasks are not sequential in nature and that monitoring is a continuous process.

Tasks to Manage Risk	
Task	**Name/Description**
1	Inventory controls.
2	Monitor operational alignment with risk tolerance thresholds.
3	Respond to discovered risk exposure and opportunity.
4	Implement controls.
5	Report IT risk response plan progress.

Part I—Risk Management and Information Systems Control Theory and Concepts
Domain 2—Risk Response
D. Risk Response Process Details

Steps Associated With Task 1

The following overview and table describe the steps to inventory controls.

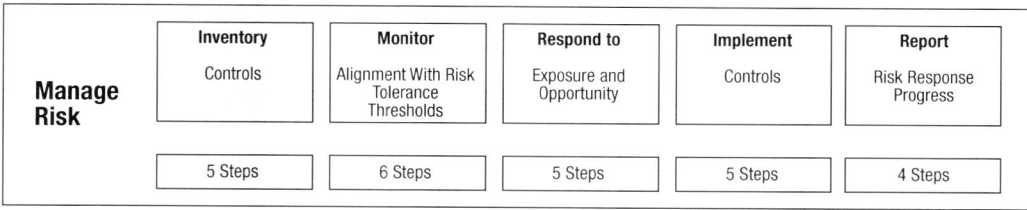

	Steps to Inventory Controls
Step	**Action**
1	Inventory the controls in place across the risk focus areas. **Rationale:** Manages and enables risk and controls to be tracked and measured to be in line with risk tolerance
2	Classify controls, and map them to specific IT risk statements and aggregations of IT risk. **Examples:** Predictive, preventive, detective and corrective controls
3	Develop tests for control design and control operating effectiveness.
4	Identify procedures and technology used to monitor the operation of controls. **Examples:** Monitoring of controls when IT is involved or the automation of enterprise monitoring processes
5	Partition operational controls into the following categories: • Controls deployed in line with expectations with no known operating deficiencies • Controls deployed in line with expectations with known operating deficiencies • Controls deployed beyond expectations with no known operating deficiencies **Note:** This third category of controls may not be justified and could indicate opportunity for cost reduction while maintaining the same level of risk.

Steps Associated With Task 2

The following overview and table list the steps to monitor operational alignment with risk tolerance thresholds.

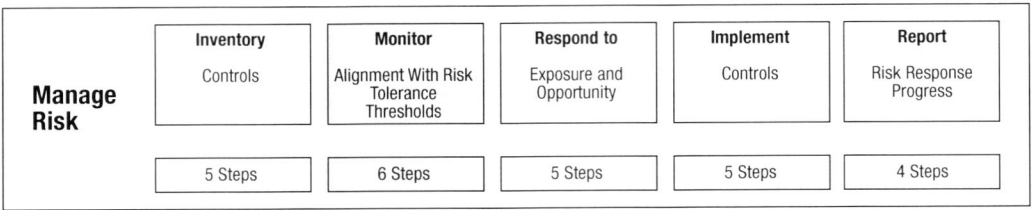

Step	Steps to Monitor Operational Alignment With Risk Tolerance Thresholds
	Action
1	Ensure that each business line accepts accountability for: • Operating within its individual and portfolio risk tolerance levels • Embedding monitoring tools into key operating processes • Monitoring control performance (e.g., control self-assessment) • Measuring variance from thresholds against objectives
2	Periodically test control design and operating effectiveness for key risk issues.
3	Obtain buy-in from management on indicators that will function as key risk indicators (KRIs).
4	When implementing KRIs: • Set thresholds and checkpoints (e.g., weekly, daily, continuously). • Configure where to send notifications (e.g., line management, senior management, internal audit) so that the recipients can respond or adjust their plans.
5	Integrate KRI data into ongoing performance indicator reporting.
6	Ensure that there is a detailed examination of areas of residual risk outside of tolerance thresholds (e.g., request risk analysis).

Steps Associated With Task 3

The following overview and table describe the steps to respond to discovered risk exposure and opportunity.

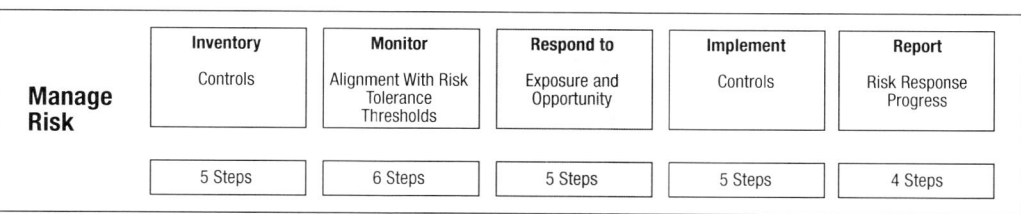

Steps to Respond to Discovered Risk Exposure and Opportunity	
Step	**Action**
1	Emphasize projects that are expected to reduce the potential likelihood and magnitude of adverse events/losses, and balance them with projects that enable the seizing of strategic business opportunities.
2	Hold cost-benefit discussions regarding the contribution of new or existing controls toward operating within risk tolerance.
3	Select candidate controls based on: • Specific threats • The degree of risk exposure • Probable loss • Mandatory requirements specified in internal and/or external standards
4	Monitor changes to the underlying business operational risk profiles.
5	Adjust the rankings of risk response projects.

Steps Associated With Task 4

The following overview and table describe the steps to implement controls.

Note: The design and implementation of information systems controls is addressed in more detail in Domain 4—Information Systems Control Design and Implementation.

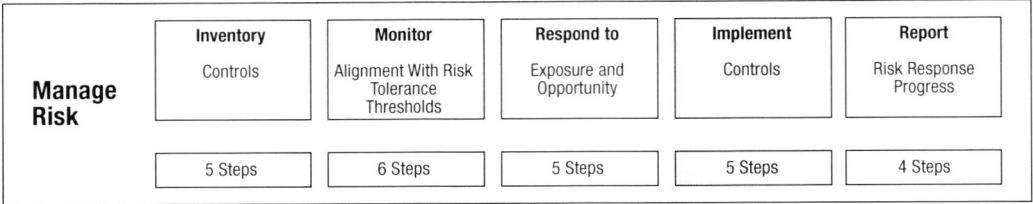

Steps to Implement Controls	
Step	**Action**
1	Take appropriate steps to ensure the effective deployment of new controls and adjustments to existing controls.
2	Communicate with key stakeholders early in the process.
3	Before relying on the control: • Conduct pilot testing. • Review performance data to verify operation against design.
4	Map new and updated operational controls to monitoring mechanisms that will: • Measure control performance over time • Prompt management corrective action when needed
5	Identify and train staff on new procedures as they are deployed.

Steps Associated With Task 5

The following overview and table describe the steps to report IT risk response plan progress.

Note: Monitoring of risk responses, particularly information systems controls, is covered in more detail in Domain 5—Information System Monitoring and Maintenance.

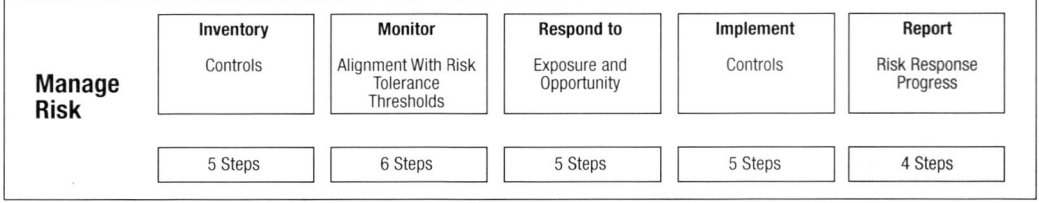

Steps to Report Risk Response Plan Progress	
Step	**Action**
1	Monitor risk response plans at all levels.
2	Ensure the effectiveness of required risk responses.
3	Determine whether acceptance of residual risk has been obtained.
4	Ensure that committed risk responses are owned by the affected process owner(s) and that deviations are reported to senior management.

PHASE 3—REACT TO RISK EVENTS

Introduction

Phase 3 of the risk response process requires reacting to risk events to ensure that measures for seizing immediate opportunities or limiting the magnitude of loss from events are activated in a timely manner and are effective.

Tasks Associated With Phase 3

The following table lists the tasks to react to risk events.

	Tasks to React to Risk Events
Task	**Name/Description**
1	Maintain incident response plans.
2	Monitor risk.
3	Initiate incident response.
4	Communicate lessons learned from risk events.

Steps Associated With Task 1

The following overview and table describe the steps to maintain incident response plans.

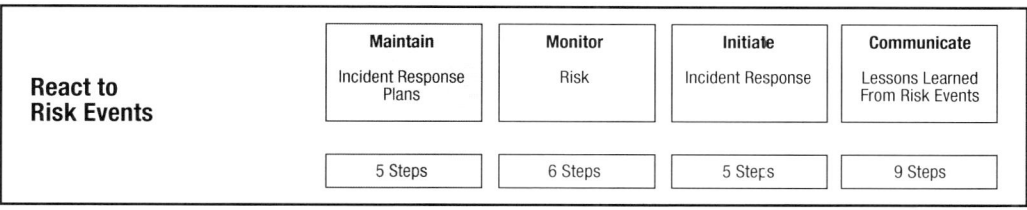

	Steps to Maintain Incident Response Plans
Step	**Action**
1	Prepare for the materialization of threats through plans that document the specific steps to take when a risk event may cause an operational, developmental and/or strategic business impact (i.e., IT-related incident) or has already caused a business impact.
2	Maintain open communication about risk acceptance, risk management activities, analysis techniques and results available to assist with plan preparation.
3	When developing action plans, consider how long the enterprise may be exposed and the time it may take to recover.
4	Based on the potential or known impact, define pathways of escalation across the enterprise, from line management to executive committees.
5	Verify that incident response plans for highly critical processes are adequate.

Steps Associated With Task 2

The following overview and table describe the steps to monitor risk.

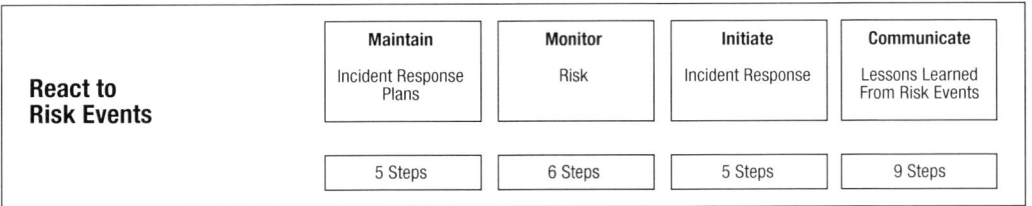

Steps to Monitor Risk	
Step	**Action**
1	Monitor the environment.
2	When a control limit has been breached, either escalate to the next step or confirm that the measure is back within limits.
3	Categorize incidents (e.g., loss of business, policy violation, system failure, fraud, lawsuit), and compare actual exposures against acceptable thresholds.
4	Communicate business impacts to decision makers.
5	Continue to take action and drive desired outcomes.
6	Ensure that policy is followed and that there is clear accountability for follow-up actions.

Steps Associated With Task 3

The following overview and table describe the steps to initiate incident response.

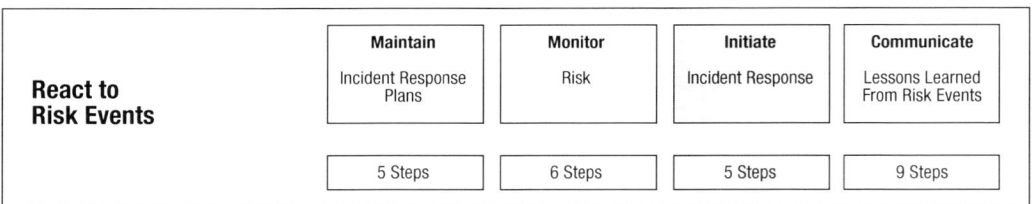

Steps to Initiate Incident Response	
Step	**Action**
1	Take action to minimize the impact of an incident in progress.
2	Identify the category of the incident, and follow the steps in the response plan.
3	Inform all stakeholders and affected parties that an incident is occurring.
4	Identify the amount of time required to carry out the plan, and make adjustments, as necessary, for the situation at hand.
5	Ensure that the correct action is taken.

Steps Associated With Task 4

The following overview and table describe the steps to communicate lessons learned from risk events.

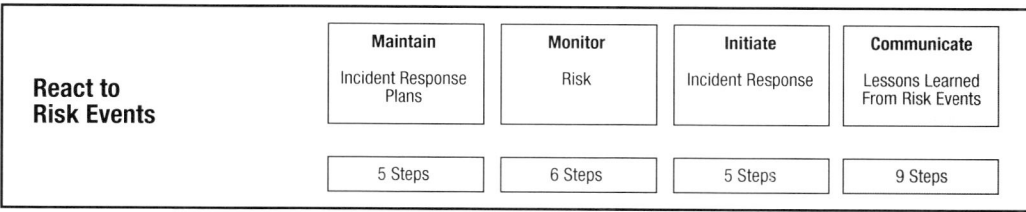

Steps to Communicate Lessons Learned From Risk Events	
Step	Action
1	Examine past adverse events/losses and missed opportunities.
2	Determine whether there was a failure stemming from lack of awareness, capability or motivation.
3	Research the root cause of similar risk events and the relative effectiveness of actions taken then and now.
4	For behavioral incidents, determine the extent of any underlying problems. **Example:** A serious systemic problem vs. an isolated case that could be managed through staff training or greater documentation of procedures
5	Identify tactical corrections; potential investments in projects; or adjustments to overall risk governance, evaluation and/or response processes.
6	To identify and correct the underlying root causes of operations and service delivery incidents and service levels (e.g., defects, rework), integrate with the: • Service center • Incident response process • Problem management process
7	Identity the root cause of incidents through open communication across business and IT functions.
8	Request additional risk analysis as needed.
9	Communicate root cause, additional risk response requirements and process improvements to risk governance processes and appropriate decision makers.

Summary

This domain is an important function of the risk management process through the determination of a cost-benefit analysis and recommendation of:
- Prioritization of risk responses, including IS control implementation efforts
- Appropriate risk response
  - Necessary level of control

The response strategies selected throughout this domain will be implemented during the next domains and will also be expanded in more detail in Domain 4—Information Systems Control Design and Implementation.

E. SUGGESTED RESOURCES FOR FURTHER STUDY

Suggested Resources for Further Study

In addition to the resources cited throughout this manual, the following resources are suggested for further study:
- ISACA:
  - COBIT 5, 2012 (*A Business Framework for the Governance and Management of Enterprise IT*)
  - *COBIT 5: Enabling Processes*, 2012
  - *COBIT 5 Implementation*, 2012
  - *Enterprise Value: Governance of IT Investments, The Val IT Framework 2.0*, 2008
  - *The Risk IT Framework*, 2009
  - *The Risk IT Practitioner Guide*, 2009
- International Organization for Standardization (ISO), ISO 27005:2011, *Information technology—Security techniques—Information security risk management*, Switzerland, 2011
- Project Management Institute (PMI), *A Guide to the Project Management Body of Knowledge (PMBOK)*, 4th Edition, USA, 2008

Domain 3—Risk Monitoring

A. CHAPTER OVERVIEW

Introduction

This chapter provides information on monitoring risk and communicating information to the relevant stakeholders to ensure the continued effectiveness of the enterprise's risk management strategy.

Learning Objectives

As a result of completing this chapter, the CRISC candidate should be able to:
- Explain the principles of risk ownership.
- List common risk and compliance reporting requirements, tools and techniques.
- Discuss various risk assessment methodologies.
- Differentiate between key performance indicators (KPIs) and key risk indicators (KRIs).
- Describe data extraction, aggregation, and analysis tools and techniques at a high level.
- Distinguish between various types of processes to review the enterprise's risk monitoring process.
- Name various standards, frameworks and practices related to risk monitoring.

Inputs From Other Domains

In the previous domains—*Risk Identification, Assessment and Evaluation; and Risk Response*—risk was identified and prioritized and risk response activities were chosen and implemented to ensure that risk was within acceptable limits.

Relevance

This phase of risk management concerns itself with monitoring and reporting on risk, taking into consideration risk events, threat events and vulnerability events. Risk monitoring ensures that current and emerging risk is within the risk tolerance levels of the enterprise.

This requires the determination of key risk indicators (KRIs) and risk reporting schedules to help ensure that:
- Senior management, operational managers, auditors, regulators, business continuity planners and security staff are aware of current risk and that
- Risk responses effectively mitigate the risk facing the enterprise.

Outputs to Other Domains

Risk monitoring provides input into the risk response and information systems control design and monitoring processes.

Information systems control monitoring can be seen as a distinct subset of risk monitoring.

Contents

This chapter contains the following sections:

Section	Starting Page
A. Chapter Overview	97
B. Task and Knowledge Statements	99
C. Essentials of Risk Monitoring	100
1. Key Terms and Concepts	100
2. Key Risk Indicator Selection	101
3. Data Extraction, Validation, Aggregation and Analysis	104
4. Risk Monitoring	106
5. Process Capability Models	108
6. Threat Analysis	110
7. Risk Reporting	111
D. Suggested Resources for Further Study	116

B. TASK AND KNOWLEDGE STATEMENTS

Introduction

This section describes the task and knowledge statements for Domain 3, which focuses on developing and implementing risk responses to ensure that risk factors and events are addressed in a cost-effective manner and in line with business objectives.

Task Statements

The following table describes the task statements for Domain 3 that the CRISC candidate must know how to perform.

No.	Task Statement (TS)
TS3.1	Collect and validate data that measure key risk indicators (KRIs) to monitor and communicate their status to relevant stakeholders.
TS3.2	Monitor and communicate key risk indicators (KRIs) and management activities to assist relevant stakeholders in their decision-making process.
TS3.3	Facilitate independent risk assessments and risk management process reviews to ensure that they are performed efficiently and effectively.
TS3.4	Identify and report on risk, including compliance, to initiate corrective action and meet business and regulatory requirements.

Knowledge Statements

The following table describes the knowledge statements for Domain 3. The CRISC candidate must have a good understanding of each of the areas delineated by the knowledge statements. These statements are the basis for the exam.

No.	Knowledge Statement (KS) Knowledge of:
KS3.1	Standards, frameworks and leading practices related to risk monitoring
KS3.2	Principles of risk ownership
KS3.3	Risk and compliance reporting requirements, tools and techniques
KS3.4	Key performance indicators (KPIs) and key risk indicators (KRIs)
KS3.5	Risk assessment methodologies
KS3.6	Data extraction, validation, aggregation and analysis tools and techniques
KS3.7	Various types of reviews of the organization's risk monitoring process (e.g., internal and external audits, peer reviews, regulatory reviews, quality reviews)

C. ESSENTIALS OF RISK MONITORING

Introduction

This section provides information on:
- Risk indicators
- Key risk indicators (KRIs)
- Key performance indicators (KPIs)

1. KEY TERMS AND CONCEPTS

Key Performance Indicator (KPI)

A measure that determines how well a process is performing in enabling the goal to be reached.

A KPI is a lead indicator of whether a goal will likely be reached, and a good indicator of capabilities, practices and skills. It measures an activity goal, which is an action that the process owner must take to achieve effective process performance.

Example: A KPI may indicate that an error rate of five percent is acceptable. An error rate higher than five percent would be unacceptable and require escalation and some form of response.

Key Risk Indicator (KRI)

A subset of risk indicators that:
- are highly relevant
- possess a high probability of predicting or indicating important risk

KRIs are the prime risk monitoring indicators for the enterprise.

Risk Indicator

A metric capable of showing that the enterprise is subject to, or has a high probability of being subject to, a risk that exceeds the defined risk appetite

They are used to measure levels of risk in comparison to defined risk thresholds and alert the enterprise when a risk level approaches a high or unacceptable level of risk. The purpose of a risk indicator is to set in place tracking and reporting mechanisms that alert staff to a developing or potential risk.

Risk indicators are specific to each enterprise, and their selection depends on a number of parameters in the internal and external environment, including, but not limited to the:
- Strategic focus of the enterprise
- Size and complexity of the enterprise
- Type of market in which the enterprise operates (e.g., highly regulated)

Comparing KPIs and KRIs

A KPI indicates the threshold or level at which the performance is not acceptable, and where action must be taken to address the measured item.

A KRI indicates the threshold at which performance is still within acceptable limits but may require some response to address the problem before it exceeds the KPI level.

An example is a KPI that indicates that an error level of five percent is unacceptable. The KRI will trigger when the error rate is at four percent. This will indicate the trend that the error rate is climbing and may soon exceed the KPI level. This will alert the risk team to a potential developing problem and to initiate a response to control the problem before it crosses the KPI threshold.

2. KEY RISK INDICATOR SELECTION

Introduction

A common mistake when implementing KRIs—other than selecting too many KRIs—includes choosing KRIs that:
- Are not linked to specific risk
- Are incomplete or inaccurate due to unclear specifications
- Are difficult to measure
- Are difficult to aggregate, compare and interpret
- Provide results that cannot be compared over time

The selection and maintenance of appropriate KRIs is critical to the ongoing success of the risk monitoring process.

This section provides an overview of:
- KRI benefits
- Factors influencing the selection of KRIs
- Criteria for KRI effectiveness
- Example of KRI reliability vs. sensitivity
- KRI maintenance

KRI Benefits

The selection of the right set of KRIs benefits the enterprise by:
- Providing an early warning (forward-looking) signal that a high risk is emerging to enable management to take proactive action (before the risk actually becomes a loss)
- Providing a backward-looking view on risk events that have occurred, enabling risk responses and management to be improved
- Enabling the documentation and analysis of trends
- Providing an indication of the enterprise's risk appetite and tolerance through metric setting (that is, KRI thresholds)
- Increasing the likelihood of achieving the enterprise's strategic objectives
- Assisting in continually optimizing the risk governance and management environment

Factors Influencing the Selection of Key Risk Indicators

The following table describes some factors to be considered when selecting key risk indicators.

Factors Influencing the Selection of Key Risk Indicators	
Factor	**Description**
Stakeholders	Select risk indicators with the involvement of relevant stakeholders to ensure greater buy-in and ownership. Risk indicators should be identified for all stakeholders and should not focus solely on IT, but should include measurement of the operational and strategic risk. IT-based metrics should be aligned as much as possible with other metrics used in the organization to report to stakeholders.
Balance	Make a balanced selection of risk indicators, covering: • Lag indicators (indicating risk after events have occurred) • Lead indicators (indicating which controls are in place to prevent events from occurring) • Trends (analyzing indicators over time or correlating indicators to gain insights)
Root cause	Ensure that selected indicators drill down to the root cause of events and not just the symptoms. The relation of a risk indicator to a root cause is not necessarily a one-to-one relationship. Therefore, it is important to map the unique root cause to a single or predefined set of indicators and vice versa to avoid false conclusions about the most appropriate risk response.

Criteria for KRI Effectiveness

The following table describes the criteria to be considered when selecting KRIs.

Criteria for KRI Effectiveness	
Criterion	**Description**
Impact	Indicators of risk with high business impact are more likely to be KRIs.
Effort	For different indicators that are equivalent in sensitivity, the one that is easier to measure and maintain is preferred.
Reliability	The indicator must possess a high correlation with the risk and be a good predictor or outcome measure.
Sensitivity	The indicator must be representative of risk and capable of accurately indicating risk variances.
Repeatable	A KRI must be repeatable in that it can be measured on a regular basis to show trends and patterns in activity and results.

> **Note:** The complete set of KRIs should also balance indicators for risk, root causes and business impact.

Part I—Risk Management and Information Systems Control Theory and Concepts
Domain 3—Risk Monitoring
C. Essentials of Risk Monitoring

KRI Optimization

To ensure accurate and meaningful reporting, KRIs will need to be optimized to ensure that: 1) the right data are being collected and reported on, and 2) that the KRI thresholds are set correctly. KRIs that are reporting on the data points that cannot be controlled by the enterprise, or are not alerting management at the correct time to an adverse condition, must be adjusted (optimized) to be more precise, more relevant or more accurate.

The following table describes a few examples in which KRIs may need to be optimized.

Examples in Which KRIs Should Be Optimized	
Metric Criterion	**Description**
Sensitivity	Management has implemented an automated tool to analyze and report on access control logs based on severity; the tool generates excessively large amounts of results. Management performs a risk assessment and decides to configure the monitoring tool to report only on alerts marked "critical."
Timing	Management has implemented strong segregation of duties (SoD) within the enterprise resource planning (ERP) system. One monitoring process tracks system transactions that violate the defined SoD rules before month-end processing is completed so that suspicious transactions can be investigated before reconciliation reports are generated.
Frequency	Management has implemented a key control that is performed multiple times a day. Based on a risk assessment, management decides that the monitoring activity can be performed weekly because this will capture a control failure in sufficient time for remediation.
Corrective action	Automated monitoring of controls is especially conducive to being integrated into the remediation process. This can often be achieved by using existing problem management tools, which help prioritize existing gaps, assign problem owners and track remediation efforts.

Example: Reliability vs. Sensitivity vs. Frequency vs. Corrective Action

Example: A smoke detector can be used to illustrate the difference between reliability, sensitivity, frequency and corrective action:
- **Reliability**—The smoke detector will sound an alarm every time there is smoke.
- **Sensitivity**—The smoke detector will sound an alarm when a specified threshold of smoke density (number of particles per cubic foot) is reached.
- **Frequency**—The smoke detector will operate continuously rather than on a periodic basis.
- **Corrective action**—The smoke detector will alert personnel to a risk condition early enough to enable an effective response.

KRI Maintenance

Since the enterprise's internal and external environments are constantly changing, the risk environment is also highly dynamic and the set of KRIs needs to be changed over time.

Each KRI is related to the risk appetite and tolerance levels of the enterprise. KRI trigger levels should be defined at a point that enables stakeholders to take appropriate action in a timely manner.

3. DATA EXTRACTION, VALIDATION, AGGREGATION AND ANALYSIS

Introduction

Because KRIs often rely on information from diverse sources, it is important that the risk practitioner understands the basic concepts related to data extraction, validation, aggregation and analysis.

This section provides an overview of the tasks related to data extraction, validation, aggregation and analysis:
- Requirements gathering
- Data access
- Data validation
- Data analysis
- Reporting and corrective action

> **Note:** Variations of these phases exist, but the steps within them and the principles used are generally the same.

Requirements Gathering

It is important to first understand the risk to be monitored, prepare a detailed plan and define the project's scope. In the case of a monitoring project, this step should involve process owners, data owners, system custodians and other process stakeholders.

Data Access

As part of the data access step, management identifies which data are available and how they be acquired in a format that can be used for analysis. There are two options for data extraction:
- Extracting data directly from the source system(s) after system owner approval
- Receiving data extracts from the system custodian (IT) after system owner approval

The recommended course of action is direct extract, especially since this risk monitoring generally involves management monitoring its own controls rather than auditors/third parties monitoring management's controls. If it is not feasible to get direct access, a data access request form should be submitted to the data owner(s) that details the appropriate data fields to be extracted. The request should specify the method of delivery for the file (i.e., posting on a dedicated server, via e-mail, or on CD/DVD). Most of the data analysis tools can handle any delimited text file, fixed length text file or spreadsheet.

Data Validation

Data validation ensures that extracted data are ready (in the correct format) and accurate enough for analysis. One objective is to perform data quality tests to ensure data are valid, complete and free of errors or duplication. This may also involve reformatting data that have been gathered from different sources to make the data suitable for comparative analysis.

There is a risk that the data could become corrupted; either through data manipulation, in storage, or during data transfer. As most of the data used for risk monitoring is the result of a download, report, transfer or other form of operation, it is imperative to validate data before they are analyzed. Omitting the step of data validation may compromise the results of the analysis. Whereas specific data validation techniques may vary depending on the tool, one should consider the concepts in the following table every time data are extracted.

Data Validation Practices

The following table provides an overview of common data validation concepts and practices.

Concepts to Consider When Validating Data	
Checking for …	**Helps …**
Validity	Ensure that data match definitions in the table layout.
Control totals	Ensure that the data are complete.
Ranges	Ensure that extracted data contain only the data requested, e.g., if data are requested for the first quarter of the year (January, February and March), one should not be provided data for the entire year.
Missing items	Identify missing data, such as gaps in sequence or blank fields/records.
Duplicates	Identify and confirm the validity of duplicates. For example, many alarms may report a problem repeatedly. The data analyst may want to eliminate duplicate alarm reporting or set thresholds that will only report on certain alarm conditions if the alarm occurs more than ten times in a short time period. Such reporting thresholds may be called clipping levels.
Reliability	Determine the confidence that an analyst may have in the integrity of the data—Were the data extracted directly from a source? Were the data computed from other values?
Reasonableness	Make certain assumptions about the information; e.g., if the average number of transactions per month is 2,400, and the report from one month is substantially different from normal volumes, this may be an indication that a data source file is missing or duplicated. A data value that is substantially different from normal values (a statistical anomaly, for example) may need further validation. Doing statistical analysis of data to determine data norms, standards of deviation and outliers, may help the analyst identify exceptional or erroneous data. **Note:** This test requires familiarity with the data used for monitoring.
Relationships (sequencing)	Identify table fields that relate to each other. For example, in an invoice table, the due date should always be later than the invoice date. To test relationships, use a filter to compare one field to another.
Orphan records	Record, in a transaction or detail table, that the record has no match in a master table. **Example:** There are two purchase order (PO) tables—one for header records and the other for line items. Every PO header record should match to at least one line; one or more line item records should match to one header record. Orphan records on either side could mean that orders are not properly completed and processed.

Note: Statistical analysis helps assess data validity through a variety of techniques.

Data Analysis

Analysis can involve a simple set of steps or can be a complex combination of commands and other functionality. Data analysis must be designed to achieve the stated objectives from the project plan. Although this may be applicable to any monitoring activity, it is beneficial to keep transferability and scalability in mind. This may include robust documentation, use of software development standards and naming conventions.

After the data extracts have been validated (this process can be automated), the enterprise should develop the logic to be used for data analysis. The logic should then be executed and reviewed for errors.

At this stage, it may be necessary to troubleshoot testing issues and refine the logic to ensure that the outcome is valid and accurate. This step also includes formatting the output for reporting purposes. The output should be reviewed again to ensure that it is providing the correct results. If necessary, the logic and output must be further refined and the previous steps repeated. The monitoring process can then be set up to be run on a repeatable basis at the appropriate levels of sensitivity, range, time and frequency.

> **Note:** The final step in data analysis, which involves making logical conclusions about the data, remains an important task for the CRISC.

Reporting and Corrective Action

Reporting structure and distribution depends on the requirements of the business and on regulatory demands. The format of the report may be determined by the monitoring objectives and the technology being used. Reporting procedures include the manner in which reports are distributed, and who should get the reports so that they are directed to the right people and in the right format, (e.g., dashboards, workflows). Similar to the data analysis stage, reporting may also identify areas in which changes to the sensitivity of the reporting parameters or the timing and frequency of the monitoring activity may be required.

Risk Monitoring Capabilities

Risk monitoring tasks can range from *ad hoc* queries performed on demand, to scheduled, repeatable monitoring processes, or even to solutions that integrate risk monitoring processes into strategic, risk management and performance management processes. Each approach has different challenges and benefits.

4. RISK MONITORING

Introduction

Risk monitoring provides timely information on the actual status of the enterprise with regard to risk. This includes information such as:
- The risk profile of the enterprise; i.e., the overall portfolio of (identified) risk to which the enterprise is exposed
- KRIs to support management reporting on risk
- Event/loss data
- The root cause of loss events
- Options to mitigate risk (cost and benefit calculations)

This section provides information on:
- Internal and external risk monitoring sources

Part I—Risk Management and Information Systems Control Theory and Concepts
Domain 3—Risk Monitoring
C. Essentials of Risk Monitoring

Risk Monitoring Sources

A key factor in the value of the risk monitoring process is to ensure that risk reporting is accurate, timely and complete. If the data used to generate the reports are biased or incomplete, management will have an incorrect understanding of the true risk levels and the appropriate risk responses may not be implemented where required.

To ensure that the reports are correct and complete, it is important to gather input from all available sources.

The following sources of risk monitoring information should not be considered all-inclusive. In determining the propriety of any specific source risk practitioners should apply their professional judgment:
- Suppliers or manufacturers of hardware, software or applications
- Antivirus/antispam/content filters
- Devices placed in the demilitarized zone (DMZ)
- Mail servers
- Communication software
- Trade unions
- Computer emergency response team (CERT) alerts/blogs/newsgroups
- Governmental advice (US Department of Homeland Security)
- Outcomes from conferences (including DEF CON®)
- Newspapers
- Online news
- Amnesty International
- ArcSight, Archer
- Bloomberg
- *Citicus.com*, *www.limesurvey.org*
- Greenpeace
- iDefense Labs Vulnerability Contributor Program (VCP)
- *Nimsoft.com*, *www.curasoftware.com/Pages/default.asp*
- Red Cross
- Reuters
- Risk & Opportunity Management, Software Engineering Institute (SEI), Carnegie Mellon University (CMU), *www.sei.cmu.edu/risk/research/index.cfm*
- RSA FraudAction
- SANS: @Risk Critical Vulnerability List
- Symantec™ DeepSight™
- The SysAdmin, Audit, Network, Security [SANS™] Institute
- TippingPoint DVLabs
- United Nations
- US Computer Emergency Readiness Team (US-CERT), *www.us-cert.gov*
- US-CERT National Cyber Alert System, *www.us-cert.gov/cas/alldocs.html*
- US National Institute of Standards and Technology (NIST) National Vulnerability Database
- The World Factbook, US Central Intelligence Agency (CIA)
- World Health Organization (WHO)
- Swiss Confederation, Reporting and Analysis Centre for Information Assurance (MELANI), *www.melani.admin.ch*

Specific, nongeneric sources are listed in alphabetical order.

5. PROCESS CAPABILITY MODELS

Value of Using Process Capability Models

Process capability models are an excellent tool to measure the maturity or development of the processes and operational procedures of the enterprise. A capability model allows the enterprise to measure and track its progress to developing, implementing and following reliable, consistent and reportable procedures.

Capability models enable the enterprise to rate itself from the least mature level (having nonexistent or unstructured processes) to the most mature (having adopted and optimized the use of good practices).

Using a capability model, management can identify:
- **The actual performance of the enterprise**—Where is the enterprise today?
- **A benchmark of what other organizations are doing**—What are the industry leading practices or international standards as represented within the model?
- **The enterprise's target for improvement**—Where does the enterprise want to be at a point in the future?

Relevance of Process Capability Models

Use of a process capability model helps determine the current risk management process capability level and allows management to determine whether it is in alignment with the desired state.

The model helps determine how to close the gap between actual and desired state and tracks process performance over time.

Management Responsibility for Risk Management Process Capability

Boards and executive management need to consider how effective their enterprises are at managing risk and should be able to answer the following related questions:
- What is the enterprise's current risk management capability level?
- What is the enterprise's desired risk management capability level?
- How does the enterprise ensure that risk prioritization reflects an enterprisewide view and is followed consistently across the enterprise?
- How does the enterprise identify which activities are necessary to reach the desired risk management process capability level?

Process Capability Levels

The levels within a process capability model are designed as profiles that allow an enterprise to identify symptoms or descriptions of its current and possible future states. In general, the purpose is to:
- Identify where enterprises are in relation to the consistent and reliable performance of certain activities or practices.
- Suggest how to identify areas for, and set priorities for, improvements.

Exhibit 3.1: Process Capability Levels

Process capability levels and related performance attributes can be summarized as shown in exhibit 3.1:

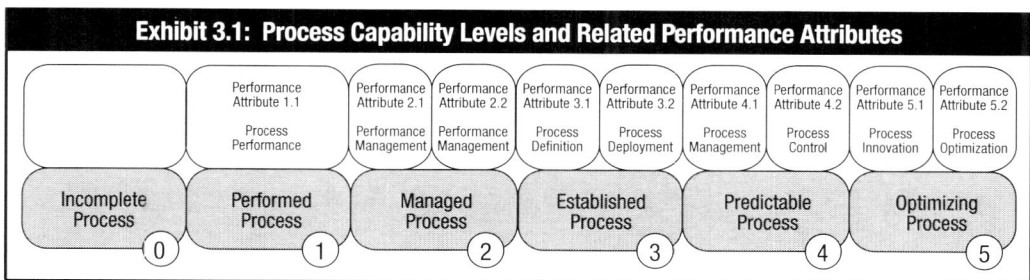

Exhibit 3.1: Process Capability Levels and Related Performance Attributes

Process Capability Levels

The following table describes the levels in the process capability levels:

COBIT 5 ISO/IEC 15504-based Capability Levels	Meaning of the COBIT 5 ISO/IEC 15504-based Capability Levels	Context
5 Optimized	The previously described predictable process is continuously improved to meet relevant current and projected business goals.	Enterprise view/Corporate knowledge
4 Predictable	The previously described established process now operates within defined limits to achieve its process outcomes.	Enterprise view/Corporate knowledge
3 Established	The previously described managed process is now implemented using a defined process that is capable of achieving its process outcomes.	Enterprise view/Corporate knowledge
2 Managed	The previously described performed process is now implemented in a managed fashion (planned, monitored and adjusted) and its work products are appropriately established, controlled and maintained.	Instance view/Individual knowledge
1 Performed	The implemented process achieves its process purpose.	Instance view/Individual knowledge
0 Incomplete	The process is not implemented or fails to achieve its process purpose. At this level, there is little or no evidence of any systematic achievement of the process purpose.	Instance view/Individual knowledge

Note: The COBIT 5 framework is available at no charge from ISACA and can be downloaded at *www.isaca.org/cobit*.

Exhibit 3.2: Current and Desired Process Capability

To make the results of a process capability model assessment easily usable in management briefings, where they should be presented as a means to support the case for future plans to improve risk governance and management efforts, the following graphic presentation method, **exhibit 3.2**, may be useful.

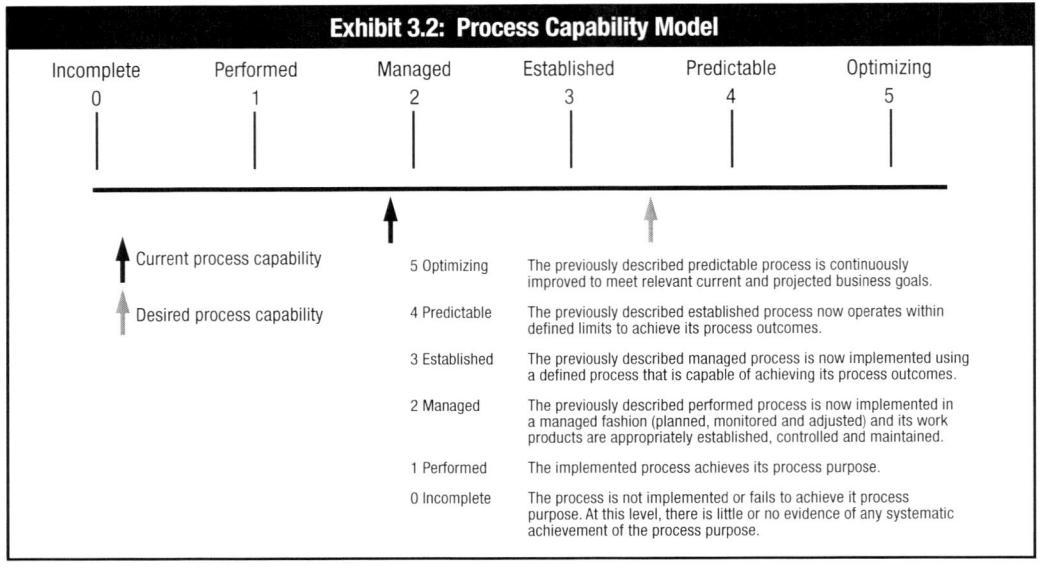

6. THREAT ANALYSIS

Importance of Periodic Threat Analysis

Technical and behavioral threats to an enterprise evolve as a result of several internal and external factors, including:
- Implementation of new technologies
- Broader network and application access to partners and customers
- The ever-growing capabilities of attackers
- Lack of staff education or attentiveness

Threats change over time due to:
- Changing market conditions
- Aging technology
- Regulations and legislation
- New connections to global systems

The risk practitioner must always be aware of changing threat levels or new and emerging threat vectors. The risk monitoring activity is a key tool for the risk practitioner to notice new threats and measure the effectiveness of the risk management program. All of these changes to threats warrant periodic reassessment of the threat landscape that an enterprise faces.

CRISC Threat Analysis Responsibilities	Internal factors such as new business units, new or upgraded technologies, changes to products and services, and changes in roles and responsibilities all represent areas in which new threats may emerge. This process of analyzing and communicating the impact on the enterprise's risk posture is critical to ensuring that stakeholders: • Are aware of potential business impact • May take actions to mitigate risk accordingly > **Note:** As new threats are identified and prioritized in terms of impact, the risk practitioner must help evaluate the ability of existing controls to mitigate risk associated with new threats and, in some cases, facilitate the: > • Modification of the technical architecture > • Deployment of a threat-specific countermeasure > • Implementation of a compensating mechanism or process until mitigating controls are developed > • Education of staff or business partners
Timing of Threat Analysis— Annual and Incremental Approaches	The enterprise should perform a threat analysis at least annually by evaluating changes in the technical and operating environments of the enterprise, particularly where external entities are granted access to organizational resources. > **Note:** The enterprise may also choose to take an incremental approach, analyzing portions of the enterprise monthly or quarterly. This is often required for regulatory compliance or compliance with industry standards such as the Payment Card Industry Data Security Standard (PCI-DSS).
Incremental Risk Assessment— Changing Asset Values and Risk Characteristics	The risk practitioner must recognize that asset values and risk characteristics can change, requiring reanalysis of risk posture. **Example:** A company can grow increasingly revenue-dependent on an application that was initially not considered to be critical to the enterprise. Asset value can increase or decrease over time in terms of real monetary value or strategic value to the enterprise. As an asset reaches the end of its useful life, its strategic value decreases. The risk associated with an asset can grow as it moves from a new product to a main revenue source. **Example:** A small database may initially contain only a few dozen personal information records; five years later, the same database may contain 10,000, representing a much higher impact if compromised.

7. RISK REPORTING

Introduction	Risk reporting, based on effective risk monitoring, helps highlight issues and enables management to make educated decisions on risk response activities. Much of the risk reporting will focus on the effectiveness of the controls and countermeasures implemented in the risk response phase. The risk practitioner must be able to report on whether the controls that were implemented are working effectively to mitigate risk and whether the residual risk to the enterprise is acceptable. The objective is to initiate corrective action and meet regulatory requirements.

Reporting Content

Risk reporting covers a broad array of information flows and may include the major types of risk communication as shown in the following table.

Types of Risk Communication	
Reporting Content	**Function**
Expectation	This is essential communication about the overall strategy the enterprise takes toward risk and drives all subsequent efforts on risk management. It sets the overall expectations from risk management. **Example:** Risk strategy, policies, procedures, awareness training, continuous reinforcement of principles, etc.
Status	This includes information on the actual status with regard to risk, such as: • The risk profile of the enterprise, i.e., the overall portfolio of (identified) risk to which the enterprise is exposed • The root cause of loss events • Thresholds for risk • Options to mitigate (cost and benefits) risk • Event/loss data • KRIs to support management reporting on risk
Risk management capability	This information allows monitoring of the state of the "risk management engine" in the enterprise and is a key indicator for good risk management. It has predictive value for how well the enterprise is managing risk and reducing exposure.
Actionable items	When actionable items and roles and responsibilities of the risk owners are included in the risk reporting matrix, the risk report overlaps with the risk response plan.
Risk register	The changes in risk must also be noted in the risk register to track ongoing mitigation activities and risk level.

Risk Reporting Criteria

To be effective and enable decision making, risk reporting has to be:
• Clear
• Concise
• Useful
• Timely
• Designed for the correct target audience
• Available on a need-to-know basis

The following table depicts the key focus areas for risk reporting.

Focus Areas for Risk Reporting Criteria	
Communication must be ...	**To ...**
Clear	Enable understanding by all stakeholders.
Concise	Focus the reader on the key points. Concise information is well structured and complete and avoids peripheral information, jargon and technical terms, except where necessary.

Risk Reporting Criteria (cont.)

Focus Areas for Risk Reporting Criteria (cont.)	
Communication must be …	**To …**
Useful	Enable decision making. Useful information is relevant and presented at the appropriate level of detail. Usefulness includes consideration of the target audience because information that may be useful to one party may not be useful to another.
Timely	Allow action at the appropriate moment to identify and treat the risk. For each risk, critical moments exist between its origination and its potential business consequence; a delay in reporting may increase the level of impact. **Example:** Communicating a potential problem too late to undertake corrective or preventive action serves no useful purpose
Designed for the correct target audience	Enable informed decisions. Information must be communicated at the right level of aggregation and adapted for the audience. Aggregation must not hide root causes of risk. **Example:** A security officer may need technical data on intrusions and viruses to deploy solutions. An IT steering committee may not need this level of detail, but it does need aggregated information to decide on policy changes or additional budgets to treat the same risk.
Available on a need-to-know basis	Ensure that information related to IT risk is known and communicated to only those parties with a genuine need. A risk register with all documented risk is not public information and should be properly protected against internal and external parties with no need for it.

Risk Reporting Channels

As risk management is an enterprisewide effort, communication flows to and from the CRISC.

The following table provides a quick overview of the most important communication channels for effective and efficient risk management including, but not limited to, those from the risk practitioner to other stakeholders. The list does not include the source and destination of the information nor the actions that should be taken on it.

Risk Reporting Channels		
Input	**Stakeholder**	**Output**
• Executive summary risk reports • Current risk exposure/profile • KPIs and KRIs	Executive management and board	• Enterprise appetite for IT risk • Key performance objectives • Risk Responsible, Accountable, Consulted and/or Informed (RACI) charts • Risk policies that express management's risk tolerance • Risk awareness expectations • Risk culture • Risk analysis request

Part I—Risk Management and Information Systems Control Theory and Concepts
Domain 3—Risk Monitoring
C. Essentials of Risk Monitoring

Risk Reporting Channels *(cont.)*

Risk Reporting Channels *(cont.)*		
Input	**Stakeholder**	**Output**
• Risk management scope and plan • Risk register • Risk analysis results • Executive summary risk reports • Integrated/aggregated risk report • KRIs • Risk analysis request	Chief risk officer (CRO) and enterprise risk committee	• Enterprise appetite for risk • Residual risk exposures • Risk action plan
• Enterprise appetite for risk • Risk management scope and plan • Key performance objectives • Risk RACI charts • Risk framework and scoring methodology • Risk register	Chief information officer (CIO)	• Residual risk exposures • Operational risk information • Business impact of the risk and impacted business units • Ongoing changes to risk factors
• Key performance objectives	Chief financial officer (CFO)	• Financial information with regard to programs and projects (budget, actual, trends, etc.)
• Risk management scope • Plans for ongoing business and risk communication • Risk culture • Business impact of the IT-related business risk and impacted business units • Ongoing changes to IT risk factors	Business management and business process owners	• Control and compliance monitoring • Risk analysis request
• Key performance objectives • Risk management plan • Risk framework and scoring methodology • Risk register • Risk culture	IT management (including security, service management)	• Residual risk exposures
• Key performance objectives • Risk responsible, accountable, consulted, informed (RACI) charts • Risk management plan • Control and compliance monitoring	Compliance and audit	• Audit findings

Part I—Risk Management and Information Systems Control Theory and Concepts
Domain 3—Risk Monitoring
C. Essentials of Risk Monitoring

Risk Reporting Channels *(cont.)*

Risk Reporting Channels *(cont.)*		
Input	**Stakeholder**	**Output**
• Key performance objectives • Risk management plan • Risk framework and scoring methodology • Risk register • Audit findings	Risk control functions	• Residual risk exposures • Risk reports
• Risk awareness expectations • Risk culture	Human resources (HR)	• Potential risk • Support on risk awareness initiatives

Reporting Results of Periodic Risk Assessment

Periodic risk assessment results should be provided to the steering committee and/or senior management for use in guiding risk management priorities and activities.

Risk Reporting Tools and Techniques

Risk reporting can range from a face-to-face meeting with stakeholders to structured e-mails and reports on risk in specific focus areas to integrated governance, risk and compliance (GRC) solutions with automated workflows and system-generated heat maps and dashboards. The tools and techniques will differ significantly between enterprises and the intended target audience.

Link to Other Domains

In the previous domains—*Risk Identification, Assessment and Evaluation and Risk Response*—risk was identified and prioritized and risk response activities were chosen and implemented to ensure that risk was within acceptable limits. In this domain, we set in place the mechanisms to monitor and report on the operation and effectiveness of the controls selected. In the next domain, we will examine the specific design and selection of the information systems related controls. In Domain 5, we will examine how to monitor and maintain those IS controls.

D. SUGGESTED RESOURCES FOR FURTHER STUDY

Suggested Resource for Further Study

In addition to the resources cited throughout this manual, the following resource is suggested for further study:
- ISACA:
  - *The Risk IT Practitioner Guide*, 2009

Part I—Risk Management and Information Systems Control Theory and Concepts
Domain 4—Information Systems Control Design and Implementation
A. Chapter Overview

Domain 4—Information Systems Control Design and Implementation

A. CHAPTER OVERVIEW

Introduction	This chapter provides information on the system development life cycle (SDLC) and project management with particular focus on IS control design and implementation.
Learning Objectives	As a result of completing this chapter, the CRISC candidate should be able to: • List different control categories and their effects. • Judge control strength. • Explain the importance of balancing control cost and benefit. • Leverage understanding of the SDLC process to implement IS controls efficiently and effectively. • Differentiate between the four high-level stages of the SDLC. • Relate each SDLC phase to specific tasks and objectives. • Apply core project management tools and techniques to the implementation of IS controls.
Inputs From Other Domains	The IS control design and implementation process is a distinct subset of *Domain 2—Risk Response*, particularly risk mitigation/reduction. The enterprise must determine the appropriate response to the risk to the enterprise, whether the risk has an internal or external source. This chapter focuses on the specific risk response required for handling risk related to information systems. Risk responses are directly driven by the risk (scenarios) that the enterprise identified and prioritized in *Domain 1—Risk Identification, Assessment and Evaluation*.
Relevance	IS control design and implementation focuses on the evaluation and selection of the information system (IS) controls that need to be designed and implemented into information systems to meet organizational risk management requirements. The selection of IS controls is based on the enterprise's risk profile, its risk appetite, risk tolerance and risk culture. The purpose of this effort is to establish timely and cost-effective solutions capable of supporting enterprise strategic and operational objectives. The COBIT 5 process BAI03 *Manage solutions identification and build* describes this process as: Establish and maintain identified solutions in line with enterprise requirements covering design, development, procurement/sourcing and partnering with suppliers/vendors. Manage configuration, test preparation, testing, requirements management and maintenance of business processes, applications, information/data, infrastructure and services. Consideration for implementing specific IS controls are cost-benefit, available solutions, ease of maintenance, existing IT architecture and strategy, and the ability to adapt to a changing risk environment. This CRISC must be knowledgeable in how to design and implement IS controls that both mitigate risk and still align with business objectives and are in compliance with the enterprise's risk appetite and risk tolerance levels.

Part I—Risk Management and Information Systems Control Theory and Concepts
Domain 4—Information Systems Control Design and Implementation
A. Chapter Overview

Outputs to Other Domains The IS control design and implementation process provides documentation of IS controls and related metrics and key performance indicators (KPIs) used in *Domain 5—Information Systems Control Monitoring and Maintenance* to enable control monitoring and maintenance.

Contents This chapter contains three primary topic areas:
- IS controls
- Design and implementation of controls through the phases of the systems development life cycle (SDLC)
- Project risk management

The chapter is formatted into the following sections.

Section	Starting Page
A. Chapter Overview	117
B. Task and Knowledge Statements	120
C. IS Controls	122
1. Key Terms and Concepts	122
2. The Control Life Cycle	125
3. Control Selection	126
4. Control Design and Development	128
5. Control Testing and Implementation	129
D. Building Control Design Into the SDLC	131
1. Introduction to the SDLC	131
2. Risk Associated With Software Development	132
E. System Development Life Cycle (SDLC) Phases	134
1. Project Initiation Phase	134
2. Project Design and Development	137
3. Project Testing	140
4. Project Implementation	144
F. Managing Project Risk	152
1. Key Terms and Concepts	152
2. Overview of Project Risk Management	153
3. Scope Management	154
4. Time Management	156
5. Budget Management	157
6. Use of Metrics to Support Resource Planning	158
G. Project Management Tools and Techniques	160
1. General Project Management Techniques	160
2. Gantt Charts	162
3. Critical Path Methodology (CPM) and Program Evaluation Review Technique (PERT)	162
H. Suggested Resources for Further Study	165

Part I—Risk Management and Information Systems Control Theory and Concepts
Domain 4—Information Systems Control Design and Implementation
A. Chapter Overview

Contents *(cont.)*

Process-specific controls for the following processes are addressed in Part II of this manual:
1. Managing the IT Strategy
2. Portfolio, Program and Project Management
3. Change Management
4. Third-party Service Management
5. Continuity Management
6. Information Security Management
7. Configuration Management
8. Problem Management
9. Knowledge Management
10. IT Operations Management

B. TASK AND KNOWLEDGE STATEMENTS

Introduction This section describes the task and knowledge statements for Domain 4, which focus on the design and implementation of IS controls that mitigate risk and are in alignment with the enterprise's risk appetite and tolerance levels to support business objectives.

Task Statements The following table describes the task statements for Domain 4 that the CRISC candidate must know how to perform..

No.	Task Statement (TS)
TS4.1	Interview process owners and review process design documentation to gain an understanding of the business process objectives.
TS4.2	Analyze and document business process objectives and design to identify required information systems controls.
TS4.3	Design information systems controls in consultation with the process owners to ensure alignment with business needs and objectives.
TS4.4	Facilitate the identification of resources (e.g., people, infrastructure, information, architecture) required to implement and operate information systems controls at an optimal level.
TS4.5	Monitor the information systems control design and implementation process to ensure that it is implemented effectively and within time, budget and scope.
TS4.6	Provide progress reports on the implementation of information systems controls to inform stakeholders and to ensure that deviations are promptly addressed.
TS4.7	Test information systems controls to verify effectiveness and efficiency prior to implementation.
TS4.8	Implement information systems controls to mitigate risk.
TS4.9	Facilitate the identification of metrics and key performance indicators (KPIs) to enable the measurement of information systems control performance in meeting business objectives.
TS4.10	Assess and recommend tools to automate information systems control processes.
TS4.11	Provide documentation and training to ensure that information systems controls are effectively performed.
TS4.12	Ensure that all controls are assigned control owners to establish accountability.
TS4.13	Establish control criteria to enable control life cycle management.

Part I—Risk Management and Information Systems Control Theory and Concepts
Domain 4—Information Systems Control Design and Implementation
B. Task and Knowledge Statements

Knowledge Statements

The following table describes the knowledge statements for Domain 4. The CRISC candidate must have a good understanding of each of the areas delineated by the knowledge statements. These statements are the basis for the exam.

No.	Knowledge Statement (KS) Knowledge of:
KS4.1	Standards, frameworks and leading practices related to information systems control design and implementation
KS4.2	Business process review tools and techniques
KS4.3	Testing methodologies and practices related to information systems control design and implementation
KS4.4	Control practices related to business processes and initiatives
KS4.5	The information systems architecture (e.g., platforms, networks, applications, databases and operating systems)
KS4.6	Controls related to information security
KS4.7	Controls related to third-party management
KS4.8	Controls related to data management
KS4.9	Controls related to the system development life cycle
KS4.10	Controls related to project and program management
KS4.11	Controls related to business continuity and disaster recovery management
KS4.12	Controls related to management of IT operations
KS4.13	Software and hardware certification and accreditation practices
KS4.14	The concept of control objectives
KS4.15	Governance, risk and compliance (GRC) tools
KS4.16	Tools and techniques to educate and train users

Note: Knowledge statements 4.4 and 4.6-4.12 are process specific and addressed in Part II of this manual.

C. IS CONTROLS

Introduction

This section provides an overview of IS controls because a thorough understanding of controls in general, and IS controls in particular, is crucial for the effective design and implementation of IS controls.

Project risk management and design and implementation of IS controls throughout the SDLC are addressed in separate chapter sections.

Process-specific controls and a practitioner level of IS control design and implementation are addressed in Part II of this manual.

1. KEY TERMS AND CONCEPTS

Control

The means of managing risk, including policies, procedures, practices, guidelines or organizational structures, which can be of an administrative, technical, management or legal nature. Controls are designed to provide reasonable assurance that:
- Business objectives are achieved.
- Undesired events are prevented or detected and corrected.

> **Note:** The relative mix and importance of process, application and general controls will be unique to each enterprise.

Defense-in-Depth

The practice of layering defenses to provide added protection

Defense in depth—also called layered defense— increases security by raising the effort needed in an attack. This strategy places multiple barriers between an attacker and an enterprise's computing and information resources.

Defense in depth can be deployed along horizontal or vertical vectors.

Examples:
- Horizontal defense in depth—controls are placed along a network path—a firewall, then a network-based intrusion detection system (IDS), an intrusion prevention system (IPS), compartmentalization or segmentation of the network, and host-based controls.
- Vertical defense in depth—controls are placed along various layers of the system—on the hardware, operating systems, database, application and user levels.

Governance of enterprise IT

A governance view that ensures that information and related technology support and enable the enterprise strategy and the achievement of enterprise objectives. It also includes the functional governance of IT, i.e., ensuring that IT capabilities are provided efficiently and effectively.

Information Security

Ensures that within the enterprise, information is protected against disclosure to unauthorized users (confidentiality), improper modification (integrity), and non-access when required (availability)

Part I—Risk Management and Information Systems Control Theory and Concepts
Domain 4—Information Systems Control Design and Implementation
C. IS Controls

Information Systems (IS)	The combination of strategic, managerial and operational activities involved in gathering, processing, storing, distributing and using information and its related technologies. **Scope Notes:** Information systems are distinct from information technology (IT) in that an information system has an IT component that interacts with the process components.
Information Technology (IT)	The hardware, software, communication and other facilities used to input, store, process, transmit and output data in whatever form
Control Categories	The following table describes control categories.

Control Categories	
Category	**Description**
Compensating controls	An alternate form of control that corrects a deficiency or weakness in the control structure of the enterprise *Compensating controls may be considered when an entity cannot meet a requirement explicitly, as stated, due to legitimate technical or business constraints, but has sufficiently mitigated the risk associated with the requirement through implementation of other controls.* **Example:** Adding a challenge response component to weak access controls can compensate for the deficiency in the access control mechanism.
Corrective controls	Remediate errors, omissions and unauthorized uses and intrusions, once they are detected **Example:** Backup restore procedures enable a system to be recovered if harm is so extensive that processing cannot continue without recourse to corrective measures.
Detective controls	Warn of violations or attempted violations of security policy and include such controls as audit trails, intrusion detection methods and checksums
Deterrent controls	Provide warnings that can deter potential compromise such as warning banners on login screens or offering rewards for the arrest of hackers
Directive controls	Directive controls mandate the behavior of an entity by specifying what actions are, or are not, permitted. **Example:** A policy is an example of a directive control.
Preventive controls	Inhibit attempts to violate security policy and include such controls as access control enforcement, encryption and authentication

Exhibit 4.1: Control Category Interdependencies

Exhibit 4.1 describes the interdependencies of different control categories.

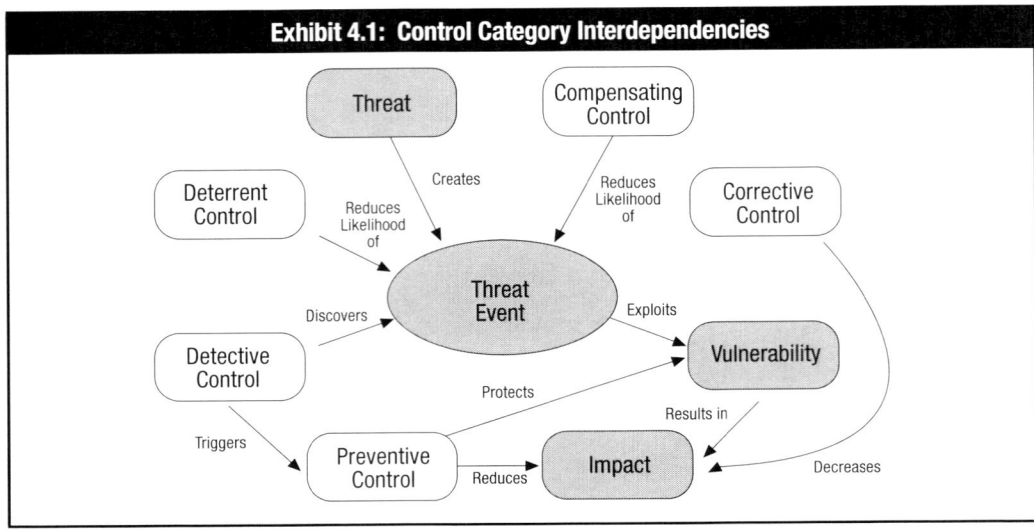

Technical and Nontechnical Control Methods

Technical controls (e.g., a firewall) must be supported by nontechnical controls such as operational controls (i.e., a configuration management process for rule changes to the firewall configuration) and managerial controls (ownership and oversight for the control functionality, awareness training for users).

Examples of Technical and Nontechnical Control Methods

An effective security management system will consist of a variety of control methods:
- Technical controls
- Nontechnical controls:
  - Managerial controls
  - Operational controls

The following table describes technical and nontechnical control methods.

Technical and Nontechnical Control Methods		
Method	**Description**	**Examples**
Technical controls	Safeguards and countermeasures that are incorporated into computer hardware, software or firmware	• Access control mechanisms • Identification and authentication mechanisms • Encryption methods • Intrusion detection software
Nontechnical controls	Management and operational controls	• Security policies • Operational procedures • Personnel, physical and environmental security

Note: Controls, such as two-factor authentication required for high-security situations, can include both automated and manual processes; for example, smart cards requiring a personal identification number (PIN).

2. THE CONTROL LIFE CYCLE

Introduction

This section provides an overview of the control life cycle as well as the related concepts and principles.

The relative mix and importance of process, application and general controls will be unique to each enterprise.

Exhibit 4.2: The Control Life Cycle

The control life cycle maps the various phases in the life of a control from the initial selection/design through development, implementation, maintenance and disposal of the system.

The CRISC must ensure that the controls are properly managed throughout each phase of the life cycle.

Exhibit 4.2 shows the phases of the control life cycle.

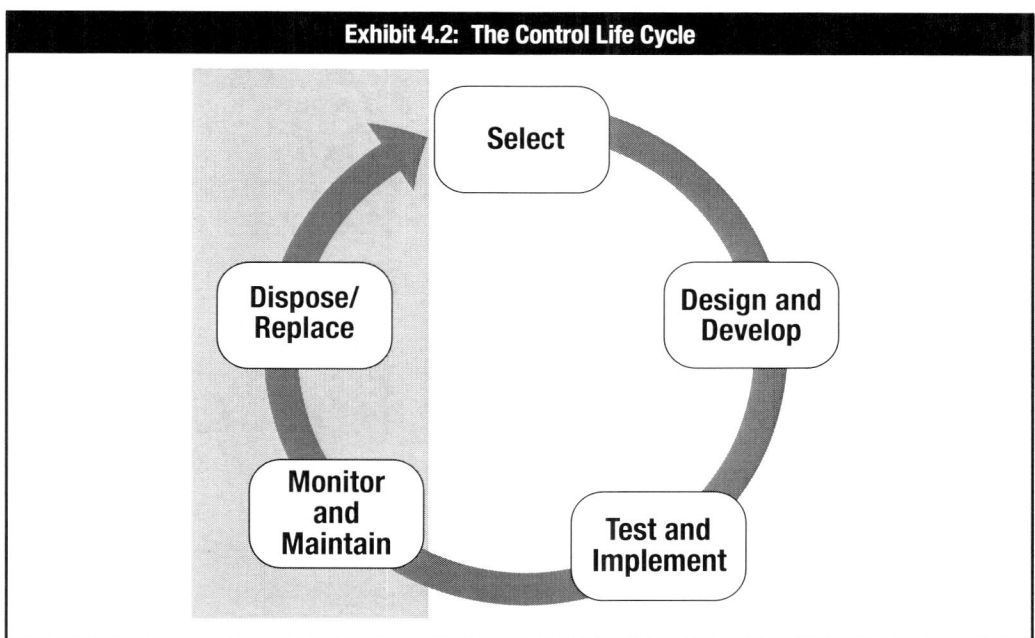

The shaded areas indicate those steps of the control life cycle that are addressed in *Domain 5—Information Systems Control Monitoring and Maintenance*.

3. CONTROL SELECTION

Introduction

Control selection does not only address specific controls, but also addresses the high-level or control architecture considerations.

Control Selection Objective

The objective of the control selection and implementation is to:
- Install and configure solutions and integrate with business process activities.
- Implement control, security and auditability measures during configuration, and during integration of hardware and infrastructural software, to protect resources and ensure availability and data integrity.
- Maintain a services catalog that must be updated to reflect the new solutions.

Excellent resources that can assist the risk practitioner in evaluating and selecting controls can be found in *COBIT 5: Enabling Processes*, specifically processes BAI02 *Manage requirements definition* and BAI03 *Manage solutions identification and build*.

The risk practitioner should ensure that the business requirements and input of all relevant stakeholders is considered in the selection of controls. Different stakeholders from different parts of the enterprise may have substantially different priorities and requirements. Selection of an ideal control for one group may be unacceptable for other areas of the enterprise.

Control Costs and Benefits

When controls or countermeasures are planned, an enterprise should consider the costs and benefits.

Cost-benefit analysis helps:
- Provide a monetary impact view of risk
- Determine the cost of protecting what is important
- Make smart choices based on potential:
  - Risk mitigation costs
  - Opportunities
  - Losses (risk exposure)

> **Note:** If the costs of specific controls or countermeasures (control overhead) exceed the benefits of mitigating a given risk, the enterprise may choose to accept the risk rather than incur the cost of mitigation. This acceptance of risk follows the general principle that the cost of a control should never exceed the expected benefit.

Part I—Risk Management and Information Systems Control Theory and Concepts
Domain 4—Information Systems Control Design and Implementation
C. IS Controls

Calculating the Benefits of a Control	Calculating the benefits to be realized from the implementation of a control can be a challenge since benefits will be both tangible and intangible. The benefits are measured against the resulting reduction in quantitative and qualitative risk and the expected exposure or impact from an incident.
	
	The calculation of risk and residual risk was covered earlier in this book; however, these are some of the factors that will affect the benefit portion of the cost-benefit calculation for IS controls:
	• Reduction in financial cost:
	  – Less impact on productivity, overtime and staff costs
	  – Lower insurance premiums
	  – Avoidance of penalties for non-compliance
	  – Fewer calls to help desk or production support
	• Impact on employee morale (less staff turnover and training)
	• More time to work on quality improvement of systems instead of repair
	• Financial savings
	  – Less system downtime
	  – Increase in productivity
	  – Reduced cost of incidents
	• Easier management of controls and consistency
	• Accurate compliance reporting

Other Factors Impacting Control Selection	The selection of the appropriate level of control is based on various critical factors: • Selection of specific IS controls should consider available control options, cost, time constraints, availability of skilled personnel and business priorities. • Some controls may place limitations on the way that the business can operate and may impact system performance, operational cost and productivity. It is important to fine-tune controls to optimize the control benefit against the adverse effect a control may have on operations.

Frameworks for Control Selection	The selection of IS controls may be influenced by international standards, guidelines and leading practices. Some of the tools that may be used to select and/or justify specific controls are:
	• International Organization for Standardization (ISO)—Standard 27001
	• Sherwood Applied Business Security Architecture Framework (SABSA)
	• ISACA—COBIT 5
	  Note: The following COBIT 5 publications are available from ISACA and can be downloaded at *www.isaca.org/cobit:*
	  – The COBIT 5 framework (complimentary PDF)
	  – *COBIT® 5: Enabling Processes* (member complimentary PDF)
	  – *COBIT® 5 Implementation* (member complimentary PDF)
	  – *COBIT® 5 for Information Security*
	• ISACA—*The Risk IT Framework*
	• Committee of Sponsoring Organizations of the Treadway Commission (COSO)—Enterprise Risk Management—Integrated Framework
	• NIST SP800-53 Recommended Security Controls for Federal Information Systems
	• Benchmarking activities

Total Cost of Ownership (TCO) for Controls	When considering costs, the TCO must be considered for the full life cycle of the control or countermeasure. This can include such elements as: • Acquisition and licensing costs • Deployment and implementation costs • Recurring maintenance costs • Testing and assessment costs • Compliance monitoring and enforcement • Inconvenience to users • Reduced throughput of controlled processes • Training in new procedures or technologies as applicable • End of life decommissioning

4. CONTROL DESIGN AND DEVELOPMENT

Introduction	Control design and implementation is a crucial part of the control life cycle. The complexity of information systems has a distinct influence on IS controls. Thus, the design has to consider "breadth" and "depth," where breadth represents the flow of information across multiple applications and depth represents the different layers on which controls can function and how those may cross-influence each other. The IS control design process includes the determination of which tools and control types are available to meet policy and business requirements, selection of the best controls, and the design and implementation of the selected controls.
Developing Controls in Depth	IS controls may be implemented at many levels in an information system. Some of the control locations may be: • Network-based controls • Application-level controls • Database controls • Operating system controls • Platform-specific controls • Physical controls It is important to design and implement controls at the correct place within the overall enterprise, system or network. A control that is placed incorrectly, or not configured properly, will provide little benefit and may in fact create a sense of false security. Controls must be selected that will be interoperable, and yet independent so that a breach of one control does not cause the failure of other controls. The controls must provide a complete framework of protection from risk so that no gaps remain that could be exploited by an attacker, or bypassed through unintentional misuse by an internal resource.
Establishing Key Performance Indicators (KPIs) and Key Risk Indicators (KRIs)	As controls are selected for implementation, criteria should also be established to determine the operational level and effectiveness of the controls. These criteria will often be based on KPIs that indicate whether a control is functioning correctly. Another measurement criterion may be based on key risk indicators (KRIs), which are used to alert monitoring personnel about trends or alarm thresholds that may indicate a potentially hazardous or marginal condition.

Part I—Risk Management and Information Systems Control Theory and Concepts
Domain 4—Information Systems Control Design and Implementation
C. IS Controls

Control Owner	The selection of a control requires the identification of a control owner. The control owner is responsible for the management of the control and for ensuring that the control is operated and maintained in the correct manner. Effective control operation includes reporting performance metrics and abnormal conditions to appropriate personnel.
Designing for Easy Control Maintenance	Design of the controls implemented must include maintenance considerations, including measurability. Effectiveness of controls cannot be evaluated unless they can be tested and measured. Further, confidence levels and sampling sizes for testing the effectiveness of these controls should closely mirror audit and regulatory compliance objectives. **Note:** The risk practitioner must ensure that the enterprise is in adequate compliance with relevant legal and regulatory requirements.
Example of Designing for Easy Control Maintenance	The design of a control includes designing the supporting processes and procedures for monitoring and reporting on the effectiveness of the control. This is covered in more detail in Domain 5, but will include monitoring actions: • That show daily affirmative reviews (review notes, sign-offs, approvals, etc.) • That have a clearly defined and documented follow-up process for anomalous/suspicious events • That use appropriate sampling sizes • For which information security controls are tested accordingly and compliance is built in

5. CONTROL TESTING AND IMPLEMENTATION

Introduction	Design and implementation of the selected IS controls requires a defined process for the secure configuration, deployment, change control and approval, as well as designing the ability to monitor and report on the effectiveness of the controls, once implemented. The actual monitoring and maintenance of the controls will be covered in Domain 5. Control testing and implementation includes: • Testing control effectiveness • Documentation • User and operator training
Testing Control Effectiveness	Control effectiveness cannot be determined by simply identifying the control category (preventive, detective, manual, automated, etc.). Control effectiveness can be assessed by its quantitative and qualitative compliance testing results. Control effectiveness must be assessed within context: While the initial test results may indicate that an automated control is highly reliable, a more comprehensive test may reveal that it is habitually circumvented. Controls can be effectively assessed only by determining how well they achieve the control objective within the environment in which they are operating. Specific considerations for designing meaningful controls may include the controls': • Design effectiveness • Operating effectiveness • Alignment with its operating environment (organization, people, processes and technology)

Control Documentation	Policy—The first control that must be implemented is policy. Policy is an administrative or managerial control that specifies the behaviors and actions that are/are not permitted. Policy proclaims the intent of management, states compliance with regulations, mandates security procedures and sets the tone for the enterprise. Without policy there is no authority for the implementation and supervision of security controls. Some examples of documentation include: • **Risk register**—Documentation of risk and the status of risk response efforts, such as design and implementation of control activities • **Control inventory**—Listing of controls • **Plan of action and milestones (POAM)**—Listing of ongoing risk response projects and tracking of the status of the projects
Training	**Awareness and Training**—A very important control. A strong and consistent awareness program heightens the ability of staff to detect, prevent and respond to many types of security incidents. Awareness should be conducted annually for all users. Staff must also be trained in the proper use of the tools that they will utilize. An IS control will not be effective if it is not configured, monitored and maintained correctly, but staff cannot be expected to properly maintain a tool for which they have not be trained in its proper use.
Control-specific Training	Users, administrators, managers, auditors and security staff may require training on the operation, maintenance, monitoring and reporting on the new or enhanced controls selected for the enterprise. The training should be practical, relevant and tailored to the needs of the audience.
Control Evaulation	At this point, the CRISC candidate should be able to evaluate the various controls available to mitigate risk, and to assist in the selection of the best control(s) for the enterprise based on cost-benefit analysis, risk levels and acceptance levels, stakeholder requirements, organizational factors and priorities. The CRISC candidate should ensure that adequate IS controls have been selected to mitigate risk, that the owner has determined that the supporting evaluation criteria (KPIs) have been set for each control and that the training needs are addressed.
The Life Cycle of Control Design	IS controls require continuous attention to risk levels, changes in the operational or threat environment and availability of new controls or control strategies. As IS systems go through the process of development and implementation, the risk manager must reevaluate and assess the risk levels to ensure that the controls are being designed, developed, tested, implemented and maintained properly. This requires the integration of risk management and IS control selection into each phase of the systems development life cycle (SDLC).

D. BUILDING CONTROL DESIGN INTO THE SDLC

Introduction

The design and deployment of IS controls will often be undertaken as a systems development project. While there are several project management techniques that can be used to manage system development projects, they can be described by the generic term "system development life cycle (SDLC)." The CRISC candidate should be familiar with the steps necessary to select, design, develop, test and deploy and maintain IS controls throughout their life cycle.

Relevance

The risk practitioner should advise on the design of appropriate IS controls to treat identified risk and oversee the implementation of IS controls, either as part of a system implementation or via the operational change management process.

> **Note:** The CRISC candidate should become involved as early as possible during the SDLC process and maintain that involvement throughout the remainder of the control life cycle. It is important to implement controls properly and ensure that they continue to operate securely throughout their operational life. The risk practitioner can add value during the initiation phases of the SDLC or even prior to the start of a project, when project ideas are generated, developed and communicated.

1. INTRODUCTION TO THE SDLC

Introduction

Companies often commit significant resources (e.g., people, applications, facilities and technology) to develop, acquire, integrate and maintain application systems that are critical to the effective functioning of key business processes.

The SDLC process governs the phases deployed in the development or acquisition of a software system and, depending on the methodology, may even include the controlled retirement of the system.

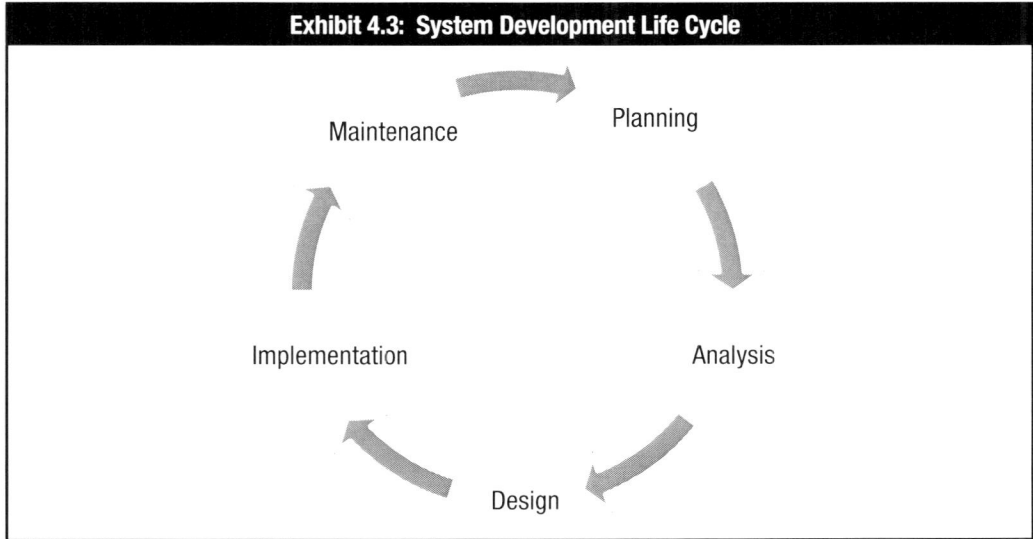

Exhibit 4.3: System Development Life Cycle

> **Note:** Typical phases of SDLC include the feasibility study, requirements study, requirements definition, detailed design, programming, testing, installation and postimplementation review. Today's SDLC also tends to include the maintenance and operational phases of the control's life cycle, up to and possibly including, its date of termination as shown in **exhibit 4.3**.

Objectives of the SDLC Process	The SDLC includes: • IT processes for managing and controlling project activity • An objective for each phase of the life cycle that is typically described with key deliverables, a description of recommended tasks and a summary of related control objectives for effective management • Incremental steps or deliverables that lay the foundation for the next phase
Relevance	The following sections contain information on the SDLC process and its relationship to the: • Achievement of business objectives • High-level SDLC process flow and architecture • Risk associated with IS/IT projects **Note:** High-level knowledge of this process is relevant to the risk practitioner's ability to design and implement IS controls.

2. RISK ASSOCIATED WITH SOFTWARE DEVELOPMENT

Introduction	This section contains information on risk associated with software design and development. The concepts can be applied to risk management, where IS controls are designed and implemented to reduce risk.
Business Risk vs. Project Risk	The following table separates and compares the risk related to designing and developing software systems along two major categories: business risk (or benefit risk) and project risk (or delivery risk).

Risk Related to Software Systems Design and Development	
Category	**Description**
Business Risk (Benefit Risk)	This relates to the likelihood that the new system may not meet the user business needs, requirements and expectations. **Example:** The business requirements that were to be addressed by the new system are still unfulfilled, and the process has been a waste of resources. Even if the system is implemented, it will most likely be underutilized and not maintained, making it obsolete in a short period of time.
Project Risk (Delivery Risk)	The project activities to design and develop the system exceed the limits of the financial or skilled technical resources set aside for the project. As a result, the project may be completed late, if ever. Project-related risk is addressed in more detail in section F, Managing Project Risk.

Root Causes of Project Delivery Risk	The foremost root cause of project risk is: • A lack of discipline in managing the software development process • Selection of a project methodology that is unsuitable to the system being developed\ In such instances: • Enterprises are not providing the infrastructure and support necessary to help projects avoid these problems. • If successful projects occur, they are not repeatable and SDLC activities are not defined or followed adequately (i.e., insufficient maturity). • With effective management, SDLC management activities can be controlled, measured and improved.

Part I—Risk Management and Information Systems Control Theory and Concepts
Domain 4—Information Systems Control Design and Implementation
D. Building Control Design Into the SDLC

CRISC Responsibilities Related to Project Risk	Merely following an SDLC management methodology does not ensure successful completion of a development project. The CRISC needs to enforce management discipline over a project to ensure that: • The project addresses the specific risk management requirements. • The project follows a defined process. • Project planning is performed, including effective estimates of resources, budget and time. • Scope creep is managed proactively. • Management tracks software design and development activities. • Senior management provides support to the software project's design and development efforts. • Periodic review and risk analysis are performed in each project phase.

E. SYSTEM DEVELOPMENT LIFE CYCLE (SDLC) PHASES

1. PROJECT INITIATION PHASE

Introduction

This section contains information on projects initiated by sponsors who gather the information required to gain approval for the project to be created.

Information often compiled into the terms of a project charter includes the:
- Objective of the project (high-level functional purpose)
- Business case and/or problem statement
- Stakeholders in the system to be produced
- Project manager and sponsor

> **Note:** Approval of a project initiation document (PID) or a project request document (PRD) is the authorization for a project to begin.

Tasks Associated With Phase 1

The following table lists the tasks to initiate the project.

Tasks to Initiate the Project		
Task	**Name/Description**	**Starting Page**
1	Conduct a feasibility study.	134
2	Define requirements.	135

1.1 Conduct a Feasibility Study

Introduction

This section contains information on conducting a feasibility study, which:
- Begins once initial approval has been given to move forward with a project
- Includes an analysis to clearly define the business needs and requirements and to identify alternatives for addressing the need

Description of a Feasibility Study

A feasibility study:
- Involves analyzing the benefits and solutions for the identified problem area
- Includes development of a business case, which:
  - States the strategic benefits of implementing the system either in productivity gains or in future cost avoidance
  - Identifies and quantifies the cost savings of the new system
  - Estimates a payback schedule for the cost incurred in implementing the system or shows the projected return on investment (ROI)

> **Note:** Intangible benefits such as improved customer relations may also be identified; however, quantify the benefits wherever possible. Nonfunctional requirements must also be known to avoid surprises in later project phases.

Main Components of a Feasibility Study

Within the feasibility study, the following are typically addressed:
- Definition of a time frame for the implementation of the required solution
- Determination of an optimum alternative risk-based solution for meeting business needs and general information technology (IT) resource requirements (e.g., whether to modify, develop or acquire a system)

Part I—Risk Management and Information Systems Control Theory and Concepts
Domain 4—Information Systems Control Design and Implementation
E. System Development Life Cycle (SDLC) Phases

Factors to Consider During the Feasibility Study

The following table describes factors to consider during the feasibility study to assist both in determining whether the project should be undertaken and in determining whether to develop or acquire a system or whether to update or replace an existing system.

Feasibility Study Consideration Factors	
Factor	**Description**
Date	The date by which the system needs to be functional, based on the business requirements
Cost	The level of effort (LOE), total cost of ownership (TCO) and anticipated ROI estimates to develop the system internally as opposed to a vendor's statement of work
Resources	Consist of the staff (availability and skill sets), consultants, licenses and hardware required to develop and implement the solution
Vendor system	The skill set of the support personnel; solvency of the vendor; system's reputation in the marketplace and with its clients; platform and compliance with security policies and regulations to which the enterprise is subject; packaging and customization design costs; consulting service costs; training costs; license characteristics (e.g., yearly renewal, perpetual) and maintenance costs
Interfaces	The other systems that will interface with the new/enhanced system through a pull, push or bidirectional relationship

Compatibility	Compatible with:
	• Strategic business plans
	• Other applications and systems in the environment
	• Alignment with the enterprise's:
	  – Information security policies
	  – Regulatory and legal requirements
	  – Existing IT infrastructure

Future requirements	The system's ability to grow as requirements change or are enhanced to meet the enterprise's strategic plan (five to seven years)

Feasibility Study Results

The completed feasibility study results should include a cost-benefit analysis report that:
• Provides the results of criteria analyzed (e.g., costs, benefits, risk, resources required and organizational impact)
• Recommends one of the alternatives/solutions and a course of action

1.2 Define Requirements

Introduction

This section contains information on defining requirements, which is concerned with identifying and specifying the requirements for the solution.

Description of Requirements

Requirements include:
• Business requirements containing descriptions of what a system should do
• Functional requirements and the use of case models describing how users will interact with a system
• Technical requirements and design specifications and coding specifications describing how the system will interact, conditions under which the system will operate and the information criteria that the system should meet

Framework for Defining Requirements	The COBIT framework defines information criteria that should be incorporated into the requirements to address issues associated with effectiveness, efficiency, confidentiality, integrity, availability, compliance and reliability. The requirements definition task of the project initiation phase also deals with issues that are sometimes called nonfunctional requirements (e.g., training or business continuity).
Defining the Requirements	The users in this process specify: • IS control needs: nonautomated and automated • How they wish to have those needs addressed by the system (e.g., access controls, regulatory requirements, management of information needs and interface requirements)
Factors to Consider When Defining Requirements	All concerned management and user groups must be actively involved in the requirements definition task to prevent problems such as expending resources on a system that will not satisfy the business requirements. User involvement is necessary to obtain commitment and full benefit from the system. Without management sponsorship, clearly defined requirements and user involvement, the benefits may never be realized.
Factors to Consider When Acquiring Software	Software acquisition should be based on various factors, such as the: • Cost differential between development and acquisition • Availability of generic software • Time gap between development and acquisition **Note:** Ensure that the feasibility study contains documentation that supports the decision to acquire the software.

Part I—Risk Management and Information Systems Control Theory and Concepts
Domain 4—Information Systems Control Design and Implementation
E. System Development Life Cycle (SDLC) Phases

2. PROJECT DESIGN AND DEVELOPMENT

Leading Practices for Design and Implementation

This section contains information on the seven leading practices for designing systems that reflect risk management strategies, which serve as checkpoints to ensure good systems design.

The following table provides a short overview of the seven leading practices and how they help mitigate risk associated with the design and implementation of IS controls.

Seven Leading Practices for System Design and Implementation		
No.	**Leading Practice**	**Description and Associated Risk**
1	Align with the business.	• Identify or create the business opportunity that makes the system worth building. • Ensure that any systems development project directly supports the enterprise in achieving one or more of its goals. • Provide sustained benefit to the enterprise by ensuring that a system continues to support the efficient exploitation of the business opportunity it was built to address.
2	Use technology to enable change.	• Investigate tasks and activities that may seem impossible at the current time, but, if IT-enabled, would fundamentally and positively change the ways that the enterprise does business. This may include: – Looking for opportunities to create a transformation or value shift in the context of the enterprise business space – Finding ways to do things that provide dramatic cost savings or productivity increases – Thinking of courses that would be least likely to be foreseen or quickly countered or copied by competitors
3	Leverage existing technology.	• Find ways to incorporate existing systems that have proven to be stable and responsive over time into the design of new systems because the design of systems embodies strategy and the purpose of strategy is to use the means available to the enterprise to best accomplish its goal. • Build new systems on the strengths of older systems, as in the principle of evolutionary process.
4	Embrace simplicity.	• Use a (simple) mix of technology and business procedures to achieve business objectives, where possible. **Rationale:** • This reduces complexity and the risk associated with the work and spreads the cost across multiple objectives. • Using a different technology or process to achieve each different project objective multiplies cost and complexity and reduces the overall probability of project success.

Leading Practices for Design and Implementation (cont.)

	Seven Leading Practices for System Design and Implementation (cont.)	
No.	Leading Practice	Description and Associated Risk
5	Remain flexible.	• Decompose the system design into separate components or objectives and, whenever possible, run the work on different objectives in parallel. • Promote flexibility by preventing the achievement of one objective from becoming dependent on the achievement of another objective (isolation). **Rationale:** Delays in the work toward one objective will not impact the progress toward other objectives. • Assign personnel to develop the systems that have applicable skills that can achieve a variety of different objectives. • Use the same development technology to achieve several different objectives, making it much easier to shift personnel from one objective to another, as needed, because they use the same skill sets. • Ensure that the project plan foresees, and provides for, an alternative plan in case of failure or delays in achieving objectives as scheduled. • Build the design of the systems to allow for some system features to be dropped from development, if needed, and still be able to deliver substantially.
6	Build within the enterprise's capability.	• Design and implement controls that are within the enterprise's capability and can be supported over time. **Rationale:** Unrealistic goals do not—in the long run—support the enterprise.
7	Learn from failure.	• Rework and increased effort are an inadequate response to failure. Provide lessons learned to ensure that avoidable challenges during the current project are not repeated in the future. **Risk:** Avoidable failures are repeated from project to project.

2.2 System Design

Introduction

While the architecture significantly influences the IS control design, the established requirements must also be addressed. This section contains information on IS control design.

System Design Objectives

After the design has been completed, architects should be able to:
- Explain how the software architecture will satisfy the risk response requirements in the system and application.
- Outline the rationale for key design decisions.

Rationale: Choices of particular hardware and software configurations may have cost implications of which stakeholders need to be aware and control implications that are of interest to the risk practitioner.

Part I—Risk Management and Information Systems Control Theory and Concepts
Domain 4—Information Systems Control Design and Implementation
E. System Development Life Cycle (SDLC) Phases

CRISC Involvement in System Design	CRISC involvement is focused on whether: • Risk response requirements are properly communicated and documented. • An adequate system of controls is incorporated into system specifications and test plans. • Project progress is tracked to identify and correct any potential issues. Monitoring functions are built into the system, particularly for electronic commerce (e-commerce) applications and other types of paperless environments.
Completion of the System Design	After the detailed design has been completed and approved, distribute the design to the developers for coding. The key deliverables coming out of project design include: • System, subsystem, program and database specifications • Test plans • A defined and documented formal software change control process

2.3 System Development

Introduction	This section contains information on system development, which uses the detailed design developed previously, to begin coding; thus moving the system one step closer to a final physical software product.
CRISC Involvement in System Development	While the responsibilities in this system development rest primarily with the programmers and systems analysts who are building the system, the CRISC may supervise the development of specific controls to ensure that they are adequately addressed. The risk practitioner will work alongside the project team to develop, resource and execute a quality assurance plan aligned with the enterprise's quality management system to obtain the quality specified in the requirements definition and the enterprise's quality policies and procedures. Some of the tasks with which the risk practitioner can assist are listed in the COBIT 5 process practice BAI03.11 as: 1. Propose definitions of the new or changed IT services to ensure that the services are fit for purpose. Document the proposed service definitions in the portfolio list of services to be developed. 2. Propose new or changed service level options (service times, user satisfaction, availability, performance, capacity, security, continuity, compliance and usability) to ensure that the IT services are fit for use. Document the proposed service options in the portfolio.

3. PROJECT TESTING

Introduction

This section describes system testing, which:
- Is an essential part of the development process that verifies and validates that a program, subsystem or application, and the designed security controls perform the functions for which it has been designed
- Determines whether the units being tested operate without any malfunction or adverse effect on other components of the system
- Uses a variety of development methodologies and organizational requirements to provide for a large range of testing schemes or levels.

The COBIT 5 process practice BAI03.07 lists the requirements to develop a test program as: Establish a test plan and required environments to test the individual and integrated solution components, including the business processes and supporting services, applications and infrastructure.

The process practice BAI03.07 describes the activities to develop a test program as:
1. Create an integrated test plan and practices commensurate with the enterprise environment and strategic technology plans that will enable the creation of suitable testing and simulation environments to help verify that the solution will operate successfully in the live environment and deliver the intended results and that controls are adequate.
2. Create a test environment that supports the full scope of the solution and reflects, as closely as possible, real-world conditions, including the business processes and procedures, range of users, transaction types, and deployment conditions.
3. Create test procedures that align with the plan and practices and allow evaluation of the operation of the solution in real-world conditions. Ensure that the test procedures evaluate the adequacy of the controls, based on enterprisewide standards that define roles, responsibilities and testing criteria, and are approved by project stakeholders and the sponsor/business process owner.

> **Note:** Each set of tests is performed with a different set of data and under the responsibility of different people or functions. The risk practitioner can play a role is ensuring that adequate tests are performed to test the functionality of the security controls.

Part I—Risk Management and Information Systems Control Theory and Concepts
Domain 4—Information Systems Control Design and Implementation
E. System Development Life Cycle (SDLC) Phases

Testing Types

The following table describes a variety of tests that:
- Relate to the previously mentioned approaches
- Are performed based on the size and complexity of the modified system

Testing Types	
Type	**Description**
Unit testing	• Tests an individual program or module • Uses a set of test cases that focus on the control structure of the procedural design • Ensures that the internal operation of the program performs according to specification
Interface or integration testing	• Uses a hardware or software test that evaluates the connection of two or more components that passes information from one area to another • Takes unit-tested modules and builds an integrated structure dictated by design **Note:** The term "integration testing" also refers to tests that verify and validate the functioning of the application under test with other systems, in which a set of data is transferred from one system to another.
System testing	• A series of tests designed to ensure that modified programs, objects, database schema, etc., which collectively constitute a new or modified system, function properly • Often performs these test procedures in a nonproduction test/development environment by software developers designated as a test team The following table describes specific analyses that may be carried out during system testing.

Test	Description
Recovery testing	Checks the system's ability to recover after a software or hardware failure
Security testing	Verifies that the modified/new system includes provisions for appropriate access controls and does not introduce any security holes that may compromise other systems
Stress/volume testing	Tests an application with large quantities of data to evaluate its performance during peak hours
Volume testing	Studies the impact on the application by testing with an incremental volume of records to determine the maximum volume of records (data) that the application can process
Stress testing	Studies the impact on the application by testing with an incremental number of concurrent users/services on the application to determine the maximum number of concurrent users/services the application can process
Performance testing	Compares the system's performance to other equivalent systems using well-defined benchmarks

Testing Types *(cont.)*

Testing Types *(cont.)*	
Type	**Description**

Final acceptance testing	• Begins on the modified system:
	  – After the system staff is satisfied with its initial or system tests
	  – During the implementation phase
	• Incorporates the defined methods of testing into the enterprise's quality assurance (QA) methodology
	• Proactively encourages QA activities to perform adequate levels of testing on all software development projects
	
	The following table identifies the two major parts of final acceptance testing.

Type of Test	Description
QA testing (QAT)	• Focuses on the technical aspect of the application (documented specifications and the technology employed) • Verifies that the application works as documented by testing the logical design and the technology itself • Ensures that the application meets the documented technical specifications and deliverables • Is performed primarily by the IS department **Note:** The participation of the end user is minimal and on request. • Does not focus on functionality testing

User acceptance testing (UAT)	• Focuses on the functional aspect of the application
	• Supports the process of ensuring that the system is production-ready
	• Satisfies all documented requirements
	• Methods include:
	  – Definition of test strategies and procedures
	  – Design of test cases and scenarios
	  – Execution of the tests
	  – Utilization of the results to verify system readiness

Note: Because they have different objectives, do not combine QAT and UAT.

Acceptance Criteria	"Acceptance criteria" refers to the criteria that a deliverable must meet to satisfy the predefined needs of the system/business owner or user.
Use of Production Data in Integrated Test Facilities (ITFs)	Many enterprises rely on ITFs to process test data in production-like systems to confirm the behavior of the new application or modules in real-life conditions, including peak volume and other resource-related constraints. In this environment, the IS function performs tests with a set of fictitious data in which the client uses extracts of production data to cover the most possible scenarios and some fictional data for scenarios that would not be tested. **Note:** When production data are used in a test environment, scramble the data so the confidential nature of data is obscured from the tester and in case the system inadvertently discloses or leaks information due to system defect or misconfiguration. **Note:** Such data leakage can occur when the acceptance testing is done by team members who, under usual circumstances, would not have access to such production data.
Certification and Accreditation Process	On completion of acceptance testing, the final step is usually a certification and accreditation process (also called a systems authorization, see NIST SP800-37 Revision 1), which: • Includes evaluating program documentation and testing effectiveness • Results in a final decision for deploying the business application system For information security issues, the evaluation process includes reviewing: • Security plans • The risk assessments performed and test plans • The evaluation process results in an assessment of the effectiveness of the security controls and processes to be deployed **Rationale:** This process generally involves security staff and the business owner of the application and provides some degree of accountability to the business owner regarding the state of the system that needs to be accepted for deployment.
Final Test Report to Management	When the tests are completed, the risk practitioner should issue an opinion to management as to whether the system: • Meets the business requirements • Has appropriate controls implemented • Would present an acceptable level of risk to the enterprise • Is ready to be migrated to production Be sure that this report: • Specifies the deficiencies in the system that need to be corrected • Identifies and explains the risk that the enterprise is taking by implementing the new system

4. PROJECT IMPLEMENTATION

4.1 Implementation Planning

Introduction

This section contains information on implementation planning, which is a project in itself and requires a methodology and the adoption of best practices that may be based on past experiences.

Description of Support Structure

Once a project is operational, it requires an efficient support structure for the new system delivered by the project. A support structure requires:
- Setting up roles and naming people to fulfill these roles
- Providing personnel with new skills
- Distributing the workload so that the right people support the right issues
- Developing new processes while respecting the specificities of IT department requirements
- Dedicating an infrastructure for support staff

Major Challenges for Implementation

One of the major challenges is to manage implementation:
- From build to integrate to migrate
- For the phasing-out of the existing system
- For the phasing-in of the new system

Migration must be set up in a step-by-step transition of the affected services.

> **Note:** The implemented processes for a legacy environment:
> - May be different from what may be implemented with the new platform
> - Must be communicated to users and system support staff if there are any changes

4.2 End-user Training

Introduction

This section contains knowledge on developing a training plan that ensures that end users can become self-sufficient in the operation of the system.

> **Note:** End-user training and the training plan must start early in the development process.

End-user Training Requirements

End-user training is required to permit the proper operation, maintenance and support for IS controls. Users must be trained in how to follow the processes and operations associated with the controls. End users must know how to react to any problems, errors or alerts generated by the controls. Ideally, the users will understand the purpose of the controls so that they can more effectively identify control issues.

End-user Training Development

The end-user training plan must start early in the development process. The development process follows a similar SDLC process as the software development:
- Define requirements/audience.
- Develop content/delivery method.
- Deliver training.
- Evaluate effectiveness of training.
- Maintain training program as controls are modified.

4.3 Data Migration

Introduction This section contains information on the data conversions involved with data migration.

Description of Data Conversion Data conversion is required if the source and target systems utilize different:
- Field formats or sizes
- File or database structures (e.g., relational database, flat files, Virtual Storage Access Method [VSAM])
- Coding schemes
- Hardware and/or operating system (OS) platforms

The object is to convert existing data into the new required format, coding and structure while preserving the meaning and integrity of the data.

Minimizing Data Migration Risk Carefully plan the data migration, and use the appropriate methodologies and tools to minimize the risk of:
- Disruption of routine operations
- Violation of the security and confidentiality of data
- Conflicts and contention between legacy and migrated operations
- Data inconsistencies and loss of data integrity during the migration process

4.4 Fallback (Rollback) Scenario

Introduction This section contains knowledge regarding problems associated with new system implementation or changes to an existing system and how to mitigate the risk when system deployment does not go as planned. To mitigate the risk of downtime for mission-critical systems, best practices dictate that the tools and applications required to reverse the migration are available prior to attempting the production cutover.

Components have to be delivered that can back out all changes and restore data to the original applications in the case of nonfunctioning new applications.

Some or all of these tools and applications may need to be developed as part of the project.

Introduction *(cont.)*

Data Conversion Key Considerations	
Consideration	**Guidelines**
Completeness of data conversion	The total number of records from the source database is transferred to the new database (assuming the number of fields is the same).
Data integrity	The data are not altered manually, mechanically or electronically by a person, program or substitution or by overwriting in the new system. **Note:** Integrity problems also include errors due to transposition and transcription errors and problems transferring particular records, fields, files and libraries.
Storage and security of data under conversion	Data are backed up before conversion for future reference or any emergency that may arise out of data conversion program management. **Note:** An unauthorized copy or too many copies can lead to misuse, abuse, or theft of data from the system.
Data consistency	The field/record called for from the new application should be consistent with that of the original application. **Note:** This enables consistency in repeatability of the testing exercise.
Business continuity	The new application should be able to continue with newer records as added (or appended) and help in ensuring seamless business continuity.

4.5 Changeover (Go-live) Techniques

Introduction

This section contains information on changeover, which refers to an approach to shift application users from the existing (old) system to the replacing (new) system.

Description of Changeover (Go Live)

Changeover (go live) is appropriate only after testing the new system with respect to its program and relevant data.

Types of Changeover Techniques

The following, selected changeover (go-live) techniques are discussed in more detail in this section:
- Parallel changeover
- Phased changeover
- Abrupt changeover

4.5.1 Parallel Changeover

Description of Parallel Changeover

This technique involves using both systems during a period of overlap. This includes, in order:
- Running the old system
- Running both the old and new systems in parallel
- Fully changing over to the new system after gaining confidence in the working of the new system

After a period of overlap, the:
- User gains confidence and assurance in relying on the newer system
- Use of the older system is discontinued
- New system becomes totally operational

Benefits of Parallel Changeover

Parallel changeover:
- Minimizes the risk of using the newer system
- Helps identify problems, issues or any concerns that the user initially comes across in the newer system

Exhibit 4.4: Parallel Changeover

Exhibit 4.4 depicts a parallel changeover.

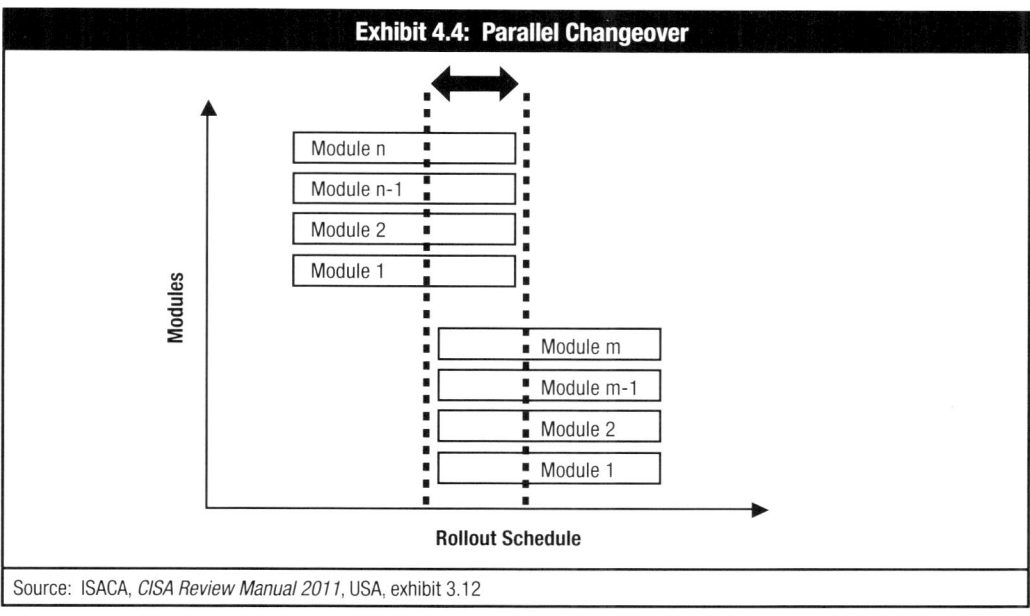

Source: ISACA, *CISA Review Manual 2011*, USA, exhibit 3.12

Note: The number (*m* and *n*, respectively) of modules in the new and old systems may be different.

4.5.2 Phased Changeover

Phased Changeover Steps

The following table describes how the changeover from the older system to the newer system takes place in a preplanned, phased manner.

Step	Phased Changeover Steps — Description
1	Module 1 of the older system is phased out and replaced by Module 1 of the newer system. **Note:** This relationship is not necessarily one to one. Modules 1-3 of the old system could be replaced by module 1 of the new system and so forth.
2	Module 2 of the older system is phased out and replaced by Module 2 of the newer system.
n-1	Module n-1 of the older system is phased out and replaced by Module n-1 of the newer system.
Alternate	This technique may be used when a system is to be deployed at multiple locations. In the phased rollout process, the changes are implemented in location after location over a period of time rather than all at one time.

Phased Changeover Risk Factors

Some of the risk factors that may exist in the phased changeover include:
- Resource challenges (both on the IT side—to be able to maintain two unique environments such as hardware, OSs, databases and code—and on the operations side—to be able to maintain user guides, procedures and policies; definitions of system terms, etc.)
- Maintaining consistency of data on multiple systems or locations
- Extension of the project life cycle to cover two systems
- Change management for requirements and customizations to maintain ongoing support of the older system

Exhibit 4.5: Phased Changeover

Exhibit 4.5 depicts a phased changeover.

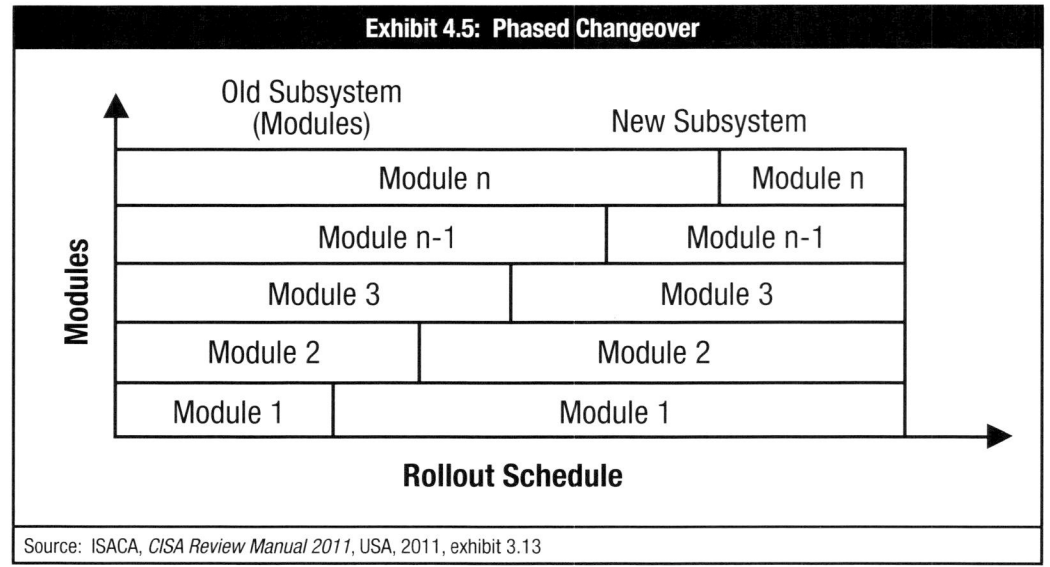

Source: ISACA, *CISA Review Manual 2011*, USA, 2011, exhibit 3.13

4.5.3 Abrupt Changeover

Description of Abrupt Changeover

In this approach, the newer system is changed over from the older system on a cut-off date and time and the older system is discontinued once changeover to the new system takes place.

Abrupt Changeover Steps

The following table describes the steps of the abrupt changeover process.

Abrupt Changeover Steps	
Step	**Description**
1	Convert files and programs; perform test runs on the test bed.
2	Install new hardware, OS, application system and migrated data.
3	Train employees or users in groups.
4	Schedule operations and test runs for go live or changeover.

Exhibit 4.6: Abrupt Changeover

Exhibit 4.6 depicts the abrupt changeover process.

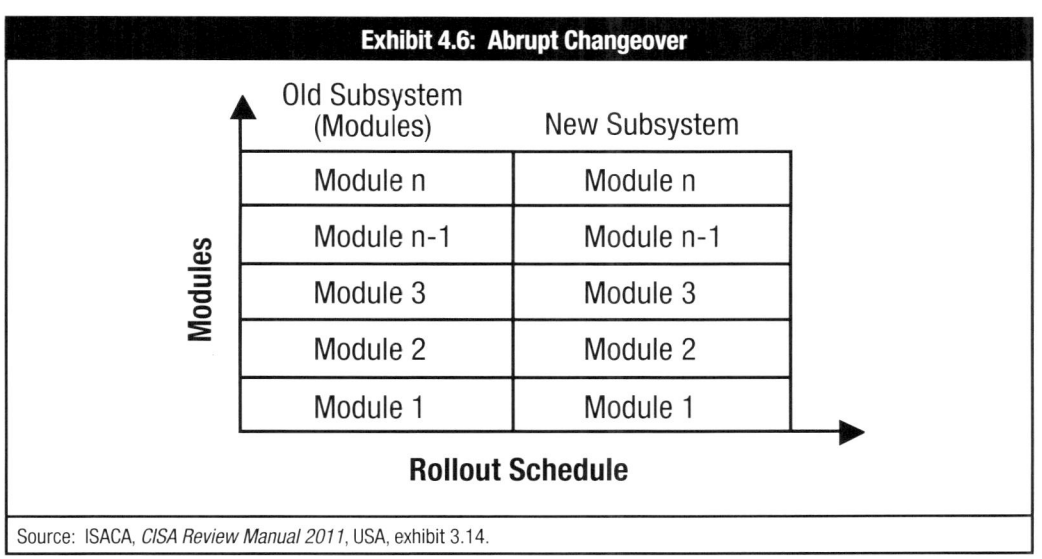

Abrupt Changeover Risk Areas

Some of the potential abrupt changeover risk areas include:
- Asset safeguarding
- Data integrity
- System effectiveness
- System efficiency
- Change management challenges (depending on the configuration items considered)
- Duplicate or missing records (duplicate or erroneous records may exist if data cleansing is not done correctly)

4.6 Postimplementation Review

Introduction

This section contains information on postimplementation review, which is:
- Used following the successful implementation of a new or extensively modified system
- Beneficial to verify that the system has been properly designed and developed and that proper controls have been built into the system

Postimplementation Review Objectives

A postimplementation review should meet the following objectives:
- Assess the adequacy of the system:
  - Does the system meet user requirements and business objectives?
  - Have controls been adequately defined and implemented?
- Evaluate the projected cost benefits or ROI measurements.
- Develop recommendations that address the system's inadequacies and deficiencies.
- Develop a plan for implementing the recommendations.
- Assess the development project process:
  - Were the chosen methodologies, standards and techniques followed?
  - Were appropriate project management techniques used?
  - Is the risk of operating the system within acceptable risk levels?

Example of Key Performance Indicators (KPIs)

A KPI for the implementation process measures the relative success of the changeover compared to desired performance objectives. Success of a changeover is often measured as a percentage of errors, number of trouble reports, duration of system outage, or degree of customer satisfaction.

The use of the KPI indicates to management whether the change control process was managed correctly, with sufficient levels of quality and testing.

Example: A KPI may be set at an error rate of 3%. If the changeover results in less than 3% errors, then the objective was satisfied. An error rate of greater than 3% would indicate an unacceptable level of errors and faults in the process.

Postimplementation Review Timing

The project development team and appropriate end users perform a post-project review jointly after the project has been completed and the system has been in production for a sufficient time period to assess its effectiveness.

4.7 Project Closeout

Introduction

This section contains information on closing a project. Projects should have a finite life: At some point, the project is closed and the new or modified system is handed over to the users and/or system support staff.

Part I—Risk Management and Information Systems Control Theory and Concepts
Domain 4—Information Systems Control Design and Implementation
E. System Development Life Cycle (SDLC) Phases

Project Closeout Steps

The following table describes the steps to close a project.

	Project Closeout Steps
Step	**Action**
1	Assign any outstanding issues to individuals responsible for remediation and identify the related budget, if applicable.
2	Assign custody of contracts, and archive or pass on documentation to those who will need it.

3	Survey the project team, development team, users and other stakeholders to:
	• Identify any lessons learned that can be applied to future projects.
	• Include content-related criteria such as:
	  – Performance fulfillment and project-related incentives
	  – Fulfillment of additional objectives
	  – Adherence to the schedule and costs
	• Include process-related criteria such as:
	  – Quality of the project teamwork
	  – Relationships to relevant environments

4	Conduct reviews in a formal process such as a post-project review in which lessons learned and an assessment of project management processes used are documented and referenced, in the future, by other project managers or users working on projects of similar size and scope.
5	Complete a postimplementation review once the project has been in use (or in "production") for some time—long enough to realize its business benefits and costs—and measure the project's overall success and impact on the business units.

Note: The project sponsor should be satisfied that the system produced is acceptable and ready for delivery.

F. MANAGING PROJECT RISK

Introduction — This section contains information on managing the risk associated with an information systems project, such as the failure of the project to deliver expected benefits on time, within budget and to meet the needs and expectations of the stakeholders.

Relevance — While the SDLC focuses on ensuring benefits realization at a technical level, project management practices—while closely related—focus on managing project-related risk through the management of project scope, time and resources.

The risk practitioner must apply project risk management techniques to ensure that needed IS controls are implemented at a reasonable cost.

> - A 2002 Gartner survey found that 20 percent of all expenditures on IT is wasted—a finding that represents, on a global basis, an annual destruction of value totalling about US $600 billion.
> - A 2004 IBM survey of Fortune 1000 CIOs found that, on average, CIOs believe that 40 percent of all IT spending brought no return to their organisations.
> - A 2006 study conducted by The Standish Group found that only 35 percent of all IT projects succeeded while the remainder (65 percent) were either challenged or failed.
>
> Source: ISACA, *Enterprise Value Governance of IT Investments: The Val IT Framework 2.0*, USA, 2009, page 7

1. KEY TERMS AND CONCEPTS

Introduction — This section introduces terms and principles related to project management as well as terms that help relate the process to other key business processes.

Portfolio — A portfolio is a grouping of "objects of interest" (investment programs, IT services, IT projects, other IT assets or resources) managed and monitored to optimize business value.

Portfolio management is distinct from project and project management in that the distinct objective is to create maximum value from a grouping of projects and programs.

Program — A structured grouping of interdependent projects that is both necessary and sufficient to achieve a desired business outcome and create value

> **Note:** These interdependent projects could include, but are not limited to, changes in the nature of the business, business processes, the work performed by people, as well as the competencies required to carry out the work, enabling technology and organizational structure.

Project — A structured set of activities concerned with delivering a defined capability (that is necessary, but not sufficient, to achieve a required business outcome) to the enterprise based on an agreed-on schedule and budget

Exhibit 4.7: Relationship Among Portfolio, Program and Project Management

Exhibit 4.7 describes the relationship among portfolio management, program management and project management.

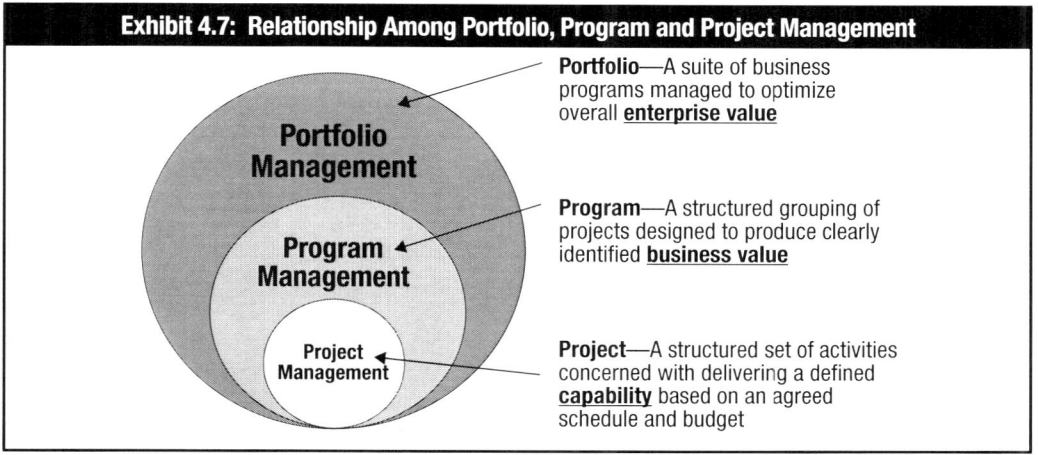

2. OVERVIEW OF PROJECT RISK MANAGEMENT

Project Management Iron Triangle

A project is bound by what can be termed the "iron triangle." As shown in **exhibit 4.8**, the iron triangle is defined by three constraints that affect project quality. A change to any one of the three constraints will force a change on the other two. In other words, an increase in scope will force a change in the allocated resources, the project schedule or both.

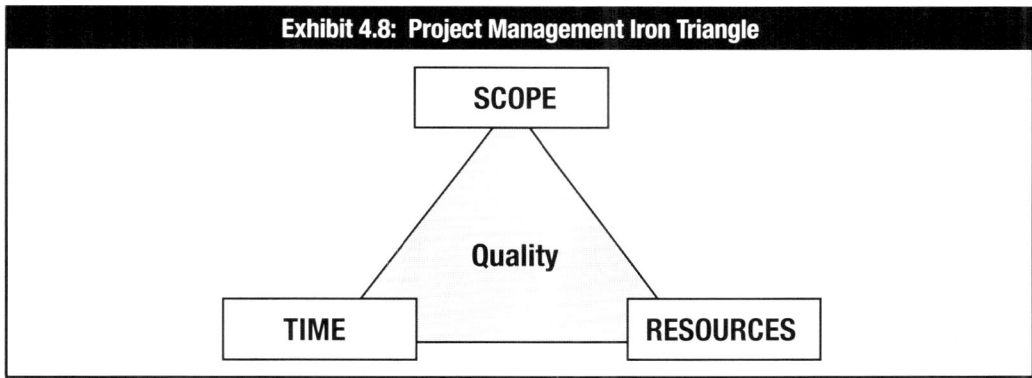

Key Project Management Phases

Key phases of the project and program management methodology are:
1. Define a program/portfolio management framework for IT investments.
2. Establish and maintain an IT project management methodology.
3. Establish and maintain an IT project monitoring, measurement and management system.
4. Build project charters, schedules, quality plans, budgets, and communication and risk management plans.
5. Ensure the participation and commitment of project stakeholders.
6. Ensure the effective control of projects and project changes.
7. Define and implement project assurance and review methods.

This section revolves around the operational aspects of project management only, specifically items 4, 5, 6 and 7.

Project Risk	There is always the risk that an IT project will not deliver the expected results and that the investment in the project has been wasted or ineffective. The cause of project failure may be due to many factors such as: • Changing requirements • Poorly defined requirements • New priorities • Lack of resources or skill • Lack of oversight for the project • Problems with new technology • Financial challenges

3. SCOPE MANAGEMENT

Introduction	Managing project deliverables requires careful control of project scope, deliverables and schedule. This control is effected through comprehensive documentation of the project, clearly defined requirements and resource management.
Project Scoping	Project scope is defined by the project owner or sponsor and is often a balance between cost, time and business needs. The scope should be clearly documented early in the project requirements phase of the project life cycle, once the feasibility studies and initial requirements gathering has been completed. It is extremely important that the scope includes identifying and addressing the security requirements in the initial documentation and budgeting for the information system. A failure to include security costs and deliverables will almost certainly lead to expensive redesign of the project later in the life cycle, or the delivery of a system that is not adequately protected. The documentation of the project deliverables should be detailed enough to provide an understanding of the complexity and amount of work required to complete the project. This can often be done through the use of a statement of work (SOW), or a work breakdown structure (WBS). This prevents confusion or disappointment when one party does not meet the expectations of the other party due to misunderstanding of the project deliverables. Any changes to the scope of the project should be subject to review and approval by a change management board (sometimes called a change control board). This is examined later in this chapter.
Earned Value	Earned value management (EVM) is a technique for measuring project performance and progress in an objective manner. Earned value has the ability to combine measurements of scope, schedule and cost in a single integrated system. EVM is notable for its ability to provide accurate forecasts of project performance problems. Using the methodology helps improve both scope definition as well as the analysis of overall project performance.

Part I—Risk Management and Information Systems Control Theory and Concepts
Domain 4—Information Systems Control Design and Implementation
F. Managing Project Risk

Project Review and Status Reporting

All projects require regular review and reporting. A project must have milestones that require the careful review of, and reporting on, the project status. These reports will indicate whether a project is on schedule or on budget and highlight any problems that may be developing with project scope, resourcing, deliverables, funding or time.

Such reviews often mandate that predefined deliverables have been produced before the next project phase is entered. Rigorous reviews enable the identification of any current or future project challenges, enable corrective action where necessary and provide management with decision points as to whether to continue, modify or abandon a project.

Managing Scope Changes

"The only thing certain in life is change." This premise certainly applies to projects and certain changes will have to be made to accommodate new requirements, adjust to changing priorities, or manage unexpected complications with integrating new technology to the existing IT architecture or vice versa.

To manage project change it is important to allow changes where necessary and—at the same time—prevent uncontrolled changes or scope creep. A formal project change management process helps achieve this while minimizing the effect that such changes may have on project schedule, budget, resources or stakeholder expectations.

A formal project change management process ensures that the change is executed according to the approval given, and that the change is documented, thoroughly tested and correctly implemented.

Change Request Approval Process

While the change management process is the overall process to change a system/software, the change approval process describes how a single change request is submitted, evaluated and approved.

The following table describes the responsibilities related to the change request approval process.

	Change Request Approval Process	
Step	**Responsibility**	**Description**
1	Stakeholder	• Initiates a formal project change request that contains clear description of the: – Requested change – Reasons for the change • Submits the change request to the project manager
2	Project manager	• Judges the impact of each change request on project activities (scope), schedule and budget • Archives copies of all change requests in the project file
3	Change advisory board	• Evaluates the change request (on behalf of the sponsor) • Decides whether to recommend the change • If accepted, instructs the project manager to update the project plan to reflect the requested change
4	Project sponsor	• Formally accepts or rejects the updated project plan recommendation by the change advisory board

4. TIME MANAGEMENT

Introduction

This section contains information on managing time by establishing and scheduling time frames.

Budgeting vs. Scheduling

Managing time involves two key efforts:
- **Budgeting**—Totaling the resource (financial, human and machine) effort involved in each task
- **Scheduling**—Establishing the sequential (or networked) relationship between tasks

Since tasks of a project reflect a self-contained, ordered group of activities, a project can be represented as a network in which tasks are shown as branches connected at nodes immediately preceding and following other tasks.

Critical Path Method (CPM)

All project schedules have a critical path, which is the set of successive activities that go from the beginning to the end of the project with the shortest possible completion time.

The essential technique for using the critical path method (CPM) is to construct a model of the project that includes the following:
1. A list of all activities required to complete the project (typically categorized within a work breakdown structure [WBS]),
2. The time (duration) that each activity will take to completion
3. The dependencies between the activities

Some activities may be performed concurrently (at the same time as other tasks); others may not be able to start before preceding activities have been completed (consecutively). Identifying such interdependencies between tasks is a key benefit of the CPM.

Exhibit 4.9: CPM

Exhibit 4.9 depicts the CPM.

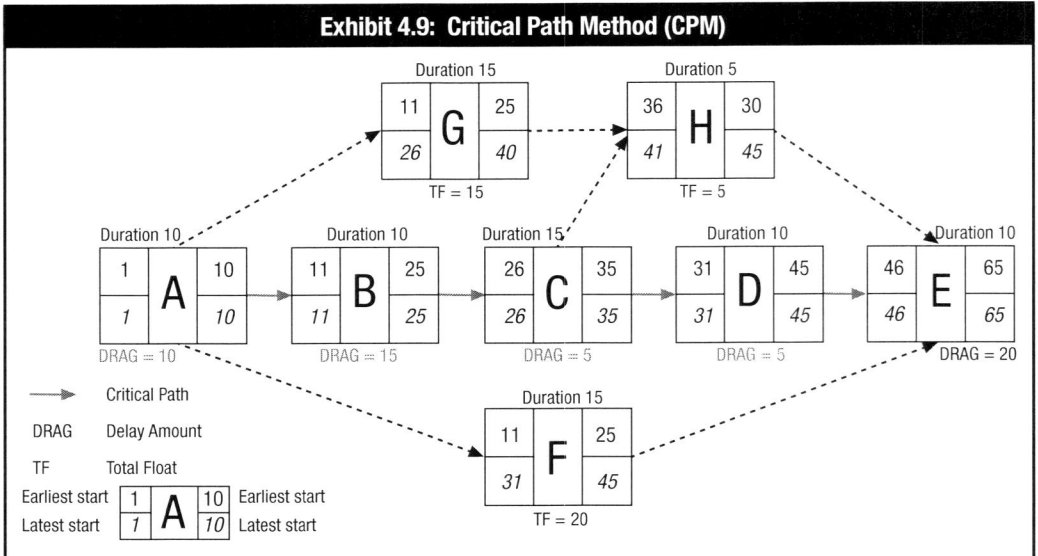

Slack (float) Time	Slack (float) time—Activities that are not part of the critical path have some slack time (float time), which is the difference between the: • Latest possible completion time of each activity that will not delay the completion of the overall project • Earliest possible completion time based on completion of all predecessor activities **Note:** Activities on the critical path have zero float time, and conversely, activities with zero slack time are on the critical path.
Drag Time	The calculation of the impact that a delay in one activity will have on the completion date of the project. The drag of an activity is equal to its duration if it does not have another activity in parallel; however, if it has parallel activities, its drag is the lesser of either the activity duration or the total float of the parallel activity.

5. BUDGET MANAGEMENT

Introduction	This section contains information on managing the project budget, which focuses on managing resources efficiently to achieve the desired outcome.
Description of Resource Usage	Resource usage is the process by which the project budget is being spent. If actual spending is in line with planned spending, resource usage must be measured and reported. Projects incur both fixed and non-fixed costs. A fixed cost such as licenses or office space will often remain relatively unchanged if the time to complete the project is extended or decreased. Salaries may be affected by overtime—more hours required than budgeted for or the cost of recruiting staff with specialized skills. **Note:** It is not sufficient to monitor only actual spending.
Example—Resource Productivity	If a task is planned to take 24 man-hours, then it is implicitly supposed that: • The resource being deployed is capable of finishing the task within the scheduled time. • At the same time, results are to be delivered at a satisfactory quality level. This assumes that every budget and project plan presupposes a certain "productivity" of resources.

Techniques for Measuring Resource Productivity

Resource usage can be checked with a technique called earned value analysis (EVA), which consists of comparing the following metrics at regular intervals during the project:
- Budget to date
- Actual spending to date
- Estimate to complete and estimate at completion

Example: A project task is budgeted to take three eight-hour working days. One work day has passed:
- From a pure time perspective, it will take another 16 hours to complete the task.
- From an earned value perspective, the resource is asked how much of the task he/she was able to complete in the eight hours:
  - If the resource performed a third of the task, he/she is in alignment with projections.
  - If the resource performed 25 percent of the task, he/she is behind projections and will need another day.
  - If the resource performed 50 percent of the task, he/she is ahead of projections and will be able to complete the task in two, instead of three, days.

6. USE OF METRICS TO SUPPORT RESOURCE PLANNING

Introduction

This section contains information on gathering metrics to support resource planning, which is required to ensure that day-to-day operations proceed smoothly while identifying and designating the technology and roles and responsibilities required for program development.

> **Note:** These metrics help ensure that resource deficiencies are detected and corrected before they impact the performance of overall security program development efforts.

Part I—Risk Management and Information Systems Control Theory and Concepts
Domain 4—Information Systems Control Design and Implementation
F. Managing Project Risk

Collecting Metrics

The following table describes the steps to collect metrics to support resource planning.

Steps for Resource Planning Metrics	
Step	**Action**
1	Develop metrics for resource utilization to support efforts to maximize program development efforts. **Rationale:** Program development activities are subject to the same staffing and organizational dependency issues as any other management process.
2	Gather historical data on resource dependencies that may affect the security program. Use this data in planning periodic activities designed to: • Identify changing security resource requirements in time to devise alternate plans • Meet control objectives that may also change in response to new requirements
3	Ensure that all personnel who take lead roles in the performance of critical security functions have a backup that can perform the given function unassisted in the absence of the primary leader. **Rationale:** Metrics that count the number of people in critical functions without a trained backup herald a call to action.
4	Ensure that lead roles are covered. **Note:** If critical elements of the security program are not managed with appropriate accountability, this metric may indicate gaps in the information security program that need to be addressed.
5	Where responsibilities for operating technologies are delegated to other departments, collect metrics that provide insight into the resource requirements and planning processes of the responsible enterprise.

G. PROJECT MANAGEMENT TOOLS AND TECHNIQUES
1. GENERAL PROJECT MANAGEMENT TECHNIQUES

Introduction	This section contains knowledge on general project management practice techniques.
Project Management Knowledge and Practices	Project management knowledge and practices are best described in terms of their component processes of initiating, planning, executing, controlling and closing a project. Overall characteristics of successful project planning are that it is a risk-based management process and iterative in nature.
Project Management Techniques and Tools for Controlling Time and Resources	Project management techniques and tools: • Provide systematic quantitative and qualitative approaches to software size estimating, scheduling, allocating resources and measuring productivity • Assist the project manager in controlling the time and resources utilized in the development of a system • Vary from a simple manual effort to a more elaborate computerized process Note: Base the approach on the project size and complexity.
Exhibit 4.10: Relationships Between Project Management Elements	Project management should pay attention to the intertwining relationships between project management elements: • Deliverables • Duration • Budget Exhibit 4.10 depicts an oversimplified and schematized complex relationship.

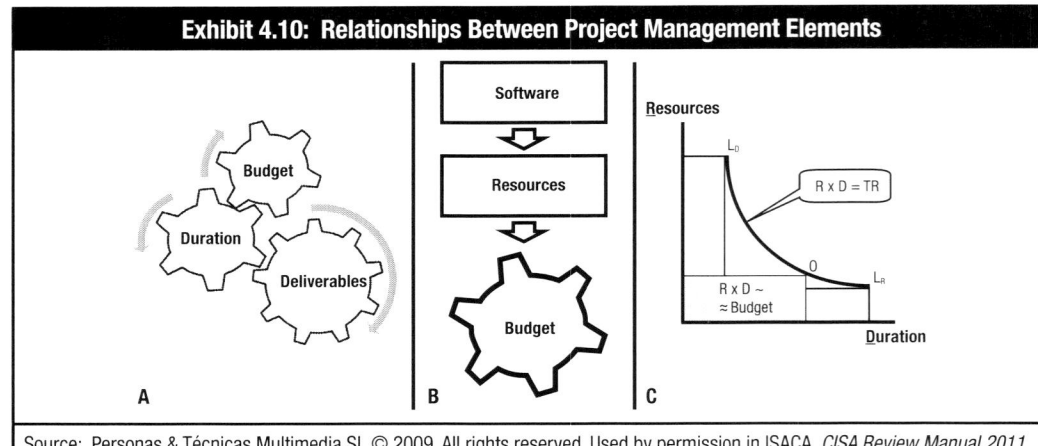

Source: Personas & Técnicas Multimedia SL © 2009. All rights reserved. Used by permission in ISACA, *CISA Review Manual 2011*, USA, 2011, exhibit 3.4

Part I—Risk Management and Information Systems Control Theory and Concepts
Domain 4—Information Systems Control Design and Implementation
G. Project Management Tools and Techniques

Project Management Elements

The following table describes the relationships between the project management elements shown in **exhibit 4.10**.

Project Management Elements	
Element	**Description**
Deliverables	• Project duration and budget must be commensurate with the nature and characteristics of the deliverables. • In general, there will be a positive correlation (growing together) between highly demanding deliverables, a long duration and a high budget. • The quality of the deliverables is an important element that is also considered during the management of time and resources: – The parameters for quality of the deliverables may be specified clearly by the project steering committee or project sponsor, or the project manager may have to elicit the parameters from user management. – Regardless of who specifies the parameters for quality, the project manager must have a clear and documented view of the quality expectations for the deliverables of the project steering committee, sponsor and users.
Budget	• Budget is deduced from the resources required to carry out the project by multiplying fees or costs by the amount of each resource. • At the beginning of the project, the required resources are estimated by using techniques of software/project size estimation. **Note:** Size estimation yields a "total resources" calculation.
Resources	• For simplification purposes, there is the assumption that resources are fixed for the duration of the project. • The curve shows resources assigned (R) × duration (D) = total resources (TR, a constant quantity); which is the classic "man × month" dilemma curve. • Any point along the curve meets the condition R × D = TR. • If any point O on the curve is chosen, the area of the rectangle will be TR, proportional to the budget. | If resources are … | Then the project will take a … | |---|---| | Few | Long time (a point close to LR). | | Many | Shorter time (a point close to LD). | LR and LD are two practical limits and result in: • A duration that is too long may not seem reasonable. • The use of too many (human) resources at once would be unmanageable.

Note: There are a few heuristics (rules of thumb) available to choose a convenient combination of assigned resources and project duration.

2. GANTT CHARTS

Introduction

This section contains knowledge on constructing Gantt charts, which aid in scheduling the activities (tasks) needed to complete a project.

Purpose of Gantt Charts

Gantt charts (depicted in **exhibit 4.11**):
- Show when an activity should begin and end along a timeline
- Show which activities:
  - Can be in progress concurrently
  - Must be completed sequentially
- Reflect the resources assigned to each task and by what percent they are allocated
- Compared to a baseline project plan, outline:
  - Which activities have been completed early or late
  - The project progress indicating whether the project is behind, ahead or on schedule
- Track the achievement of milestones or significant accomplishments for the project such as the end of a project phase or completion of a key deliverable

Exhibit 4.11: Sample Gantt Chart

Exhibit 4.11 shows a Gantt chart within a work breakdown structure.

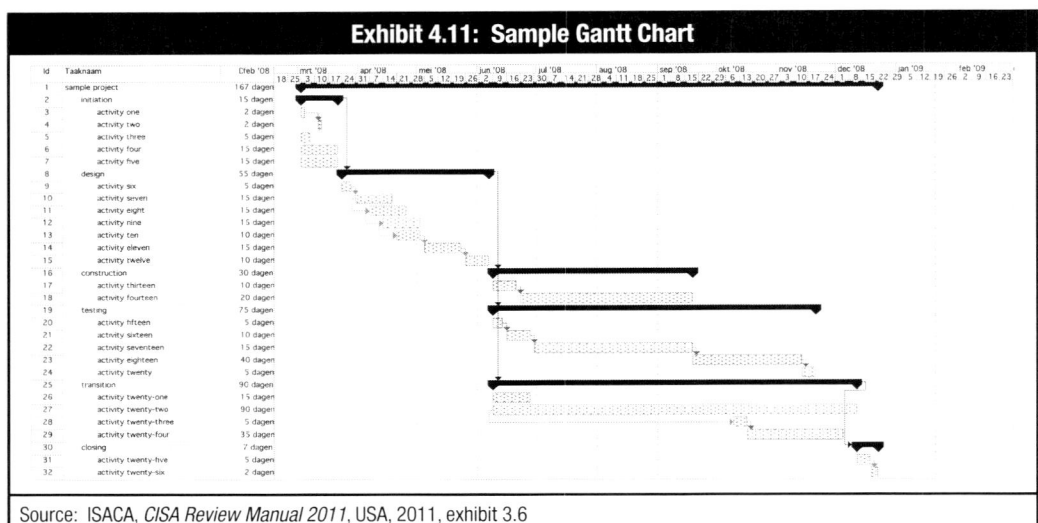

Source: ISACA, *CISA Review Manual 2011*, USA, 2011, exhibit 3.6

3. CRITICAL PATH METHODOLOGY (CPM) AND PROGRAM EVALUATION REVIEW TECHNIQUE (PERT)

Introduction

This section contains information on PERT, which is a CPM-type technique that uses three different estimates for each activity instead of using a single number (as used by CPM).

The advantage of using three different estimates for each activity is that the formula is based on the reasonable assumption that:
- The three time estimates follow a beta statistical distribution.
- Probabilities (with associated confidence levels) can be associated with the total project duration.

PERT Usage	PERT is often used in system development projects with uncertain durations (e.g., pharmaceutical research or complex software development).
Identification of All Activities and Related Events/Milestones	When designing a PERT network for system development projects, the first step is to identify all the activities and related events/milestones of the project and their relative sequence. **Note:** The risk practitioner: • Must be careful not to overlook any activity • May prepare many diagrams that provide increasingly more detailed time estimates **Note:** Some activities such as analysis and design must be preceded by others before program coding can begin. The list of activities determines the detail of the PERT network. **Example:** An event or result may be the completion of the operational feasibility study or the point at which the user accepts the detailed design.
Identification of a Critical Path Using PERT	A critical path is the: • Longest path through the project (There is only one critical path in a project.) If everything goes perfectly, this path indicates the minimum amount of time required to complete the project. • Route along which the project is shortened (accelerated) or lengthened (delayed) **Example:** In **exhibit 4.12**, the critical path is A, C, E, F, H and I.
PERT Time Estimates	The following table describes the three estimates used for completing each task's activity. | Task Activity Completion Estimates	
Estimate	Description
First	This is the most optimistic time, if everything went well.
Second	This is the most likely scenario and is based on experience attained from projects similar in size and scope.
Third	This is the pessimistic or worst-case scenario.
Calculation of the PERT Time Estimate	To calculate the PERT time estimate for each given activity, the following calculation is applied: [Optimistic + Pessimistic + 4(most likely)]/6. The three PERT estimates are: • Reduced (applying a mathematical formula) to a single number • Applied to the classic CPM algorithm

Exhibit 4.12: PERT Network-based Chart

Exhibit 4.12 illustrates the use of the PERT network management technique, in which events are points in time or milestones for starting and completing activities (arrows).

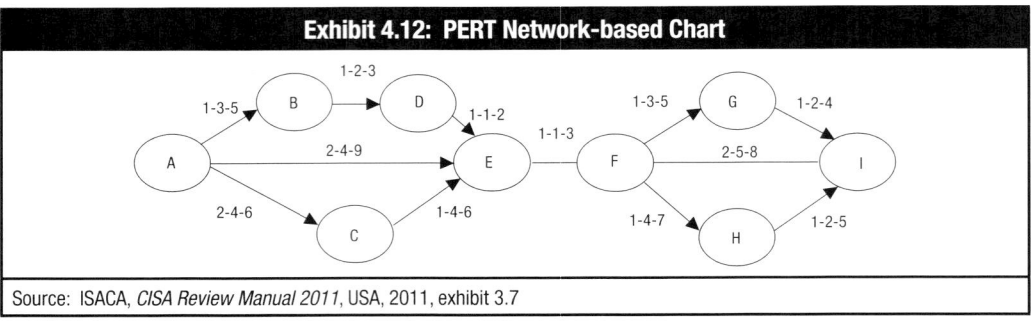

H. SUGGESTED RESOURCES FOR FURTHER STUDY

Suggested Resources for Further Study

In addition to the resources cited throughout this manual, the following resources are suggested for further study:

- ISACA:
  - *Change Management Audit/Assurance Program*, 2009
  - *CISM Review Manual 2013*, 2012
  - *COBIT 5*, 2012
  - *COBIT 5 for Information Security*, 2012
  - *ITAF: A Professional Practices Framework for IT Assurance*, 2008
  - *IT Assurance Guide: Using COBIT*, 2007
  - *Systems Development and Project Management Audit/Assurance Program*, 2009
  - *The Business Model for Information Security (BMIS)*, 2010
  - *Value Management Guidance for Assurance Professionals: Using Val IT 2.0*, 2010
- Barnier, Brian; *The Operational Risk Handbook for Financial Companies: A Guide to the New World of Performance-oriented Operational Risk*, Harriman House Ltd, UK, 2011
- Committee of Sponsoring Organizations of the Treadway Commission (COSO), *Internal Control—Integrated Framework: Guidance on Monitoring Internal Control Systems*, USA, 2009
- IEEE Computer Society, Standard 1074-2006 for Developing a Software Project Lifecycle Process, USA, 2006, www.ieee.org
- International Project Management Association (IPMA), *IPMA Competence Baseline (ICB), Version 3.0*, The Netherlands, 2006
- Krutz, Ronald L.; Russell Dean Vines; *The CISM Prep Guide: Mastering the Five Domains of Information Security Management*, Wiley, USA, 2003
- Maizlish, Bryan; Robert Handler; *IT Portfolio Management Step-by-step: Unlocking the Business Value of Technology*, John Wiley & Sons, USA, 2005
- National Institute of Standards and Technology (NIST), *Security Controls in External Environments, NIST Special Publication (SP) 800-53, Revision 3*, USA, 2009
- NIST, Special Publication 800-53, Revision 3, Online Database, USA, http://web.nvd.nist.gov/view/800-53/home
- Office of Government Commerce (OGC), *Projects in Controlled Environments 2 (PRINCE2): Managing Successful Projects With PRINCE2*, UK, 2009
- Project Management Institute (PMI), *A Guide to the Project Management Body of Knowledge (PMBOK), 4th Edition*, USA, 2008

Page intentionally left blank

Domain 5—Information Systems Control Monitoring and Maintenance

A. CHAPTER OVERVIEW

Introduction

This chapter focuses on monitoring of information systems controls to confirm that controls are well designed and continue to effectively mitigate identified risk over time. The chapter provides information on the implementation of a control monitoring process, reporting of control status to management and automated tools that are available to help enterprises achieve information systems (IS) control monitoring appropriate for their unique needs. It introduces process capability models to help enterprises determine the current state of their control environment and the steps necessary to move toward their desired state.

Learning Objectives

As a result of completing this chapter, the CRISC candidate should be able to:
- Discuss the purpose and levels of a process capability model.
- Compare different monitoring tools and techniques.
- Describe various testing and assessment tools and techniques.
- Explain how monitoring of IS controls relates to applicable laws and regulations.
- Understand the need for control maintenance.
- Establish a process for the ongoing operation and maintenance of controls.

Inputs From Other Domains

The IS control monitoring and maintenance process is a distinct subset of *Domain 2—Risk Response*, particularly risk mitigation/reduction.

Information systems control monitoring and maintenance efforts are directly based on and affected by *Domain 4—Information Systems Control Design and Implementation*. Newly implemented IS controls are captured in the controls inventory and key controls are selected for monitoring.

Relevance

Risk management relies on a controls monitoring process to ensure that IS controls remain effective and efficient over time. Monitoring requires the definition of meaningful performance indicators, systematic and timely reporting of performance, and prompt response to deviations. Monitoring makes sure that the right things are done and are in line with business directions, corporate policies and best practice standards.

The ultimate purpose of monitoring is for an evaluator to gather enough information to evaluate whether IS controls are operating effectively to mitigate risk.

Outputs to Other Domains

Reports on the effectiveness or ineffectiveness of controls are an important input into *Domain 1—Risk Identification, Assessment and Evaluation*. Control effectiveness significantly influences the enterprise's overall risk profile, particularly residual risk. As technologies or operational environments change, the effectiveness of controls may be diminished or bypassed. These changes will trigger the need to return to Domain 1 and reassess the controls and risk profile of the organization.

Part I—Risk Management and Information Systems Control Theory and Concepts
Domain 5—Information Systems Control Monitoring and Maintenance
A. Chapter Overview

Contents

This chapter contains the following sections:

Section	Starting Page
A. Chapter Overview	167
B. Task and Knowledge Statements	169
C. Process Capability Assessment	171
1. Key Terms and Concepts	171
2. The Process Dimension	173
3. The Capability Dimension	174
4. Assessment Indicators for the Risk Management Process	174
D. The Control Life Cycle	177
E. Information Systems Control Monitoring and Maintenance	179
1. The IS Control Monitoring and Maintenance Process	180
F. Identify and Assess Information	182
1. Key Terms and Concepts	182
2. Control Monitoring Tasks	183
G. Tools for Monitoring	189
1. Tools for Monitoring Transaction Data	190
2. Tools for Monitoring Conditions	191
3. Tools for Monitoring Changes	192
4. Tools for Monitoring Process Integrity	193
5. Tools for Error Management and Reporting	194
6. Continuous Monitoring	195
H. Implementing Control Monitoring Processes	198
1. Determine Monitoring Method and Frequency	198
2. Select and Implement Automated Monitoring Tools	199
3. Clarify Reporting Requirements and Expectations	201
I. Implementing Control Maintenance Processes	204
1. Determine Control Maintenance Requirements	204
2. Implement Control Maintenance Process	206
J. Suggested Resources for Further Study	208

Process-specific controls for the following processes are addressed in Part II of this manual:
1. Managing the IT Strategy
2. Portfolio, Program and Project Management
3. Change Management
4. Third-party Service Management
5. Continuity Management
6. Information Security Management
7. Configuration Management
8. Problem Management
9. Knowledge Management
10. IT Operations Management

B. TASK AND KNOWLEDGE STATEMENTS

Introduction   This section describes the task and knowledge statements for Domain 5, which focuses on monitoring and maintaining IS controls to ensure that they function effectively and efficiently.

Task Statements   The following table describes the task statements for Domain 5 that the CRISC candidate must know how to perform..

No.	Task Statement (TS)
TS5.1	Plan, supervise and conduct testing to confirm continuous efficiency and effectiveness of information systems controls.
TS5.2	Collect information and review documentation to identify information systems control deficiencies.
TS5.3	Review information systems policies, standards and procedures to verify that they address the organization's internal and external requirements.
TS5.4	Assess and recommend tools and techniques to automate information systems control verification processes.
TS5.5	Evaluate the current state of information systems processes using a maturity model to identify the gaps between current and targeted process maturity.
TS5.6	Determine the approach to correct information systems control deficiencies and maturity gaps to ensure that deficiencies are appropriately considered and remediated.
TS5.7	Maintain sufficient, adequate evidence to support conclusions on the existence and operating effectiveness of information systems controls.
TS5.8	Provide information systems control status reporting to relevant stakeholders to enable informed decision making.

Part I—Risk Management and Information Systems Control Theory and Concepts
Domain 5—Information Systems Control Monitoring and Maintenance
B. Task and Knowledge Statements

Knowledge Statements

The following table describes the knowledge statements for Domain 5. The CRISC candidate must have a good understanding of each of the areas delineated by the knowledge statements. These statements are the basis for the exam..

No.	Knowledge Statement (KS) Knowledge of:
KS5.1	Standards, frameworks and leading practices related to information systems control monitoring and maintenance
KS5.2	Enterprise security architecture
KS5.3	Monitoring tools and techniques
KS5.4	Maturity models
KS5.5	Control objectives, activities and metrics related to IT operations and business processes and initiatives
KS5.6	Control objectives, activities and metrics related to incident and problem management
KS5.7	Security testing and assessment tools and techniques
KS5.8	Control objectives, activities and metrics related to information systems architecture (platforms, networks, applications, databases and operating systems)
KS5.9	Control objectives, activities and metrics related to information security
KS5.10	Control objectives, activities and metrics related to third-party management
KS5.11	Control objectives, activities and metrics related to data management
KS5.12	Control objectives, activities and metrics related to the system development life cycle
KS5.13	Control objectives, activities and metrics related to project and program management
KS5.14	Control objectives, activities and metrics related to software and hardware certification and accreditation practices
KS5.15	Control objectives, activities and metrics related to business continuity and disaster recovery management
KS5.16	Applicable laws and regulations

Note: Knowledge statements 5.6 and 5.8-5.15 are process specific and addressed in Part II of this manual.

C. PROCESS CAPABILITY ASSESSMENT

Introduction To effectively assess process capability for an organization, the risk practitioner has to take a two-dimensional approach. In one dimension, the process dimension, the enterprise's processes are defined and classified into process categories. In the other dimension, the capability dimension, a set of process attributes grouped into capability levels is defined. The process attributes provide the measurable characteristics of process capability. Frameworks, such as COBIT 4.1 or COBIT 5 can be used to define and classify process categories. A recognized process assessment model is ISO/IEC 15504-2 *Requirements for a Process Assessment Model*, and it can be used as the basis for conducting an assessment of the capability of each process.

Relevance Process maturity levels help the risk practitioner determine the current state of process capability and, should the current state be different from the desired state, provide feedback into the risk assessment, risk response and IS control design and implementation processes.

1. KEY TERMS AND CONCEPTS

Assessment Indicator	Used to assess whether process attributes have been achieved. There are two types of assessment indicators: • Process capability indicators, which apply to capability levels 1 to 5 • Process performance indicators, which apply exclusively to capability level 1
Attribute Indicator	An assessment indicator that supports the judgment of the extent of achievement of a specific process attribute (ISO/IEC 15504:1, 3.16)
Base Practice	An activity that, when consistently performed, contributes to achieving a specific process purpose (ISO/IEC 15504:1, 3.17)
Capability Dimension	The set of elements in a process assessment model explicitly related to the Measurement Framework for Process Capability (ISO/IEC 15504:1, 3.18)
Capability Indicator	An assessment indicator that supports the judgment of the process capability of a specific process (ISO/IEC 15504:1, 3.19)
Generic Practice	An activity that, when consistently performed, contributes to the achievement of a specific process attribute (ISO/IEC 15504:1, 3.22)
Performance Indicator	A set of metrics designed to measure the extent to which performance objectives are being achieved on an ongoing basis Performance indicators can include service level agreements (SLAs), critical success factors (CSFs), customer satisfaction ratings, internal or external benchmarks, industry best practices and international standards.
Process Assessment Model	A model suitable for the purpose of assessing process capability, based on one or more process reference models (ISO/IEC 15504:1, 3.33)

Process Attribute	A measurable characteristic of process capability applicable to any process (ISO/IEC 15504:1, 3.31)
Process Attribute Rating	A judgment of the degree of achievement of the process attribute for the assessed process (ISO/IEC 15504:1, 3.32)
Process Capability	A characterization of the ability of a process to meet current or projected business goals (ISO/IEC 15504:1, 3.33)
Process Capability Indicator	The process capability is expressed in terms of process attributes grouped into capability levels. Process capability indicators are generic for each process attribute for capability levels 1 to 5 and consist of: • Generic Practices (GPs) • Generic Work Products (GWPs)
Process Capability Level	A point on the six-point ordinal scale (of process capability) that represents the capability of the process, each level building on the capability of the level below (ISO/IEC 15504:1, 3.36)
Process Capability Level Rating	A representation of the achieved process capability level derived from the process attribute ratings for an assessed process (ISO/IEC 15504:1, 3.37)
Process Outcome	An observable result of a process (ISO/IEC 15504:1, 3.44) **Note:** An outcome is an artefact, a significant change of state or the meeting of specified constraints.
Process Performance Indicator	Process performance indicators (base practices and work products) are specific for each process and are used to determine whether a process is at capability level 1. The base practices and work products for the risk management process are shown in **exhibit 5.5**. These are based on material in COBIT 4.1.
Process Purpose	The high-level measurable objectives of performing the process and the likely outcomes of effective implementation of the process (ISO/IEC 15504:1, 3.47)
Process Reference Model	A model composed of definitions of processes in a life cycle described in terms of process purpose and outcomes, together with an architecture describing the relationships among the processes (ISO/IEC 15504:1, 3.48)
Work Product	An artefact associated with the execution of a process (ISO/IEC 15504:1, 3.55)

Part I—Risk Management and Information Systems Control Theory and Concepts
Domain 5—Information Systems Control Monitoring and Maintenance
C. Process Capability Assessment

2. THE PROCESS DIMENSION

Exhibit 5.1: Process Dimension Using COBIT 4.1

Exhibit 5.1 shows a process reference model based on COBIT 4.1:

Exhibit 5.1: Process Dimension Using COBIT 4.1

- BUSINESS OBJECTIVES
- GOVERNANCE OBJECTIVES

COBIT

ME1 Monitor and evaluate IT performance.
ME2 Monitor and evaluate internal control.
ME3 Ensure compliance with external requirements.
ME4 Provide IT governance.

PO1 Define a strategic IT plan.
PO2 Define the information architecture.
PO3 Determine technological direction.
PO4 Define the IT processes, organisation and relationships.
PO5 Manage the IT investment.
PO6 Communicate management aims and direction.
PO7 Manage IT human resources.
PO8 Manage quality.
PO9 Assess and manage IT risks.
PO10 Manage projects.

INFORMATION CRITERIA
- Effectiveness
- Efficiency
- Confidentiality
- Integrity
- Availability
- Compliance
- Reliability

IT RESOURCES
- Applications
- Information
- Infrastructure
- People

MONITOR AND EVALUATE
PLAN AND ORGANISE
DELIVER AND SUPPORT
ACQUIRE AND IMPLEMENT

DS1 Define and manage service levels.
DS2 Manage third-party services.
DS3 Manage performance and capacity.
DS4 Ensure continuous service.
DS5 Ensure systems security.
DS6 Identify and allocate costs.
DS7 Educate and train users.
DS8 Manage service desk and incidents.
DS9 Manage the configuration.
DS10 Manage problems.
DS11 Manage data.
DS12 Manage the physical environment.
DS13 Manage operations.

AI1 Identify automated solutions.
AI2 Acquire and maintain application software.
AI3 Acquire and maintain technology infrastructure.
AI4 Enable operation and use.
AI5 Procure IT resources.
AI6 Manage changes.
AI7 Install and accredit solutions and changes.

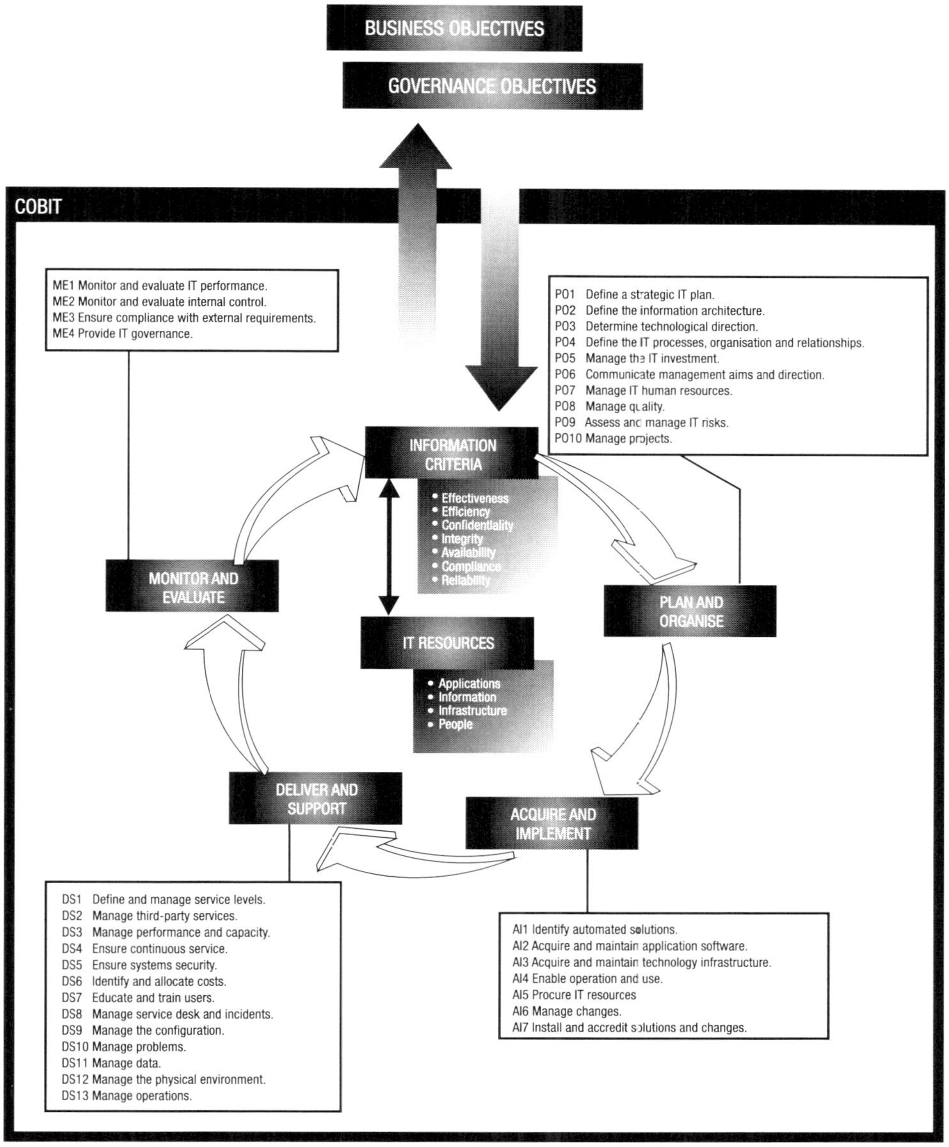

Source: ISACA, *COBIT Process Assessment Model (PAM): Using COBIT® 4.1*, USA, 2011, figure 3

3. THE CAPABILITY DIMENSION

Exhibit 5.2: Process Capability Levels

The capability dimension provides a measure of a process's capability to meet an organization's current or projected business goals for the process. The process capability is expressed in terms of process attributes grouped into capability levels. **Exhibit 5.2** describes the six capability levels of a process on the basis of the achievement of specific process attributes as per ISO/IEC 15504-2:2003.

Process Capability Level	Capability (number of attributes)	Description
Level 0	Incomplete process	The process is not implemented or fails to achieve its process purpose. At this level, there is little or no evidence of any systematic achievement of the process purpose.
Level 1	Performed process (one attribute)	The implemented process achieves its process purpose.
Level 2	Managed process (two attributes)	The previously described performed process is now implemented in a managed fashion (planned, monitored and adjusted), and its work products are appropriately established, controlled and maintained.
Level 3	Established process (two attributes)	The previously described managed process is now implemented using a defined process that is capable of achieving its process outcomes.
Level 4	Predictable process (two attributes)	The previously described established process now operates within defined limits to achieve its process outcomes.
Level 5	Optimizing process (two attributes)	The previously described predictable process is continuously improved to meet relevant current and projected business goals.

4. ASSESSMENT INDICATORS FOR THE RISK MANAGEMENT PROCESS

Exhibit 5.3: Process Performance Indicators

Assessment indicators measure the extent to which the process purpose is achieved. Full achievement of this attribute results in the process achieving its defined outcomes. Base practices for the risk management process are reflected in **exhibit 5.3**.

Exhibit 5.3: PA1.1 Process Performance

Result of Full Achievement of the Attribute	Generic Practices (GPs)	Generic Work Products (GWPs)
The process achieves its defined outcomes.	**GP 1.1.1 Achieve the process outcomes.** There is evidence that the intent of base practice is being performed.	Work products are produced that provide evidence of process outcomes, as outlined in section 3.0.

Source: ISACA, *COBIT Process Assessment Model (PAM): Using COBIT® 4.1*, USA, 2011, figure 6

Generic Work Products

The generic work products (GWPs) are those work products required to support the management of a process. Their existence, together with generic practices, provides evidence for the achievement of specific process capability attributes. The evidence includes:
- Process objectives
- Responsibilities
- Performance requirements
- Improvement plans and outcomes required at various levels of process capability

They are called "generic" because similar work products would be expected for each process. They are indicative of the types of work products and content that will be introduced to support increased process capability.

Exhibit 5.4: Generic Work Products and Relation to Capability Level

Exhibit 5.4 lists the GWPs and the capability levels at which they would be required for evidential purposes.

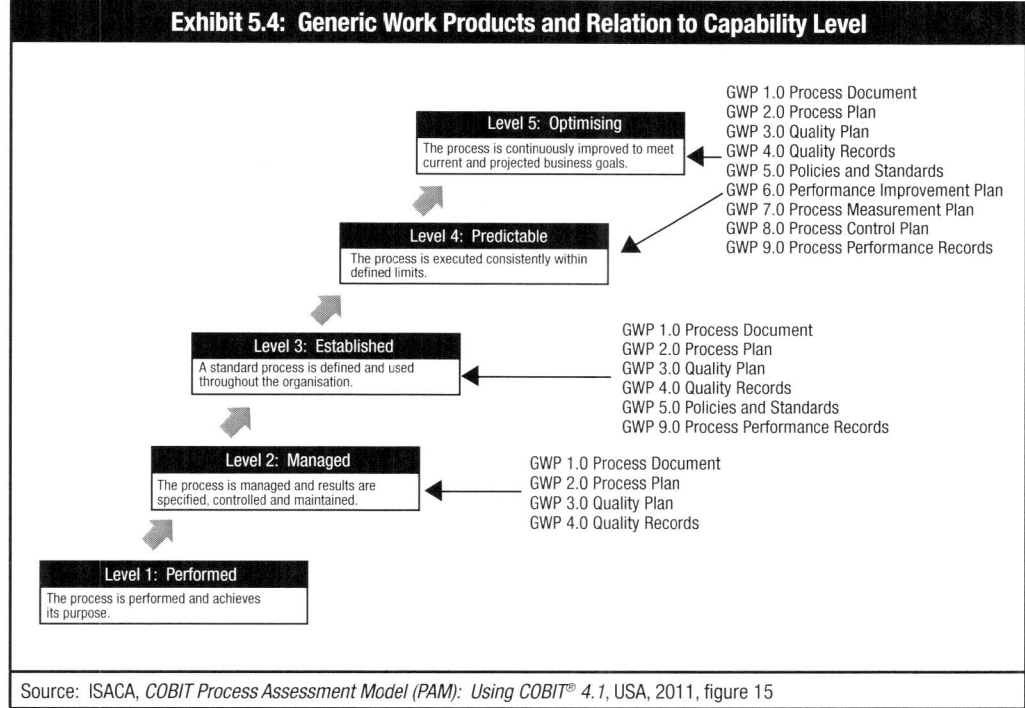

Source: ISACA, *COBIT Process Assessment Model (PAM): Using COBIT® 4.1*, USA, 2011, figure 15

Part I—Risk Management and Information Systems Control Theory and Concepts
Domain 5—Information Systems Control Monitoring and Maintenance
C. Process Capability Assessment

Exhibit 5.5: Assess and Manage IT Risks Process

Exhibit 5.5 describes IT risk management in terms of process name, purpose and outcomes (Os), based on COBIT 4.1 as well as:
- A set of base practices (BPs), providing a definition of the tasks and activities needed to accomplish the process purpose and fulfill the process outcomes
- A number of input and output work products (WPs) and characteristics associated with each WP

Exhibit 5.5: Process Dimension and Process Performance Indicators

Process ID	P09		
Process Name	Assess and Manage IT Risks		
Purpose	Satisfy the business requirement of analysing, communicating and managing IT risks and their potential impact on business processes and goals.		
Outcomes (Os)	**Number**	**Description**	
	P09-01	An IT risk management framework is established that is aligned to the organisation's (enterprise's) risk management framework.	
	P09-02	Risk remediation action plans are defined and communicated.	
Base Practices (BPs)	**Number**	**Description**	**Supports**
	P09-BP1	Determine risk management alignment (e.g., assess risk).	P09-01
	P09-BP2	Understand relevant strategic business objectives.	P09-01
	P09-BP3	Understand relevant business process objectives.	P09-01
	P09-BP4	Identify internal IT objectives related to risk management, and establish the risk context.	P09-01
	P09-BP5	Identify events associated with these objectives.	P09-01
	P09-BP6	Assess risk associated with the events.	P09-01
	P09-BP7	Evaluate risk responses.	P09-01
	P09-BP8	Prioritise and plan control activities.	P09-02
	P09-BP9	Approve and ensure funding for risk action plans.	P09-02
Work Products (WPs)			
Inputs			
Number	**Description**		**Supports**
P01-WP1	Strategic IT plan		P09-01, 02
P01-WP2	Tactical IT plan		P09-01, 02
P01-WP3	IT project portfolio		P09-01, 02
P01-WP4	IT service portfolio		P09-01, 02
P010-WP2	Project risk management plan		P09-01, 02
DS2-WP3	Supplier risks		P09-01, 02
DS4-WP1	Contingency test results		P09-01, 02
DS5-WP5	Security threats and vulnerabilities		P09-01, 02
ME1-WP3	Historic risk trends and events		P09-01, 02
ME4-WP5	Enterprise appetite for IT risks		P09-01, 02
Outputs			
Number	**Description**	**Input to**	**Supports**
P09-WP1	Risk assessment	P01, DS4, DS5, DS12, ME4	P09-01, 02
P09-WP2	Risk reporting	ME4	P09-01, 02
P09-WP3	IT-related risk management guidelines	P06	P09-01
P09-WP4	IT-related risk remedial action plans	P04, AI6	P09-02

Source: ISACA, *COBIT Process Assessment Model (PAM): Using COBIT® 4.1*, USA, 2011

D. THE CONTROL LIFE CYCLE

Introduction

This section provides an overview of the control life cycle as well as the related concepts and principles.

The relative mix and importance of process, application and general controls will be unique to each enterprise.

Exhibit 5.6: The Control Life Cycle

The control life cycle maps out the various phases in the life of a control from the initial selection/design, through development, implementation, maintenance and disposal of the system.

The risk practitioner must ensure that the controls are properly managed throughout each phase of the life cycle.

Exhibit 5.6 shows the phases of the control life cycle.

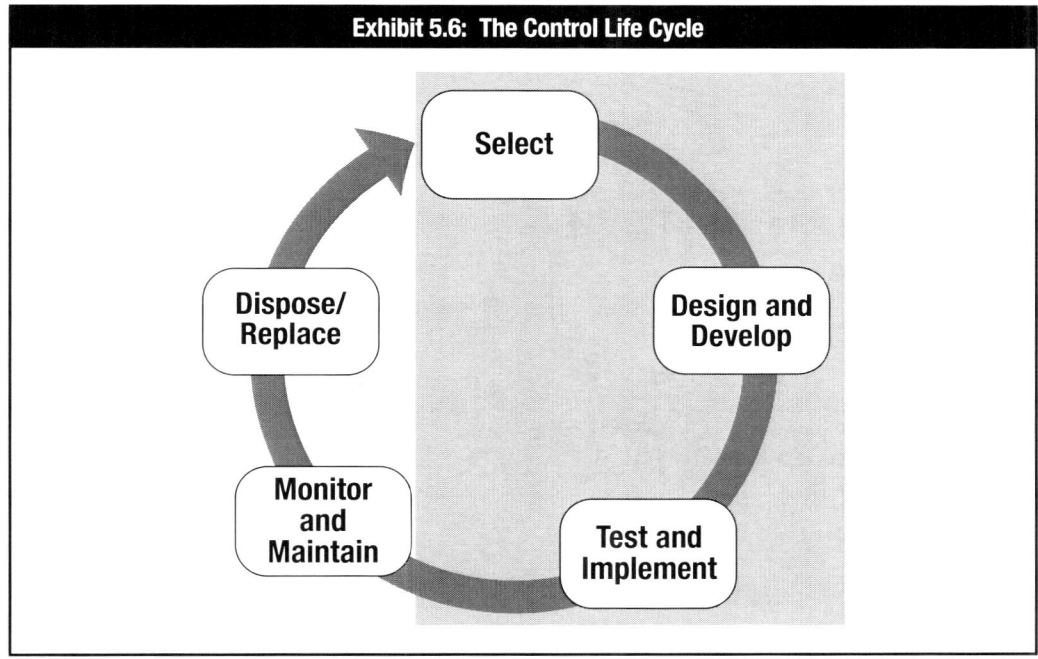

The shaded areas indicate those steps of the control life cycle that are addressed in *Domain 4—Information Systems Control Monitoring and Maintenance.*

Control Monitoring and Maintenance Within the Control Life Cycle

As **exhibit 5.6** shows, the control life cycle is a continuous process that begins with the selection of an appropriate response to identified risk.

Once a control has been selected, it must be designed into the systems and processes of the enterprise, tested, and implemented properly. Those topics are covered in *Domain 4—Information Systems Control Design and Implementation.*

> **Note:** While this chapter focuses on the monitoring and maintenance of IS controls, many principles and concepts can be applied to other control types, such as managerial or operational controls.

Part I—Risk Management and Information Systems Control Theory and Concepts
Domain 5—Information Systems Control Monitoring and Maintenance
D. The Control Life Cycle

Control Monitoring and Maintenance

This domain focuses on the last two segments of the control life cycle—Monitor and Maintain, and Dispose/Replace. Once a control has been put in place, the CRISC is concerned with continuing to monitor the control to measure its effectiveness in mitigating the risk and sustaining the risk at the level the organization deems acceptable. As environments change, business processes are modified and new threats emerge, the controls may become weak, ineffective or bypassed altogether. When risk levels reach unacceptable levels (as indicated by key performance indicators [KPIs] or key risk indicators [KRIs]) the controls must be modified or replaced to reduce the residual risk to an acceptable level once again.

When a control is no longer needed or needs to be replaced, the CRISC should ensure that the monitoring processes are updated to reflect the replacement of the control and ensure that new or compensating controls are put in place as needed.

E. INFORMATION SYSTEMS CONTROL MONITORING AND MAINTENANCE

Introduction

This section contains an overview of the IS control monitoring and maintenance process as well as the related key concepts and principles.

Relevance

Business processes and related information flows change over time to adjust to new technologies, changing market demands, new products and more complex and stringent compliance requirements. The risk practitioner must be attentive to the impact these changes may have on the effectiveness of controls.

These changes can—often quite subtly—affect the efficiency and effectiveness of controls, or even bypass them altogether. Controls that were previously effective become inefficient, redundant or obsolete and have to be removed or replaced.

Control monitoring helps maintain control effectiveness by providing timely feedback on the status of controls to ensure that controls are effectively managed throughout the control life cycle.

Control Monitoring and Maintenance Objectives

The IS monitoring and maintenance function addresses the following objectives:
- Ensure that key controls are in use, are working correctly, and are effectively mitigating the identified risk.
- Confirm that controls are being properly maintained, updates are being applied, and configurations and exceptions are handled according to policy and procedures.
- Provide assurance for management that the current risk levels are within acceptable levels.
- Ensure that management is aware of risk and provide oversight (ownership) of control activities.
- Report on the effectiveness of key controls to management and other stakeholders.
- Ensure that processes that require correction are identified and remedial actions are assigned and completed.
- Where applicable, implement preventive maintenance practices to ensure that controls continue to work as intended.

Optimizing Controls

At some point, the addition of controls begins to detract from the efficiency and profitability of a process without adding an equitable level of risk mitigation.

While controls provide a benefit, there are related costs—both tangible and intangible (opportunity costs) to the implementation and operation of controls. Controls place a burden on the enterprise—a control has a cost to implement (i.e., purchase, install), a cost to maintain (i.e., reset passwords, annual maintenance fees), a cost in productivity (i.e., slower performance), a cost to monitor (i.e., data gathering, analysis and reporting); therefore, as seen in Domains 2 and 4, the selection of controls should be done carefully with due consideration of the effect of the control on the enterprise.

Controls should only be implemented if there is a clear corresponding risk, which the control helps mitigate. As seen in *Domain 2—Risk Response* and *Domain 4—IS Control Design and Implementation*, risk response selection mechanisms are frequently based on cost-effectiveness. The purpose of Domain 5—IS Control Monitoring and Maintenance is to ensure that the cost-benefit analysis calculations remain valid, that residual risk levels are aligned with expectations and that the enterprise has an appropriate level of control.

An enterprise that attempts to implement too many controls—controls that are perceived as not justifiable, too aggravating, restrictive, confrontational or invasive—will often see their residual risk increase since employees, customers and managers ignore or bypass the controls.

Likewise, more monitoring is not necessarily better monitoring, which is why an emphasis on key controls is important; it allows an enterprise to focus its resources on the controls that matter the most.

1. THE IS CONTROL MONITORING AND MAINTENANCE PROCESS

Introduction

This section contains the IS control monitoring and maintenance process.

IS Control Monitoring and Maintenance Process Phases

The following table describes how monitoring and maintenance of IS controls ties into the overall risk management process.

IS Control Monitoring and Maintenance Process Phases		
Phase	Description	For details, refer to...

1	Prioritize risk:
	• Understand and prioritize risk to organizational objectives.
	• Identify the significant application components and flow of information through the system.
	• Understand the functionality of the application by:
	  – Reviewing the application system documentation
	  – Interviewing appropriate personnel
		Domain 1—Risk Identification, Assessment and Evaluation

2	Identify controls: • Identify key controls across the internal control systems that address the prioritized risk. • Identify the application control strengths. • Evaluate the impact of the control weaknesses. • Develop a testing strategy by analyzing the accumulated information.	Domain 2—Risk Response Domain 4—Information Systems Control, Design and Implementation

Part I—Risk Management and Information Systems Control Theory and Concepts
Domain 5—Information Systems Control Monitoring and Maintenance
E. Information Systems Control Monitoring and Maintenance

IS Control Monitoring and Maintenance Process Phases *(cont.)*

	IS Control Monitoring and Maintenance Process Phases *(cont.)*	
Phase	**Description**	**For details, refer to…**
3	Identify operational control information: • Identify information that will persuasively indicate whether the internal control system is operating effectively. • Observe and test users performing procedures.	Domain 5—Information Systems Control Monitoring and Maintenance, F. Identify and Assess Information
4	Implement monitoring: • Develop and implement cost-effective procedures to evaluate the data about control operations.	Domain 4—Information Systems Control, Design and Implementation Note: The same core principles apply as to IS control implementation. Domain 5—Information Systems Control Monitoring and Maintenance, H. Implementing Control Monitoring Processes
5	Report results.	Domain 3—Risk Monitoring, C. Essentials of Risk Monitoring Domain 5—Information Systems Control Monitoring and Maintenance, G. Tools for Monitoring **Note:** The same core principles apply to risk reporting and control reporting.
6	Maintain IS controls: • Test and deploy vendor patches as quickly as possible. – Maintain secure configuration of controls according to proscribed baselines.	Part II, Chapter 7, Configuration Management

F. IDENTIFY AND ASSESS INFORMATION

Introduction

This section describes the identification and assessment of information to enable the monitoring of IS controls.

This requires considerable thought because there are various ways in which controls operate that impact how they can be monitored.

1. KEY TERMS AND CONCEPTS

Continuous Monitoring	Note: According to paragraph 101 of volume II of the Committee of Sponsoring Organizations of the Treadway Commission (COSO) *Internal Control—Integrated Framework: Guidance on Monitoring Internal Control Systems* (USA, 2009): *Some control monitoring tools perform what is often referred to as "continuous controls monitoring." These tools complement normal transaction processing by checking every transaction, or selected transactions, for the presence of certain anomalies (e.g., identifying transactions that exceed certain thresholds, analyzing data against predefined criteria to detect potential controls issues such as duplicate payments, electronically identifying segregation of duties issues). Many of these tools serve more as highly effective control activities (detecting individual errors and targeting them for correction before they become material) than they do as internal control monitoring activities. Regardless, if they operate with enough precision to detect an error before it becomes material, they can enhance the efficiency and effectiveness of the whole internal control system and may be key controls whose operation should be monitored.*
Direct Information	The process in which the risk practitioner directly gathers the data needed to monitor system operations or controls without the assistance of another individual. This is the most trustworthy data because the analyst has direct access to the data source and thereby prevents any alteration or deletion by another person.
Indirect Information	The process in which the risk practitioner relies on another individual or process to provide data. This could lead to alteration of the data prior to submission to the analyst.
Key Control	The controls that provide the best gauge of the maturity and effectiveness of the control environment and the risk management effort. A key control is one that, if breached, would have the largest impact on the business, often by affecting more than one system, department or product line. **Note:** The ultimate definition of key controls, as well as the relative mix and importance of process, application, and general controls, will be unique to each enterprise. There are several sources of information on controls that can be used as tools in this process including frameworks, such as those provided by ISACA, ITIL or the International Organization for Standardization (ISO).
Ongoing Monitoring	The gathering of control data as part of normal processes in which control data may be supplied to the analyst on a routine scheduled basis
Separate Evaluation	The gathering of data through a separate or on demand process in which the control data is not supplied automatically or as a part of normal scheduled business processes.

Part I—Risk Management and Information Systems Control Theory and Concepts
Domain 5—Information Systems Control Monitoring and Maintenance
F. Identify and Assess Information

Exhibit 5.7: Direct and Indirect Information

Exhibit 5.7 provides a quick reference on the relative confidence that an evaluator can have in the information provided for analysis and the use of direct vs. indirect information for monitoring.

	Direct Information	Indirect Information
Ongoing Monitoring	• Typically most persuasive • Expecially valuable in high-risk areas	• Can enhance monitoring efficiency • Provides support to direct information
Separate Evaluation	• Primarily used to revalidate conclusions reached through ongoing monitoring	• Typically least persuasive • Can help scope other separate evaluation procedures

Source: ISACA, *Monitoring Internal Control Systems and IT: A Primer for Business Executives, Managers and Auditors on How to Embrace and Advance Best Practices*, USA, 2010, figure 8

2. CONTROL MONITORING TASKS

Considerations for Monitoring IS Controls

Throughout the IS control monitoring process, the risk practitioner must consider the following questions:
- How significant or meaningful is the risk that is being addressed?
- How directly does the IS control address a defined risk when considered against other potential controls?
- What is the acceptable fault tolerance of the control, or is the control working or not working? What is the nature of the control? Is it manual or automated; is it detective or preventive?
- Does the effectiveness of a control potentially depend on other controls (e.g., aspects of security or application development)?

Key Control Elements

A critical first step in this process is clearly defining the elements that make up a given key control. Although many key controls may have only a single element, others may have multiple elements.

The purpose of this exercise is to ensure that all elements of a key control are ultimately being considered when determining monitoring options.

The purpose of determining control elements is that it can often be found that a control may appear to be set correctly, but in practice is not working effectively (i.e., a firewall may be configured correctly, but if it is possible for users to circumvent or bypass the firewall then the firewall is not working effectively to mitigate the risk). The purpose of risk monitoring is not just to ensure that a control is merely in place or set correctly, but also to ensure that the control is working effectively in its real-world environment.

> **Example:** A gate across a private driveway, for example, may be set to open only to people with the correct card, but on examination it is found that people can drive around the gate and still gain access if they do not have the correct card. A control review that only checked that the gate works correctly with the correct card would not identify the fact that the risk of unauthorized access is not being mitigated by the control.

Examples of Control Elements

The following table describes examples of single- and multiple-element IS controls.

Example IS Control Elements	
Control	**Description**
Network IDs are established only after department head approval.	This control consists of a single element: the signature of the department head.
System settings require users to change passwords every 60 days.	This control likely has multiple elements: It is based on the system parameter being set at 60 days and the systems successfully enforcing the control parameters over time. **Consideration:** The password control system should be checked to see if the setting is set correctly to 60 days and, in addition, a proper control review would confirm that passwords HAVE been changed in the past 60 days and that the users are not able to bypass the controls. **Note:** When defining a monitoring plan, it is important to consider both aspects of the control.
All programming changes are subjected to comprehensive testing based on the significance of the change.	This control has multiple, complex elements: • The competence of the people defining the test plans • The determination of the "significance" of a change • The determination of "comprehensive" for the tests being performed • The completion of the tests prior to implementation **Consideration:** This control seems to be quite challenging to monitor because the selection criterion (significant change) and the related control attribute (comprehensive testing) are not measurable.

Part I—Risk Management and Information Systems Control Theory and Concepts
Domain 5—Information Systems Control Monitoring and Maintenance
F. Identify and Assess Information

Examples of Control Elements *(cont.)*	The following table describes examples of single- and multiple-element IS controls.

Example IS Control Elements *(cont.)*			
Control	**Description**		
Access controls grant system privileges on a need-to-know basis.	Access controls always have at least two elements, as described in the following table.		
	Element	**Description**	
	Programming logic	The system configuration, settings and processes within the application or infrastructure resource that: • Identify and authenticates users/other resources • Allow or disallow activity based on certain constraints, such as user ID or a media access control (MAC) address	
	Provisioning	The process to grant, maintain, change or revoke access privileges, based on changes in the environment or the user/resource	

Tasks Associated With IS Control Monitoring	The following table lists the tasks associated with planning the implementation of IS control monitoring.

Tasks Associated With IS Control Monitoring	
Task	**Task Name/Description**
1	Determine whether a specific control is already being monitored.
2	Determine the types of information available for monitoring.
3	Determine the feasibility of using automated tools.

Task 1— Determine Whether a Specific Control Is Already Being Monitored

Determine whether the monitoring functions are already in place to measure the effectiveness of a control. Other business processes or activities may be a good source of control data and metrics and make additional monitoring unnecessary.

Management must determine whether the control monitoring and reporting processes are adequate to provide for reasonable assurance that the enterprise is compliant with regulations, whether the risk is being managed effectively and whether the risk level is acceptable.

In *COBIT 5 for Information Security*, the process MEA01 *Monitor, evaluate and assess performance and conformance* describes the process of monitoring IS controls as follows:
- **MEA01 Process Description**—Collect, validate and evaluate business, IT and process goals and metrics. Monitor that processes are performing against agreed-on performance and conformance goals and metrics and provide reporting that is systematic and timely.
- **MEA01 Process Purpose Statement**—Provide transparency of performance and conformance and drive achievement of goals.

Part I—Risk Management and Information Systems Control Theory and Concepts
Domain 5—Information Systems Control Monitoring and Maintenance
F. Identify and Assess Information

Enabling a Monitoring Program	According to the *COBIT 5 for Information Security* process MEA01 *Monitor, evaluate and assess performance and conformance,* the process of setting up an IS control monitoring process requires the risk practitioner to engage with stakeholders to establish and maintain a monitoring approach to define the objectives, scope and method for measuring business solution and service delivery and contribution to enterprise objectives. Integrate this approach with the corporate performance management system. IS controls must align with the IT security and related policies of the enterprise. The control monitoring function will ensure that IT security requirements are being met; the standards are being followed and staff is complying with the policies, practices and procedures of the organization. The steps to monitoring IS controls are: 1. Identify and confirm information security stakeholders. 2. Engage with stakeholders and communicate the information security requirements and objectives for monitoring and reporting. 3. Align and continually maintain the information security monitoring and evaluation approach with the IT and enterprise approaches. 4. Establish the information security monitoring process and procedure. 5. Agree on a life cycle management and change control process for information security monitoring and reporting. 6. Request, prioritize and allocate resources for monitoring information security.
Data Analytics—Reviewing the Effectiveness of IS Controls	As described in ISACA's 2011 white paper *Data Analytics—A Practical Approach,* data analytics (DA) involves processes and activities designed to obtain and evaluate data to extract useful information. The results of DA may be used to identify areas of key risk, fraud, errors or misuse; improve business efficiencies; verify process effectiveness; and influence business decisions. The types of control monitoring and analytics may be listed as: • *Ad hoc* • Centralized/repeatable • Continuous monitoring (CM) Each of these types are examined in more detail in the following sections.

Part I—Risk Management and Information Systems Control Theory and Concepts
Domain 5—Information Systems Control Monitoring and Maintenance
F. Identify and Assess Information

***Ad hoc* Reporting**	*Ad hoc* DA is a one-use process—a starting point that may be used to help identify patterns or potential risk areas within a business system. It is typically used for an initial investigation as a way to begin to understand the business processes while becoming familiar with the data.

Ad hoc DA may be difficult to repeat if the steps performed are not well documented or if there are a number of complicated steps to perform. The documentation can serve as a starting point to move up the maturity scale for all repeatable DA and CM.

Typical *ad hoc* DA may begin with exploratory data mining, benchmarking/trending or data quality testing, but it may also include specific management-directed requests from any business area or function. The results of ad hoc DA vary from the increased knowledge of systems and processes to system improvements and cost savings.

The use of *ad hoc* DA often requires an enterprise to rely on selected skilled individuals to perform the testing; such reliance may increase risk, due to attrition and the lack of knowledge transfer.

Ad hoc DA inevitably leads to the next maturity stage: repeatable DA. Repeatable DA is predefined and scripted; it is designed to perform the same tests on similar data (e.g., data from a different time period) on a scheduled basis. The benefits include consistency, efficiency and more effective corrective actions. |
| **Centralized/Repeatable Data Analytics (DA)** | The next step up the maturity scale from repeatable DA is a centralized approach for the development, storage and operation of repeatable DA. In this approach, a central repository is established for repeatable DA programs and standard data files; standards for DA development are documented and DA applications are set up and scheduled to run against the centralized data on a regular basis or on demand.

Data to be analyzed may either be pushed to the repository or extracted directly from different sources as needed, and the analytic results themselves are stored in the repository. This centralized approach has the following advantages:
• The process is more consistent, efficient and repeatable.
• The results are more reliable and consistent.
• The chance of multiple variations being scattered across individual machines is reduced.
• The potential negative impact on the performance of production applications is minimized.
• Data security is improved.
• Backups are more easily performed.
• The use of a centralized repository increases the overall performance of workstations.
• Access to analytics and results is available to more people, increasing productivity and improving the use of supporting reference materials, analytic sample logic and source data. |
| **Continuous Monitoring (CM)** | CM marks the highest point on the maturity scale. At this stage, analytics are fully automated and running at regularly scheduled intervals and may be embedded directly into a production system. A continuous run of analytics enables the immediate identification of potential exception transactions. Many commercially available CM packages include sophisticated web interfaces, email notifications, workflows, remediation tracking, dashboards and/or heat maps.

The purpose of continuous monitoring is to continuously monitor, benchmark and improve the IT control environment and control framework to meet organizational objectives. |

Exhibit 5.8: Continuous Risk Assessment and Continuous Control Assessment	The relationship between continuous risk assessment and continuous control assessment is shown in **exhibit 5.8**.  Continuous control assessment enables more accurate and effective risk assessment.
Task 2—Determine the Types of Information Available for Monitoring	Work with the stakeholders to define, periodically review, update and approve performance and conformance targets within the performance measurement system: • Define information security performance targets consistent with overall IT performance standards. • Communicate information security performance and conformance targets with key due diligence stakeholders. • Evaluate whether the information security goals and metrics are adequate, i.e., specific, measurable, achievable, relevant and time-bound.
Issues Related to IS Monitoring	The effectiveness of IS monitoring is dependent on the: • Timeliness of the reporting • Skill of the data analyst • Quality of monitoring data available • Quantity of data to be analyzed Steps to IS control monitoring are: 1. Collect and analyze performance and conformance data relating to information security and information risk management (e.g., information security metrics, information security reports). 2. Assess the efficiency, appropriateness and integrity of collected data.
Task 3—Determine the Feasibility of Using Automated Tools	Determine the feasibility of using automated monitoring tools, which may add value to the monitoring process in situations in which the information that supports a conclusion that a control is in place and operating resides directly or indirectly in electronically stored data. A control to verify that only authorized changes can be made to system applications, configurations or files may use a tool that compares historical file sizes to current file sizes to detect any changes and alert the administrator to a potentially unauthorized change. Monitoring tools can focus on many, but not all, dimensions of internal control.

G. TOOLS FOR MONITORING

Software-based Monitoring Tools	The following list describes variations in software tools: • Some tools focus on a very narrow set of control issues; others cover a broader range. (i.e., there are many types of firewalls—some that provide much more comprehensive controls than others) • Software vendors use varying names to describe what their tools provide. • Software tools that do the same basic thing can be called by different names, and conversely, tools with the same basic names can do entirely different things (i.e., there are many different types of VPNs depending on the vendor supplying the tool). • Some tools are designed to focus on monitoring or auditing internal controls; others are integrated into IS operating processes, but provide essential information that is relevant to monitoring. • There are several software-based tools that can assist the enterprise in gathering, analyzing and reporting on control data. Many of these tools will draw data from multiple systems and aggregate the data into common files or databases. The tool will then correlate activity on different systems and seek to discover related events (or activity) that span more than one system.
Considerations for Selecting Monitoring Software	Other things to consider when selecting monitoring software tools include: • Many tools are likely to be considered multipurpose in that they are designed to provide a form of operational process or control. • Depending how they are used, these tools can also provide direct or indirect information to management that controls are in place and operating. • The presence of a tool should not be an assumption that the tool is being used. • To be considered effective for controls monitoring, there needs to be a focus on communicating results/issues from these tools to those individuals responsible for acting on findings.
Limitations of Monitoring Tools	Although automated monitoring tools can be highly effective in a number of situations, they are not without limitations and generally cannot: • Determine the propriety of the accounting treatment afforded individual transactions because this must be determined based on the underlying substance of the transaction itself • Address whether an individual transaction was accurately entered into the system; rather, they can deal only with whether the transactions met internal standards for acceptable transactions (for example, it was within a valid range of values) • Determine whether all relevant initial transactions were entered into systems in the proper period because this is typically dependent on human activity
Types of Monitoring Tools	There are various types of software tools that can be used to perform different types of control monitoring. These tools are organized into the following groups based on the focus of the tool as it relates to monitoring internal control: 1. Transaction data 2. Conditions 3. Changes 4. Processing integrity 5. Error management **Reference:** For details on each of these tools, see the applicable examples in this section.

1. TOOLS FOR MONITORING TRANSACTION DATA

Tools for Monitoring Transaction Data

Tools used in this fashion tend to deal with:
- IS controls designed to focus on the integrity of transaction processing
- Processed data, which confirms the IS controls related to the completeness of processed data

The following table lists different uses of tools focusing on transaction data.

> **IMPORTANT:** Tools cannot evaluate whether all transactions were entered into a system, but they can evaluate the completeness of transactions being moved from one system to another or whether transactions are likely to be duplicated.

Tools for Monitoring Transaction Data	
Tool Use	**Description**
Comparing transaction data against defined rule sets	The intention is to identify instances in which the controls over a process or system are not working as intended. This category of tools for monitoring information can be used to: • Highlight exceptions and/or anomalies • Analyze unusual trends in activities, values and volumes • Compare balances or details between two systems or between distinct parts of a process
Ad hoc reporting	Tools in this category can take the form of *ad hoc* programs that are run periodically against the defined: • Population (either a sample of the data population or the entire population) • Programs that are implemented into a processing environment to continuously monitor a specific set of transactions
Data correlation across multiple sources	Given their ability to be used in a variety of scenarios and to correlate data and information from multiple sources, these tools can be used in multiple situations to: • Highlight significant manual or unusual automated journal entries so that financial management can ensure that such entries were proper and approved. • Search for unusual or duplicate payments so that management can ensure that controls over disbursements are working effectively. • Analyze unusual activity as part of management's fraud prevention activities. • Determine the frequency of supervisory overrides. • Validate that all transactions fall within a specific control range.

Part I—Risk Management and Information Systems Control Theory and Concepts
Domain 5—Information Systems Control Monitoring and Maintenance
G. Tools for Monitoring

2. TOOLS FOR MONITORING CONDITIONS

Tools for Monitoring Conditions

Software tools in this category:
- Examine specific settings or parameters that control how an application or infrastructure resource is configured
- Compare the configuration information to either baseline information, a prior analysis or both to determine whether they are consistent with the enterprise's expectations
- May operate periodically (frequently described as scanning-based)
- Can be embedded in the process as either software or hardware (frequently described as agent-based)

Rationale: This increases the speed and effectiveness of the monitoring process while simultaneously allowing it to be performed on a more frequent or even continuous basis.

The following table describes where tools for monitoring conditions can best be applied.

Application of Tools for Monitoring Conditions	
Situation	**Example**
A large number of conditions across different systems	Numerous parameter settings that affect internal control in an integrated enterprise resource planning (ERP) system or multiple different "instances" or versions of the same ERP package that support different business units
Few parameters across many systems	Analysis of security parameters across relevant servers in a global network
Complex conditions or a high volume of records	A large number of users of a single application or multiple applications to be evaluated for appropriate segregation of responsibilities as defined by application access rights

These types of tools deal with integrity controls. There are also dimensions of these controls that support access and authorization-related controls. This is particularly true for tools that analyze security-related parameters because these parameters are relevant to ensuring that only those authorized to utilize resources can gain access.

3. TOOLS FOR MONITORING CHANGES

Tools for Monitoring Changes

The following table describes an example of change-monitoring tools. They can be considered an extension of the tools that focus on "conditions."

Examples of Tools for Monitoring Changes	
Topic	**Description**
Tool focus	These tools: • Usually operate on a continuous basis (i.e., they are agent-based) • Are specifically designed to identify and report changes to critical resources, data or information, thereby making it possible to verify that changes are appropriate and authorized Change within IT resources is pervasive, continuous and unavoidable, so the purpose of these controls is to make sure that the changes are authorized and implemented correctly.
Preventive vs. detective change control detection	When controlling change is considered a key control, enterprises have some form of change control that includes both a: • Preventive control that will only permit authorized personnel to make changes • Detective control, whereby all changes are recorded, reviewed and potentially approved by someone independent of those making changes
Database structure changes	Changes that have been made to database structures can be logged by the database environment itself and then subsequently reviewed by management.

Considerations for Monitoring Changes

The following table provides examples of considerations for monitoring changes.

Examples of Tools for Monitoring Changes	
Topic	**Description**
Control mechanism for changes to programming or ERP configuration	Certain ERP environments provide their own control mechanism for controlling programming changes or changes to the ERP configuration. To the extent that this exists, the reports of these changes can be used both as a detective control and a means by which management monitors controls.
Recording changes	Although this approach to controlling change works effectively, it is dependent on the ability of the resource being changed to provide an effective means to record changes and, thereby, ensure that any changes can be reviewed and approved. The risk practitioner will need to ensure that the recording logs cannot be altered or deleted by users or administrators.
Essential considerations	• Not all IT resources provide the capability for recording changes. • In large IT environments, the number of individual resource components that would need to be analyzed on a detective basis can be overwhelming. • Using native logging capabilities of some resources may affect system performance unacceptably. • In certain high-risk areas, management may not wish to rely onuse the native features of certain resources because of the potential impact that such features could have on systems performance and the simplicity of turning off these features.

Part I—Risk Management and Information Systems Control Theory and Concepts
Domain 5—Information Systems Control Monitoring and Maintenance
G. Tools for Monitoring

Considerations for Monitoring Changes (cont.)

Examples of Tools for Monitoring Changes	
Topic	**Description**
Category application with appropriate conditions	When these conditions are present, tools in this category can: • Become mechanisms for both control and monitoring • Identify changes that have been made to infrastructure resources, databases, application programs, and security rights and permissions • Provide visibility for all changes so that, ultimately, they can be validated independently, which is, in essence, monitoring that the underlying change control process is working • Provide alerts when certain types of mission-critical changes are being made so that there is transparency throughout the enterprise and necessary actions can be taken on a timely basis • Provide a verifiable audit history for direct information of control functionality over time As with tools that focus on conditions, these largely focus on integrity and authorization-related controls.

4. TOOLS FOR MONITORING PROCESS INTEGRITY

Tools for Monitoring Process Integrity

The following table describes an example of automated tools to evaluate processing integrity that:
• Are designed to verify and monitor the completeness and accuracy of the various situations
• May occur in the overall IT process

Tools for Monitoring Process Integrity	
Topic	**Description**
Tool focus	Automated tools focus on balancing and controlling data as they progress through processes and systems.
Activity performance	Tools in this category can: • Reconcile financial totals and/or transaction/record counts from one file or database to another file or database within the same or between different application or operating systems (OSs) • Ensure data file, record and field accuracy as data move across systems and processes • Monitor information from source systems and/or data warehouses to the general ledger (GL)
Design to maintain an audit trail	Because these systems operate independently of transaction processing, they can be designed to maintain an audit trail of key information for monitoring or trending studies.
Control mechanism	Within an enterprise, these tools are typically considered a control mechanism. Depending on how management uses them, these tools could be considered mechanisms for monitoring controls as they relate to information processing.

5. TOOLS FOR ERROR MANAGEMENT AND REPORTING

Tools for Error Management and Reporting

Error management tools are designed to detect transactions that do not meet defined criteria so that they can be corrected and reprocessed. Considerations are presented in the following table.

Examples:
- An automotive parts supplier may receive a technically valid electronic data interface message describing an authorized shipping schedule; however, the message may have invalid order identification.
- A telecommunications provider may receive message information from its telephone switching systems on customers whose information has not made it through the process of being added to the billing system.

The fact that invalid transactions are rejected is frequently considered an application control. These systems frequently capture the transactions in an area where they can later be reprocessed after correcting the cause of the error.

Management's monitoring of the volume and resolution of activity in these systems or accounts provides both direct and indirect information that the control is working effectively.

> **IMPORTANT:** These capabilities typically are part of an existing information system rather than an add-on vendor solution.

Considerations for Error Management Tools	
Topic	**Description**
Best approach	Determining the best approach for an enterprise should be driven by the importance of the control and the related risk that the control is designed to mitigate. There are trade-offs with respect to these solutions that need to be considered carefully by management in the process of determining whether tools of this nature will work in their environment.
Parameter specifics	Many controls that are built into systems are controlled through the configuration of a specific set of parameters within both IT infrastructure resources and application systems. A significant risk during error handling is that an error is not managed in the same way as a normal transaction. The CRISC must ensure that separation of duties, monitoring or other controls are in place for personnel handling errors just as they are for normal transactions.

6. CONTINUOUS MONITORING

Introduction

Continuous monitoring:
- Is becoming important in today's electronic business (e-business)
- Provides a method to collect evidence on system reliability while normal processing takes place
- Allows for monitoring the operation on a continuous basis
- Gathers selective monitoring evidence through the computer system

> **Note:** If the selective information collected by the computer technique is not deemed serious or material enough to warrant immediate action, store the information in separate monitor files for verification by the monitor at a later time.

Relevance

Continuous monitoring techniques are important tools:
- When they are used in time-sharing environments that process a large number of transactions, but leave a scarce paper trail
- Because they improve the security of a system by permitting IS monitors to evaluate operating controls on a continuous basis without disrupting the enterprise's usual operations
- Because they report a system misuse in a timely fashion

Results:
- This reduces the time lag between the misuse of the system and the detection of that misuse.
- IS monitoring allows management to gain greater confidence in a system's reliability due to the realization that failures, improper manipulation and lack of controls will be detected on a timely basis.

Benefits

The continuous monitor approach:
- Cuts down on paperwork
- Reduces monitoring cost and time
- Leads to the conduct of an essentially paperless monitoring process
- Allows real-time reporting on significant errors, trends or other irregularities that may require immediate action
- Improves accountability of local management for the enforcement and monitoring of controls

Selection of Continuous Monitoring Tools and Techniques

The selection and implementation of continuous monitoring tools depends, to a large extent, on the complexity of an enterprise's computer systems and applications and the IS monitor's ability to understand and evaluate the system with and without the use of continuous monitoring techniques.

The risk practitioner must recognize that continuous monitoring techniques are not a cure for all control problems and that the use of these techniques provides only limited assurance that the information processing systems examined are operating as they were intended to function. Continuous monitoring does not replace the need for periodic reevaluation of controls or the requirement for audits.

Automated Monitoring Techniques

The following table describes five types of automated evaluation techniques applicable to continuous monitoring.

Types of Automated Evaluation for Continuous Monitoring	
Technique	**Description**
Systems Control Audit Review File and Embedded Audit Modules (SCARF/EAM)	The technique involves embedding specially written monitor software in the enterprise's host application system so that the application systems are monitored on a selective basis.
Snapshots	This technique: • Takes what may be termed as "pictures" of the processing path that a transaction follows, from the input to the output stage • Tags transactions by applying identifiers to input data and recording selected information about what occurs for the subsequent review
Monitor hooks	This technique: • Embeds hooks in application systems to function as red flags if a suspicious condition occurs • Induces IS monitors to act before an error or irregularity gets out of hand
Integrated test facility (ITF)	In this technique: • Dummy entities are set up and included in a monitor's production files. • The IS monitor can make the system either process live transactions or test transactions during regular processing runs and have these transactions update the records of the dummy entity. • The operator enters the test transactions simultaneously with the live transactions that are entered for processing. • The monitor then compares the output with the data that have been independently calculated to verify the correctness of the computer-processed data.
Continuous and intermittent simulation (CIS)	In this technique: • During a process run of a transaction, the computer system simulates the instruction execution of the application. • This simulator follows the table below as each transaction is entered to decide whether the transaction meets predetermined criteria.

If the transaction …	The simulator …
Meets the predetermined criteria,	Monitors the transaction.
Does not meet the predetermined criteria,	Waits until it encounters the next transaction that meets the criteria.

Part I—Risk Management and Information Systems Control Theory and Concepts
Domain 5—Information Systems Control Monitoring and Maintenance
G. Tools for Monitoring

Exhibit 5.9: Continuous Monitoring Tools—Advantages and Disadvantages

Exhibit 5.9 describes the relative advantages and disadvantages of the various continuous monitoring tools.

	SCARF/EAM	**ITF**	**Snapshots**	**CIS**	**Audit Hooks**
Complexity	Very high	High	Medium	Medium	Low
Useful when:	Regular processing cannot be interrupted.	It is not beneficial to use test data.	An audit trail is required.	Transactions meeting certain criteria need to be examined.	Only select transactions or processes need to be examined.

Exhibit 5.9: Continuous Monitoring Tools—Advantages and Disadvantages

H. IMPLEMENTING CONTROL MONITORING PROCESSES

Introduction

The section describes the implementation of monitoring processes.

The monitoring process requires an enterprise to design and implement ongoing monitoring procedures and/or separate evaluations that are needed to gather and analyze persuasive information to support conclusions about the effectiveness of internal control.

1. DETERMINE MONITORING METHOD AND FREQUENCY

Factors to Determine Monitoring Method and Frequency

The management of an enterprise uses a unique judgment regarding monitoring methods and frequency after factoring in:
- Objectives
- Risk
- Controls
- The persuasiveness of information that is available about its controls

Controls Already Monitored

Certain controls may already be monitored as part of an existing process.

In these situations, activities may be limited to ensuring that the frequency of monitoring is sufficient for management's purpose.

Questions to Determine Control Monitoring Method

The following table outlines the questions that need to be asked and answered to determine monitoring method.

Questions to Determine Control Monitoring Method	
Question	Description
What is the nature of the control?	• The answer depends on the specific elements of the control, as discussed previously. • Generally, automated controls require less frequent monitoring of their automated aspects. • Once the control is verified to be working properly, automated aspects are unlikely to change—assuming that effective change management controls are in place.

How will monitoring be used?	• A properly designed and executed monitoring program helps support both internal needs to:
	  – Ensure that objectives and external assertions are being achieved
	  – Provide persuasive information that an internal control operated effectively at a point in time or during a particular period
	• This duality of purpose ultimately should play a role in decisions made as to the frequency and method of evaluation and to who should be the evaluator.

Part I—Risk Management and Information Systems Control Theory and Concepts
Domain 5—Information Systems Control Monitoring and Maintenance
H. Implementing Control Monitoring Processes

Questions to Determine Control Monitoring Method *(cont.)*

Questions to Determine Control Monitoring Method *(cont.)*	
Question	**Description**
Who is most appropriate to evaluate a given control?	• The answer provides needed input to the decision about how monitoring should best be performed. • There are several criteria that must be considered that relate to the competence and objectivity of the "evaluator" of an internal control. **Example:** If management decides that the chief information officer (CIO) should evaluate a control, it is most likely that a decision will also be made that the CIO's monitoring would be part of an ongoing management process, not a separate evaluation.
What is the mix of direct and indirect information?	<table><tr><th>When monitoring using …</th><th>Then …</th></tr><tr><td>Direct information,</td><td>Perform monitoring on whatever schedule is deemed sufficient by management.</td></tr><tr><td>Indirect information,</td><td>Apply additional separate evaluations as they may be required more frequently.</td></tr></table> The specifics for each monitoring procedure need to be developed for both ongoing and separate evaluations. This activity focuses largely on the development of activities and reporting responsibilities and the training and change management activities for those impacted.
Can the evaluation be an ongoing process?	• To the extent that controls can be monitored through an ongoing process, it is more likely that control issues will be highlighted on a timely basis. • Not all monitoring activities, however, lend themselves to being easily implemented as part of an ongoing process. • This decision ultimately needs to be made by management for each unique situation.

2. SELECT AND IMPLEMENT AUTOMATED MONITORING TOOLS

Acquisition of Monitoring Tools

The selection and implementation of monitoring tools should follow the same basic technology acquisition and implementation processes that an enterprise uses for any of its business systems.

Part I—Risk Management and Information Systems Control Theory and Concepts
Domain 5—Information Systems Control Monitoring and Maintenance
H. Implementing Control Monitoring Processes

Selection Criteria for Monitoring Tools

The following table provides several criteria to consider when determining whether and how to use a given tool.

> **IMPORTANT:** Each of the criteria listed below must be considered in terms of the benefit of automating the monitoring process compared to alternatives.

Criteria for Tool Use	
Criterion	**Description**
Sustainability	• It is important that monitoring software be able to change at the same speed as technology applications and infrastructure to be effective over time. • If monitoring tools do not change at the same pace as the underlying applications, there is a risk that new or changed controls in the application will not be monitored.
Scalability	• Monitoring tools have to be able to keep up with the growth of an enterprise and meet anticipated growth in process, complexity or transaction volumes.
Customizability	• Many software products come with rules that have been defined by the vendor and may not meet management's business requirements. • For software to be effective, it must be customizable to the specific needs of an enterprise. • Ideally, the ability to customize should be built into the software so that the end users can adapt the software—the cost of custom programmer changes to parameters or rule sets can be prohibitive and impact scalability.
Ownership	• Someone in the enterprise must "own" the tool and must be able to effectively identify organizational changes and opportunities that may require modifications to the tool. • Effort takes time and a certain level of expertise for the enterprise to maximize the benefits from its investment in the tool.
Impact on performance	• Because monitoring tools must operate in a manner that coexists with existing systems and data, it is important to understand the impact of a monitoring tool on base system's performance and capacity. • Embracing a monitoring tool that adversely impacts business processing capability is not sustainable in most enterprises.
Usability of existing tool	• If an operating tool currently in use can be leveraged to perform monitoring activities, there are fewer considerations to be made than when an enterprise is acquiring a new tool. Tools that may integrate or be readily interoperable with other tools may be a preferred choice.
Tool complexity	• Tool complexity and challenging user interfaces may affect its usage. How much training would be required by the administrators or users of the tool?
Transferability	• This refers to the ease of shifting the use of the tool from one user (group) to another.
Cost-benefit (return on investment [ROI])	• Enterprises must understand the total cost of ownership (TCO) when purchasing monitoring tools. • Software licensing costs are only a very minor part of the total cost of software ownership.

3. CLARIFY REPORTING REQUIREMENTS AND EXPECTATIONS

Clarify Reporting Requirements and Expectations

Because the controls being monitored often cross organizational objectives and business functions, one of the most important aspects of this activity is clearly defining how exceptions should be reported and to whom.
Report exceptions or deficiencies identified in the monitoring process to the individual:
- Who is in a position to take corrective actions
- With overall responsibility for:
  - Controls in a given area
  - A given set of objectives

> **Note:** This requires clear ownership for risk and control monitoring, The risk practitioner must ensure that there is a declared owner for risk and for ensuring that the controls are being operated and monitored effectively. Consideration must also be given to maintaining the confidentiality of some reporting metrics and ensuring that only authorized personnel are advised of control monitoring results.

Considerations for Correcting Deficiencies

It is worth noting that the process of correcting deficiencies may be considered a management activity rather than an element of internal control. Classification aside, corrective actions should be undertaken when control deficiencies are severe. It is crucial that:
- The right people receive the information necessary to enable corrective action.
- Management retains sufficient oversight to gain assurance that the corrective action has been taken.

Considerations for Prioritization of Corrective Actions

The following table describes several factors that may influence an enterprise's prioritization of identified exceptions and deficiencies as described in the Committee of Sponsoring Organizations of the Treadway Commission (COSO) 2009 *Internal Control—Integrated Framework: Guidance on Monitoring Internal Control Systems* (USA).

Monitoring Factors and Considerations	
Factor	**Consideration**
Errors	The likelihood that the deficiency will result in an error or other adverse event: • The fact that a deficiency has been identified means that there is at least some likelihood that an error could occur. • The greater that likelihood, the greater the severity of the control deficiency will be.
Other controls	• The effectiveness of other, compensating controls: • The effective operation of other controls may prevent or detect an error that results from an identified deficiency before that error can materially affect the enterprise. • The presence of such controls, when monitored, can provide support for reducing the severity of a deficiency.
Deficiencies and organizational objectives	The potential effect of a deficiency (vulnerability) on organizational objectives: • As the effect of an identified deficiency increases, its severity increases.

Monitoring Factors and Considerations *(cont.)*	
Factor	**Consideration**
Effect of deficiencies on other objectives	The potential effect of the deficiency on other objectives: • Beyond consideration of the factors listed previously, enterprises may consider the effect of a deficiency on their overall operating effectiveness or efficiency. **Example:** An identified deficiency may prove to be immaterial in relation to the financial reporting objective, but it may cause inefficiencies that warrant correction in relation to operational objectives.
Multiple deficiencies	The aggregating effect of multiple deficiencies: • When multiple deficiencies affect the same or similar risk factors, their mutual existence (the aggregated effect of the deficiencies) increases the likelihood that the internal control system may fail, thus increasing the severity of the identified deficiencies (sometimes known as the "perfect storm").

Considerations for Prioritization of Corrective Actions *(cont.)*

Prioritizing Deficiencies

Determining who prioritizes the deficiencies is a matter of judgment. Enterprises likely will consider the:
- Size and complexity of the enterprise
- Nature and importance of the underlying risk
- Experience and authority of the people involved in the monitoring process
- Regulatory or compliance requirements

The prioritization of identified deficiencies should be performed by appropriately competent and objective personnel that hold positions of responsibility within the enterprise.

Determining Where Controls Are Needed

Those responsible for implementing controls should focus on root causes of problems when:
- Analyzing a control failure resulting from monitoring
- Identifying where controls are needed

Introduction to the Cause-and-effect Diagram

A cause-and-effect diagram:
- Is generally used to explore all the potential or real causes (or inputs) that result in a single effect (or output)
- Can help identify root causes and areas where there may be problems
- Assists in comparing the relative importance of the different causes to ensure effective prioritization and coordination of remediation efforts

Note: Arranging causes according to their level of importance or detail results in a depiction of relationships and hierarchy of events.

Part I—Risk Management and Information Systems Control Theory and Concepts
Domain 5—Information Systems Control Monitoring and Maintenance
H. Implementing Control Monitoring Processes

Building a Cause-and-effect Diagram	The following table outlines the steps to successfully build a cause-and-effect diagram.

Steps to Build a Cause-and-effect Diagram	
Step	**Action**
1	Agree on the effect or problem statement before beginning.
2	Identify major categories of failures. **Note:** This may be equipment, policies, procedures and people for an IT process, but it can include other categories as well, depending on the circumstances of the environment.
3	Link the potential or observed control failures to the categories.
4	Discuss the control failure points with the project team.
5	Revise the monitoring process and repeat testing as necessary.

Cause-and-effect Diagram Process Considerations	When following the cause-and-effect diagram process, there are additional considerations, including the need to: • Be concise. • Think carefully about what could be causes and add them to the tree for each node. • Pursue each line of causality back to its root cause. • Consider grafting relatively empty branches onto others or splitting up branches that become crowded. • Identify which root causes are most likely to merit further investigation.
Exhibit 5.10: Cause-and-effect Diagram Based on Software Change Failure	**Exhibit 5.10** depicts an example of a cause-and-effect diagram based on the failure of a software change.

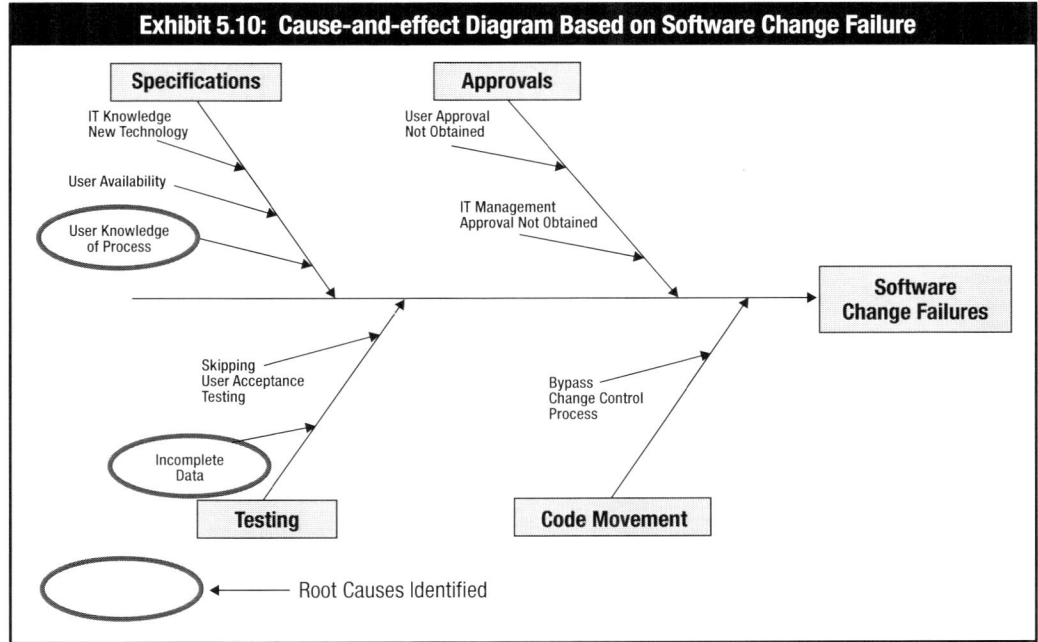

Source: ISACA, *Monitoring Internal Control Systems and IT: A Primer for Business Executives, Managers and Auditors on How to Embrace and Advance Best Practices*, USA, 2010, figure 10

I. IMPLEMENTING CONTROL MAINTENANCE PROCESSES

Introduction This section describes the implementation of control maintenance processes.

The maintenance process requires an enterprise to update, maintain, adjust and reconfigure controls according to changing risk levels and emerging threats or vulnerabilities. The maintenance of controls is done either on a regular basis to keep the controls operating correctly—including applying patches, new rules etc., and on an exception basis to respond to incidents or vulnerabilities discovered through control monitoring.

1. DETERMINE CONTROL MAINTENANCE REQUIREMENTS

Control Maintenance Objectives Effective risk management requires that the IS controls be maintained over time to ensure that they continue to provide effective risk reduction and protect the enterprise from changing or emerging risk.

Control Maintenance Activities Control maintenance includes activities such as:
- Patch management
- Configuration management
- Exception management
- Documentation

Maintenance of Vendor Supplied Controls Some controls deployed across the enterprise will be provided by hardware and software vendors. Examples of these controls may include:
- Intrusion Detection and Intrusion Prevention Systems (IDS/IPS)
- Firewalls
- Routers
- Access Control Systems
- Anti-malware (Antivirus) Systems

Maintenance of these controls starts with an inventory of what controls the enterprise has, and where the controls are located.

The enterprise is responsible for ensuring that relevant vendor released patches are deployed as quickly as possible across the enterprise. However all changes must be tested (as much as possible) prior to deployment since changes may not work properly or have sometimes disabled other associated systems or processes.

Vendor patches may be released in order to repair a vulnerability in the system, update malware or attack signatures, or provide enhanced functionality.

A critical challenge with deploying patches across multiple devices is to ensure that all systems are updated, that the rollout of the patch does not congest networks or unduly impact production activity.

Part I—Risk Management and Information Systems Control Theory and Concepts
Domain 5—Information Systems Control Monitoring and Maintenance
I. Implementing Control Maintenance Processes

Maintenance of IS Control Configurations	Many information system controls have customized settings that regulate the operation of the control according to the requirements of the enterprise's security policy.
	IS control configurations are not static, but will often require tuning or modification to meet changing risk scenarios or emerging threats. The effectiveness of IS controls is maintained through adjusting the control configuration settings; however, it is critically important to ensure that any changes to control configuration are subject to a change control or configuration management procedure that will prevent unauthorized or untested changes and will help detect any errors in the new configuration prior to implementation.
	These control configuration settings are often found in access control lists (ACLs), configuration tables, or in settings that can only be altered by a person with administrator-level access.
	The risk practitioner needs to ensure that the changes to these settings (the configuration of the device), are carefully tested, controlled and documented.
	Without proper controls over changes to control configurations, changes are made without proper approval or documentation and the enterprise cannot determine whether the control is operating correctly or not. Poor management of the rules may impact the performance of the control.
Exception Handling	It is not uncommon to encounter a situation where a user or process requires an exception to a policy that is being enforced by an IS control. In such an instance, the exception must be reviewed, approved and documented prior to granting the exception. The exception may be required for business reasons or to enable testing of a new application. The exception should be evaluated for its potential impact on risk levels prior to approval and implementation.
	Exceptions to the rules that govern the operation of control should only be made when necessary, and as infrequently as possible. Too many exceptions may be an indicator that the policy is not appropriate or that the control is not set up correctly or does not align with business priorities.
	When an exception is required, the request should be submitted formally, in writing, and then require approval by a manager with the appropriate level of authority.
	Exceptions to policy (and the needed changes to the configuration of the IS control) must be documented and reviewed on a periodic basis to ensure that the exception is still required, and removed once it is no longer needed.
Maintenance of Control Documentation	Current and concise IS control documentation is an important input into the risk management process. The documentation should reflect the current configuration of individual controls as well as the control architecture. This is why it is common to hear the documentation described as "living"—they are always being kept alive and accurate.

2. IMPLEMENT CONTROL MAINTENANCE PROCESS

Control Maintenance Process

Change management is the process that controls changes to IS controls and production systems. The objective of change management is to ensure that unauthorized changes are not permitted, authorized changes are reviewed for impact on risk prior to implementation, all changes are logged and validated to ensure that the change was implemented properly in the authorized manner, and that the effect of the change is monitored for effectiveness.

The flow chart below outlines a process to manage changes to control configuration.

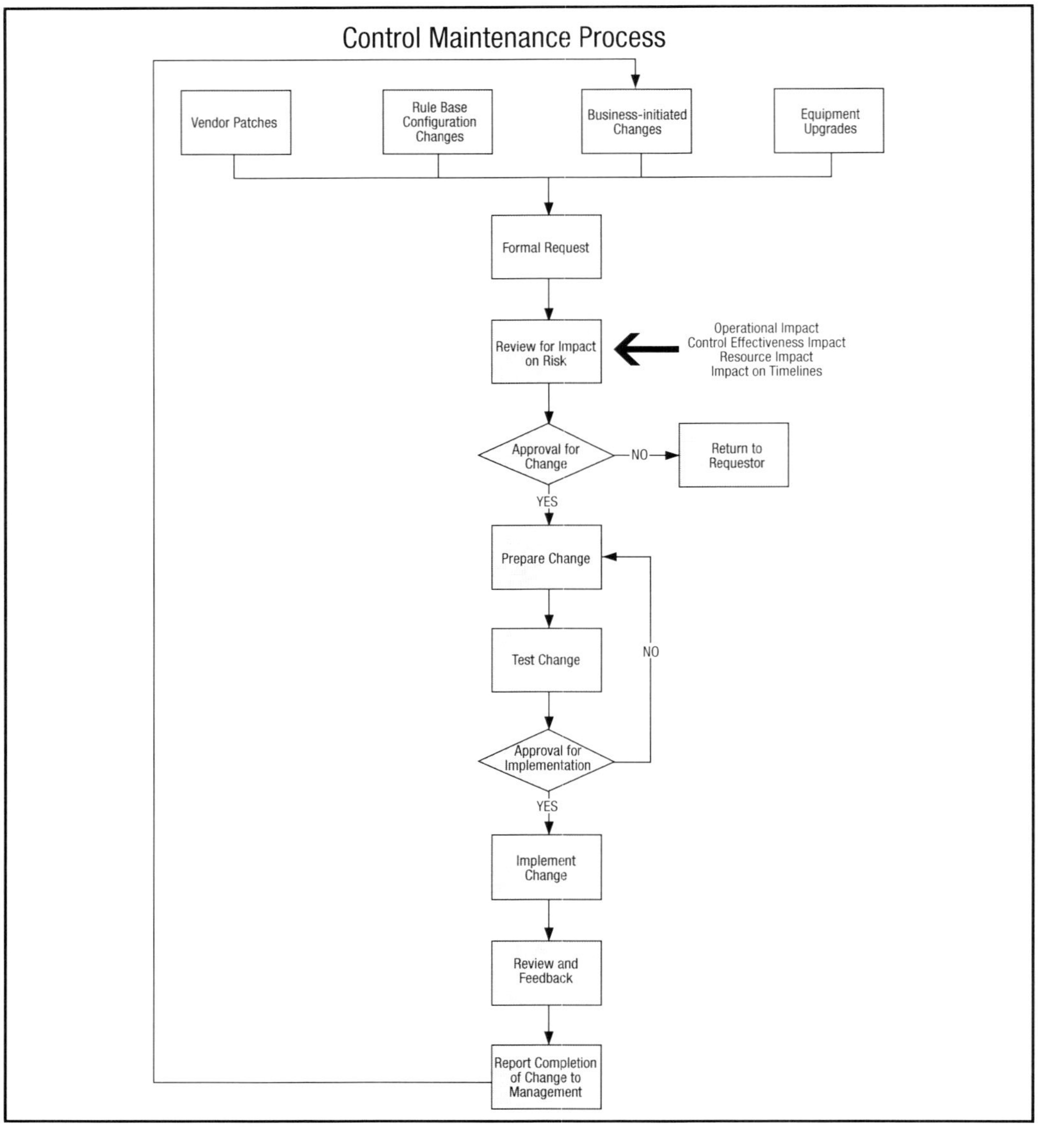

Control Maintenance Process

The Configuration Management Database (CMDB) and all other relevant documentation should be updated to reflect the changes made.
- Establishing a formal control maintenance process should avoid the problem of undocumented or improper changes to controls. The formal process mandates testing, review and documentation of all changes both prior to, and following the implementation of the proposed change.

Monitoring and Maintenance of IS Controls

During the previous phases of the risk management process, IS controls were designed and implemented. It is important that regular reviews of those controls be conducted to ensure that the IS controls are meeting their objectives and effectively mitigating the risks—as intended when the controls were selected and installed.

Reviews may be conducted on a regular scheduled basis or on an "as-needed" basis—perhaps as mandated by an incident, an emerging threat or as a result of a KRI threshold being reached.

The results of the IS control reviews should be documented and reported to management. The results may be used for further risk assessment, audit, investigation or compliance reporting as required.

Periodic Control Testing

IS control maintenance is a subset of overall risk monitoring and should be conducted in conjunction with other control monitoring. An effective risk response on one that integrates all forms of control—administrative, technical (typical IS controls) and physical control measures into an interoperable and complete response solution.

IS controls may be network based, application based, hardware based, or user-based. The tests may be conducted by internal review and testing teams (including security, audit, and IT staff) or external parties (such as penetration testing specialists, compliance auditors, or external auditors).

The purpose of the control testing is to ensure that the controls are:
- Implemented as designed
- Operating correctly
- Achieving the desired result

As issues are discovered in the testing process, the controls must be maintained and adjusted to address the issues and resolve outstanding vulnerabilities.

J. SUGGESTED RESOURCES FOR FURTHER STUDY

Suggested Resources for Further Study

In addition to the resources cited throughout this manual, the following resources are suggested for further study:
- ISACA:
  - *COBIT 5*, 2012
  - *COBIT and Application Controls: A Management Guide*, 2009
  - *Data Analytics—A Practical Approach*, 2011
  - *Electronic Discovery*, 2011
  - *Monitoring Internal Control Systems and IT: A Primer for Business Executives, Managers and Auditors on How to Embrace and Advance Best Practices*, 2010
- National Institute of Standards and Technology (NIST), *Guide for Assessing the Security Controls in Federal Information Systems and Organizations, Building Effective Security Assessment Plans*, Special Publication 800-53A, Revision 1, USA, 2010

Part II—Risk Management and Information Systems Control in Practice

OVERVIEW

Part II Introduction

Part II of the *CRISC Review Manual 2013* provides information on how the risk management theory and concepts that were introduced in Part I apply to specific processes.

It particularly addresses those knowledge statements from Domains 1, 4 and 5 that relate to risk, controls, control objectives, activities and metrics related to selected business and IT processes.

Part II Contents

Part II contains the following chapters:

Part II Chapter	Starting Page
Overview	209
1. Managing the IT Strategy	211
2. Portfolio, Program and Project Management	228
3. Change Management	253
4. Third-party Service Management	267
5. Continuity Management	281
6. Information Security Management	299
7. Configuration Management	319
8. Problem Management	333
9. Knowledge Management	349
10. IT Operations Management	369

Individual Chapter Content

Each chapter first identifies the knowledge statements addressed within the chapter. It then provides an overview of the specific process as well as:
- An explanation of its importance to achieving business objectives
- Process-related key concepts and principles
- A high-level process overview
- Examples of common risk factors and common vulnerabilities
- Examples of generic risk scenarios
- Selected key risk indicators (KRIs)
- Examples of common key control activities supporting the process
- Suggested IS control metrics for monitoring
- A description of the practitioner's perspective
- Suggested reading materials and references

Individual Chapter Structure	Each chapter contains the following sections: A. Chapter Overview B. Related Knowledge Statements C. Key Terms and Concepts D. Process Overview E. Risk Management Considerations F. Information Systems Control Design, Monitoring and Maintenance G. The Practitioner's Perspective H. Suggested Resources for Further Study

Part II—Risk Management and Information Systems Control in Practice
1. Managing the IT Strategy
A. Chapter Overview

1. Managing the IT Strategy

A. CHAPTER OVERVIEW

Learning Objectives

The CRISC candidate should have a general understanding of the process for managing the IT strategy, how IT strategy integrates into the enterprise strategy and how managing the IT strategy relates to risk management.

Introduction

This chapter provides an overview of the process for aligning the IT strategic technological direction with risk management and:
- Explains its importance in achieving business objectives
- Outlines a high-level process overview
- Introduces related key concepts
- Presents examples of common risk
- Lists selected key risk indicators (KRIs)
- Provides examples of common key control activities supporting the process
- Describes the practitioner's perspective
- Offers suggested reading materials and references

Contents

This section contains the following topics:

Topic	Starting Page
A. Chapter Overview	211
B. Related Knowledge Statements	212
C. Key Terms and Concepts	213
D. Process Overview	214
E. Risk Management Considerations	218
F. Information Systems Control Design, Monitoring and Maintenance	221
G. The Practitioner's Perspective	224
H. Suggested Resources for Further Study	226

B. RELATED KNOWLEDGE STATEMENTS

Contents

The following table lists the applicable knowledge statements from the CRISC job practice.

No.	Knowledge Statement (KS) — Knowledge of:
KS1.8	Threats and vulnerabilities related to business processes and initiatives
KS1.22	Threats and vulnerabilities associated with emerging technologies
KS2.5	Organizational risk management policies
KS4.4	Control practices related to business processes and initiatives
KS5.5	Control objectives, activities and metrics related to IT operations and business processes and initiatives

C. KEY TERMS AND CONCEPTS

Introduction	This section introduces terms and principles related to the determination of the IT strategy as well as terms that help relate the process to other key business processes.
Governance	Governance ensures that enterprise objectives are achieved by evaluating stakeholder needs, conditions and options; setting direction through prioritization and decision making; and monitoring performance, compliance and progress against plans. In most enterprises, governance is the responsibility of the board of directors under the leadership of the chairperson.
Information Systems (IS)	The combination of strategic, managerial and operational activities involved in gathering, processing, storing, distributing and using information and its related technologies IS is distinct from IT in that IS has an IT component that interacts with the process components.
Information Technology (IT)	The hardware, software, communication and other facilities used to input, store, process, transmit and output data in whatever form
IT Strategic Plan	A long-term plan—i.e., three- to five-year horizon—in which business and IT management cooperatively describe how IT resources will contribute to the enterprise's strategic objectives (goals)
IT Tactical Plan	A medium-term plan—i.e., six- to 18-month horizon—that translates the IT strategic plan direction into required initiatives, resource requirements and ways in which resources and benefits will be monitored and managed
Risk Appetite	The amount of risk, on a broad level, that an entity is willing to accept in pursuit of its mission
Risk Tolerance	The acceptable level of variation that management is willing to allow for any particular risk as the enterprise pursues its objectives
Strategic Planning	The process of deciding on the enterprise's objectives, on changes in these objectives, and the policies to govern their acquisition and use

D. PROCESS OVERVIEW

Introduction

This section introduces the process of managing the IT strategy and provides information about its:
- Relevance to organizational objectives and/or risk management.
- Process description
- Process goal
- Process-specific management practices
- Process-specific activities

Relevance

Enterprises spend significant amounts of money every year on IT resources, including applications, infrastructure, information and people.

However, studies have shown that many IT investments fail to meet organizational expectations and that the positive, expected results of these investments are not being realized.

The risk practitioner has a role in ensuring that the IT strategy is closely aligned with the overall business strategy and that the direction in which the enterprise is headed is reflected in the IS investment strategy and subsequent project selection.

Process Description

Managing the IT strategy provides a holistic view of the current business and IT environment, the future direction, and the initiatives required to migrate to the desired future environment. The process leverages enterprise architecture building blocks and components, including externally provided services and related capabilities to enable a nimble, reliable and efficient response to strategic objectives.

Process Goal

The goal of managing the IT strategy is to:
- Align strategic IT plans with business objectives
- Clearly communicate the objectives and associated accountabilities so they are understood by all
- Identify, structure and integrate IT strategic options with the business plans

Management Practices

Key management practices of managing the IT strategy are:
1. Understand enterprise direction.
2. Assess the current environment, capabilities and performance.
3. Define the target IT capabilities.
4. Conduct a gap analysis.
5. Define the strategic plan and road map.
6. Communicate the IT strategy and direction.

1. Understand enterprise direction.

Consider the current enterprise environment and business processes, as well as the enterprise strategy and future objectives. Consider also the external environment of the enterprise (industry drivers, relevant regulations, basis for competition).

Related activities include:
1. Develop and maintain an understanding of enterprise strategy and objectives, as well as the current enterprise operational environment and challenges.
2. Develop and maintain an understanding of the external environment of the enterprise.
3. Identify key stakeholders and obtain insight on their requirements.
4. Identify and analyze sources of change in the enterprise and external environments.
5. Ascertain priorities for strategic change.
6. Understand the current enterprise architecture and work with the enterprise architecture process to determine any potential architectural gaps.

2. Assess the current environment, capabilities and performance.	Assess the performance of current internal business and IT capabilities and external IT services, and develop an understanding of the enterprise architecture in relation to IT. Identify issues currently being experienced and develop recommendations in areas that could benefit from improvement. Consider service provider differentiators and options and the financial impact and potential costs and benefits of using external services. Related activities include: 1. Develop a baseline of the current business and IT environment, capabilities and services against which future requirements can be compared. Include the relevant high-level detail of the current enterprise architecture (business, information, data, applications and technology domains), business processes, IT processes and procedures, the IT organization structure, external service provision, governance of IT, and enterprisewide IT-related skills and competencies. 2. Identify risk from current, potential and declining technologies. 3. Identify gaps between current business and IT capabilities and services and reference standards and best practices, competitor business and IT capabilities, and comparative benchmarks of best practices and emerging IT service provision. 4. Identify issues, strengths, opportunities and threats in the current environment, capabilities, and services to understand current performance. Identify areas for improvement in terms of IT's contribution to enterprise objectives.
3. Define the target IT capabilities.	Define the target business and IT capabilities and required IT services. This should be based on the understanding of the enterprise environment and requirements; the assessment of the current business process and IT environment and issues; and consideration of reference standards, best practices and validated emerging technologies or innovation proposals. Related activities include: 1. Consider validated emerging technology or innovation ideas. 2. Identify threats from declining, current and newly acquired technologies. 3. Define high-level IT objectives/goals and how they will contribute to the enterprise's business objectives. 4. Define required and desired business process and IT capabilities and IT services and describe the high-level changes in the enterprise architecture (business, information, data, applications and technology domains), business and IT processes and procedures, the IT organization structure, IT service providers, governance of IT, and IT skills and competencies. 5. Align and agree with the enterprise architect on proposed enterprise architecture changes. 6. Demonstrate traceability to the enterprise strategy and requirements.
4. Conduct a gap analysis.	Identify the gaps between the current and target environments and consider the alignment of assets (the capabilities that support services) with business outcomes to optimize investment in and utilization of the internal and external asset base. Related activities include: 1. Identify all gaps and changes required to realize the target environment. 2. Consider the high-level implications of all gaps. Consider the value of potential changes to business and IT capabilities, IT services and enterprise architecture, and the implications if no changes are realized. 3. Assess the impact of potential changes on the business and IT operating models, IT research and development capabilities, and IT investment programs. 4. Refine the target environment definition and prepare a value statement with the benefits of the target environment.

5. Define the strategic plan and road map.	Create a strategic plan that defines, in cooperation with relevant stakeholders, how IT-related goals will contribute to the enterprise's strategic goals. Include how IT will support IT-enabled investment programs, business processes, IT services and IT assets. Direct IT to define the initiatives that will be required to close the gaps, the sourcing strategy and the measurements to be used to monitor achievement of goals. Then prioritize the initiatives and combine them in a high-level road map. Related activities include: 1. Define the initiatives required to close gaps and migrate from the current to the target environment, including investment/operational budget, funding sources, sourcing strategy and acquisition strategy. 2. Identify and adequately address risk, costs and implications of organizational changes, technology evolution, regulatory requirements, business process reengineering, staffing, insourcing and outsourcing opportunities, etc., in the planning process. 3. Determine dependencies, overlaps, synergies and impacts among initiatives, and prioritize the initiatives. 4. Identify resource requirements, schedule and investment/operational budgets for each of the initiatives. 5. Create a road map indicating the relative scheduling and interdependencies of the initiatives. 6. Translate the objectives into outcome measures represented by metrics (what) and targets (how much) that can be related to enterprise benefits. 7. Formally obtain support from stakeholders and obtain approval for the plan.
6. Communicate the IT strategy and direction.	Create awareness and understanding of the business and IT objectives and direction, as captured in the IT strategy, through communication to appropriate stakeholders and users throughout the enterprise. Related activities include: 1. Develop and maintain a network for endorsing, supporting and driving the IT strategy. 2. Develop a communication plan covering the required messages, target audiences, communication mechanisms/channels and schedules. 3. Prepare a communication package that delivers the plan effectively using available media and technologies. 4. Obtain feedback and update the communication plan and delivery as required.
Risk Associated With Maintenance of the IT Strategy Development	The risk associated with the development of IS strategy primarily stems from the difficulty of learning how to understand the true requirements of the business and aligning the IS strategy effectively with those requirements. It is common to find that the business cannot properly describe its needs or requirements, and IT analysts are not always adept at helping the enterprise determine or draw out its requirements. This leads to misalignment and ineffective solutions that serve to frustrate both parties and undermine the benefits that should have been realized by developing a strong strategic plan. One risk is that many strategic plans are narrow and concerned with current problems. While they are considered "strategic," they are really tactical or midrange plans based on solving "known" problems. However, enterprises change at a rapid pace, and strategic plans must be flexible enough to adapt to those changes. A failure to remain agile at the strategic level can result in projects that are solutions for yesterday's business needs and do not reflect the priorities and requirements of the enterprise today or in the future.

The Development of a Strategic Plan

The development of a strategic plan requires input from several sources, especially from top management.

A strategic plan needs to consider the operating environment (context) of the enterprise several years into the future. The operational context will consider political influences, proposed regulations, staffing issues, new technology, competition, financial constraints and economic forecasts. A strategic plan cannot be considered strategic unless it starts out by describing the world in which the enterprise will operate some time into the future.

Other sources of information that the risk practitioner will search for in the review of the strategic plan are mission statements, press releases and sources that indicate the culture of the enterprise. The organizational culture may influence how quickly the enterprise may want to deploy new technology, whether the enterprise is willing to push the edge of systems development or wait to follow the lead of other enterprises, and whether the enterprise is committed to growth or stability.

The risk practitioner should also ensure that the development of the IS strategic plan considers:
- Employee input
- Human resources and employee retention and development
- Customer expectations
- Competitive overview
- Legal requirements
- Market conditions and analysis
- Supply chain
- Vendors
- Emerging technology

E. RISK MANAGEMENT CONSIDERATIONS

Introduction

This section discusses risk management practices related to managing the IT strategy. The following points are addressed:
- Risk factors
- Generic threats and common vulnerabilities
- Generic risk scenarios
- Key risk indicators (KRIs)

> **Note:** The lack of strategic planning for technological direction is, in itself, a risk. An absence of strategic planning poses the risk of:
> - Improper oversight and governance of IT investments
> - Purchase of noncompliant or noninteroperable technologies
> - Purchase of unnecesary licenses, equipment, training and software
> - Inadequate forecasting of bandwidth, memory and processing needs
> - Lack of technology investment

Risk Factors

Examples of factors affecting the IT strategy are:
- **Enterprise size and complexity**—Size, complexity, geographic range, etc., all affect the ability to implement and maintain an IS strategic plan:
  - Organizational culture and values
  - Infrastructure: insourced, facilities managed, outsourced
  - Organizational structure: employees, consultants, hybrid
  - Operational environment: regional, international, competitive, regulated, government, military, not-for-profit
  - Control environment: mature, immature, resilient, *ad hoc*
  - Applications source: in-house, purchased off the shelf, customized
  - IT environment: centralized, decentralized, cloud, virtual, outsourced
- **Risk management capabilities**—A defined risk culture and risk management program with effective identification, management and evaluation of risk
- **Portfolio management capability**—The diligence with which business projects are approved, managed and implemented

Generic Threats and Common Vulnerabilities

Some threats and vulnerabilities are common to most organizations. These threats and vulnerabilities should be considered by the risk practitioner in determining the risk to the enterprise related to IT strategy. These common threats and vulnerabilities include:
- IT program selection
- New technologies
- Technology selection
- IT investment decision making
- Accountability over IT
- Integration of IT within business processes
- State of infrastructure technology
- Aging of application software
- Architectural agility and flexibility

Each of these vulnerabilities is discussed in more detail on the following page.

IT Program Selection	The risk related to IT program selection is based on the challenge of aligning the IT program with the objectives, mission and direction of the business. Enterprises often find a gap between the strategic plan of the enterprise as compared to management of the IT program. The IT plan is frequently driven more by technology and procedures than by strategic planning and business requirements. The management of an IT program must seek to be more closely aligned with the business and to have an enterprisewide approach. When IT projects are based on a project-by-project approach, it is common to find that the IT resources are wasted on duplicate efforts or are not compatible with the equipment, systems and processes of other parts of the organization.
New Technologies	New technologies present a considerable risk to the organization. The organization may not have the skilled staff to implement and administer the new technology or equipment properly, the technology may not be suited for the purposes for which it was designed, the organization may be slow to implement new technologies and lose a competitive position, or the new technology may not integrate with existing systems or data.
Technology Selection	Enterprises invest significant financial resources in technology. The organization may purchase the wrong technology from the wrong vendor at the wrong price. The technology may not be suited for the purpose, up to date, interoperable or may be too expensive to maintain and operate.
IT Investment Decision Making	The purchase and development of the IT program should involve the business owners, users and auditors, not just the IT department.
Accountability over IT	IT is part of business and the business must accept a portion of the accountability for ensuring that business functional requirements, priorities and operational (environmental) issues are communicated to the IT department.
Integration of IT Within Business Processes	An ideal world would see IT integrated into business processes. The users could use IT systems easily, in a standard, repeatable manner, without needing to do extraordinary or *ad hoc* processes. Systems and data would be integrated together and user reports, processes and functions would be readily available to the user community.
State of Infrastructure Technology	Old, obsolete or unpatched systems or equipment may pose a serious risk to the enterprise. Older systems may be more vulnerable to failure or attack, or may be insufficient to support business and user needs.
Aging of Application Software	Older applications may be expensive; difficult to maintain; poorly documented; or not written to modern, secure coding standards.
Architectural Agility and Flexibility	A stiff or inflexible architecture can limit the ability of the business and IT to adapt to new business requirements or processes. Changing market conditions or regulatory requirements may require the enterprise to revise business processes. An inflexible architecture may make such changes difficult.

Generic Risk Scenarios	Generic risk scenarios related to the IT strategy are: • Technological acquisitions that are inconsistent with strategic plans • Uncontrolled acquisition, use and proliferation of IS assets • Inappropriate IT infrastructure to meet organizational requirements • Increased costs due to uncoordinated and unstructured acquisition plans • Licensing violations • Regulatory requirements that are not met • Difficulty in determining compliance due to unclear requirements or incomplete evidence • Failure to maximize the use of emerging technological opportunities to improve business and IT capability • Technical incompatibilities or maintenance issues • Incompatibilities between technology platforms and applications • Increased support, replacement and maintenance costs • Lack of skilled resources • Reliance on unsupported technology
KRIs	Examples of KRIs are the: • Level of standardization of systems with the overall IT architecture • Regularity of review/update of system documentation and the technology infrastructure plan • Level of system availability • Level of business unit management perception of IT service delivery • Level of completion of scheduled maintenance tasks • Percent of applications operating on nonstandard, unpatched or unapproved operating platforms • Number of requests for exceptions to policies or standards • Number of hardware/software purchase approvals without IT technology board review • Number of unapproved changes to systems, configurations, networks or applications • Satisfaction levels of users to deployed projects

F. INFORMATION SYSTEMS CONTROL DESIGN, MONITORING AND MAINTENANCE

Introduction

This section provides an overview of common controls relating to managing the IT strategy. The following points are addressed:
- Key control activities
- IS control metrics
- Monitoring practices

Key Control Activities

Key control activities for the following aspects of planning the technological direction are described in this section:
- Managing the IT strategy
- The technology infrastructure plan
- Monitoring future trends and regulations
- Technology standards objectives
- The IT architecture board responsibilties
- Enterprise security architecture planning

Key Control Activities Related to Managing the IT Strategy

Managing the IT strategy includes the following:
- Follow up on market evolutions and relevant emerging technologies.
- Identify the latest developments in IT that could have an impact on the success of the business.
- Establish the appropriate technological risk appetite (e.g., pioneer, leader, early adopter, follower).
- Identify what is needed in terms of technological direction for business systems architecture, migration strategies and contingency aspects of infrastructure components.
- Identify security needs and ensure that security requirements are addressed throughout the IT systems development life cycle and enterprise IT security plan.

Key Control Activities Related to Technology Infrastructure Plan

The technology infrastructure plan includes:
- A clear linkage to the IS strategic and tactical plans
- A clear linkage to the IT strategic and tactical plans
- Ongoing assessments of the current vs. planned information systems, resulting in a migration strategy or road map to achieve the future state
- Transitional and other costs, complexity, technical risks, future flexibility, value and product/vendor sustainability
- The need to identify changes in the competitive environment, economies of scale for IS staffing and investments, and improved interoperability of platforms and applications
- Ensuring that controls are in place to provide the information necessary for compliance with regulations

Key Control Activities Related to Monitoring Future Trends and Regulations	Monitoring future trends and regulations includes the following: • Assign adequately skilled staff members to routinely monitor technological developments, competitor activities, infrastructure issues, legal requirements and regulatory environment changes, and provide relevant information to senior management (including consulting third-party experts to obtain their opinions and to confirm findings and proposals of internal staff). • Ascertain that the IT department maintains membership in vendor user groups, subscribes to technical journals and maintains a research budget. • Evaluate new technologies in the context of their potential contribution to the realization of broader business goals and targets using established criteria, e.g., return on investment (ROI) or the ability to achieve market leadership. • Ensure that the enterprise's legal counsel monitors legal and regulatory conditions in all relevant geographic locations and informs the IT steering committee of any changes that may impact the technology infrastructure plan. • Monitor ongoing and evolving security conditions, threats, exposures and attack methods, and ensure that the security of information and information systems is adequate to meet industry best practices. • Ensure staff maintains professional proficiency (i.e., trained on new technology, continuing education)
Key Control Activities Related to Technology Standards Objectives	Technology standards have the following common objectives: • Corporate technology standards are approved by the IT architecture board and communicated throughout the enterprise by using a technology forum. • Management establishes and maintains an approved list of vendors and system components that conform to the technological infrastructure plan and technology standards. • A process is established to prevent the acquisition of nonconforming systems or applications. • Technology guidelines are put in place to effectively support the enterprise's technological solutions. • Monitoring and benchmarking processes are implemented to enforce compliance with the standards—e.g., by measuring noncompliance with technology standards. • Technology standards are updated as part of a periodic review of the technological infrastructure plan and stakeholders are involved in the development and approval of migration strategies and change plans, taking into consideration impacts on personnel and operations. • The IT department's recruiting and training practices are aligned with the technology standards. • Technology solutions are interoperable, consistent and provide defense-in-depth (layered defense) across the enterprise.
Key Control Activities Related to IT Architecture Board Responsibilities	Those fulfilling the role of an IT architecture board should: • Provide architecture guidelines and advice on their application. • Agree on, and formally document, the board's role and authority, including the overall IT architecture design and alignment with the information architecture. • Put a process in place to monitor and benchmark instances of noncompliance to technology standards. • Meet regularly and take minutes that include actions, assignments of responsible parties, time lines and tasks. • Participate in enterprise security architecture planning.
Key Control Activities Related to Enterprise Security Architecture Planning	The role of an enterprise security architecture planning is to: • Review IT projects to ensure that security requirements are identified in all IT systems and projects. • Ensure interoperability of security controls. • Provide support for audit and compliance. • Review security control reports to ensure that they are effectively mitigating the risk.

Metrics for Process Monitoring

The following table provides examples of metrics for monitoring the IT strategy. Because strategy management itself does not generally involve high reliance on IS controls, these metrics are process metrics.

Note: This content will not be tested in the CRISC exam.

Examples of Metrics for Monitoring the IT Strategy					
Attribute	**Target**	**Current Period**	**Prior Period**	**Prior Period −1**	**Prior Period −2**
Percent of objectives in the IT strategy that support the enterprise strategy					
Percent of enterprise objectives addressed in the IT strategy					
Percent of initiatives in the IT strategy that are self-funding (financial benefits in excess of costs)					
Trends in ROI of initiatives included in the IT strategy					
Level of enterprise stakeholder satisfaction survey feedback on the IT strategy					
Percent of projects in the IT project portfolio that can be directly traced back to the IT strategy					
Percent of strategic enterprise objectives obtained as a result of strategic IT initiatives					
Number of new enterprise opportunities realized as a direct result of IT developments					
Percent of IT initiatives/projects championed by business owners					
Achievement of measurable IT strategy outcomes part of staff performance goals					
Frequency of updates to the IT strategy communication plan					
Percent of strategic initiatives with accountability assigned					
Percent of enterprise strategic goals and requirements supported by IT strategic goals					
Level of stakeholder satisfaction with scope of the planned portfolio of programs and services					
Percent of IT value drivers mapped to business value drivers					

G. THE PRACTITIONER'S PERSPECTIVE

Introduction

Setting the IT strategy of the enterprise is an integral part of IS's strategic planning responsibility. It is in this function that the decision whether (and how) to utilize technology to advance the enterprise's objectives is defined and integrated into an overall strategic plan. The following section provides an overview of how the five CRISC domains (listed below) relate to planning the technological direction in practice:
- Domain 1—Risk Identification, Assessment and Evaluation
- Domain 2—Risk Response
- Domain 3—Risk Monitoring
- Domain 4—Information Systems Control Design and Implementation
- Domain 5—Information Systems Control Monitoring and Maintenance

Domain 1— Risk Identification, Assessment and Evaluation

Risk Identification, Assessment and Evaluation:
- The risk identification, assessment and evaluation process is key to the determination of technological direction. In alignment with the enterprise strategy, the IT technology board or an equivalent IT management group determines the future IS direction. It is critical that the IS direction be based on the effective management of risk to information systems, data, business processes and assets of the enterprise. The enterprise must be committed to implementing a cost-effective, risk-based information systems control and investment strategy.
- The risk identification process related to technology considers the consequences of specific technology acquisitions, such as:
  – Interoperability with existing technologies
  – Obsolescence
  – Future costs of nonstandard solutions
  – Adherence to known and anticipated regulatory requirements
  – Failure to embrace new technologies
  – Security implications
  – Other tangible and intangible factors
  – Availability, integrity and confidentiality of information and information systems
- The process should include interaction with core business management and IT functions (systems development, operations, technical support, information security, etc.) to ensure alignment with business requirements.
- The risk identification, assessment and evaluation process provides the risk practitioner with a listing of critical systems, current risk, risk levels and an indication of acceptable risk levels.

Domain 2— Risk Response

Risk Response:
- The risk practitioner must create a suitable risk response plan to mitigate the risk identified during the risk identification assessment and evaluation process.
  – Risk acceptance (where the cost or impact of a risk is less than the cost to mitigate or reduce the risk effectively)
  – Risk avoidance (to choose to discontinue or not enter an area of exceptional uncontrolled risk)
  – Risk transference (to purchase insurance or spread the risk between various parties)
  – Risk mitigation (to reduce the risk level through the implementation of administrative, physical or technical controls)
- It is essential to identify risk in conjunction with the development of the IS strategy because once the strategy has been accepted and projects have started, it is expensive to initiate changes because investments in new technology have been made.
- Risk response is handled at a senior IT management level based on the available options and the level of risk deemed acceptable for the enterprise.
- Risk response must be undertaken with consideration of the level, impact and priority of the risk as well as the interdependency between different risk management activities.
- Risk response is a strategic inititative and obligation and must be sponsored at the highest level of the management team (particularly when the risk is not operational in level).

Part II—Risk Management and Information Systems Control in Practice
1. Managing the IT Strategy
G. The Practitioner's Perspective

Domain 3—Risk Monitoring	Risk Monitoring: • Key risk indicators (KRIs) are symptoms of a breakdown in the IS strategy, either due to misalignment with the business strategies, lack of a cohesive technical plan or poor internal controls. KRIs can be used to monitor IT activities, but the most telling risk monitoring process will be how well the integration of IS and business strategy progresses and that the identified risk to information and information systems is managed.
Domain 4—Information Systems Control Design and Implementation	Information Systems Control Design and Implementation: • The IS control objectives and activities described previously provide an excellent resource of best practice processes to ensure an effective technological direction management function. • The practitioner should periodically review the control design against stated strategic objectives and ensure that the risk is being managed proactively and risk management is built into strategic planning and development.
Domain 5—Information Systems Control Monitoring and Maintenance	Information Systems Control Monitoring and Maintenance: • The control metrics defined previously are best suited for a senior management audience. KRI metrics should be given to C-level management as an overview of the effectiveness of risk management efforts and IS strategic planning integration with the business. Failure to achieve adequate KRIs requires additional analysis because the issues may be with IT management, business management or some combination. • Annual evaluation of the maturity of controls provides a barometer of controls in their current state, comparison to previous periods, and the target maturity level. Target maturity levels should be agreed on by stakeholders and evaluated by senior management as part of the IT annual review. • Maturity assessment can be performed as a self-assessment, with oversight by an objective third party, peer review or independent assessment (internal audit or external provider).

H. SUGGESTED RESOURCES FOR FURTHER STUDY

Suggested Resources for Further Study

In addition to the resources cited throughout this manual, the following resources are suggested for further study:
- ISACA:
  - COBIT 5, 2012
    Note: The COBIT 5 framework is available at no charge from ISACA and can be downloaded at *www.isaca.org/cobit*
  - *Enterprise Value: Governance of IT Investments, The Val IT Framework 2.0*, 2008
  - *Implementing and Continually Improving IT Governance*, 2010
  - *The Risk IT Framework*, 2009
  - *The Risk IT Practitioner Guide*, 2009
- International Organization for Standardization (ISO):
  - ISO/IEC 27005:2011, *Information technology—Security techniques—Information security risk management*, Switzerland, 2011
  - ISO/IEC 38500:2010, *Corporate governance of information technology*, Switzerland, 2010
- Lientz, Bennet P.; Lee Larssen; *Risk Management for IT Projects: How to Deal With Over 150 Issues and Risks*, Butterworth-Heinemann, USA, 2006
- Sherwood, John; Andrew Clark; David Lynas; *Enterprise Security Architecture: A Business-Driven Approach*; CMP Books, USA, 2005

2. Portfolio, Program and Project Management

A. CHAPTER OVERVIEW

Learning Objectives

Enterprise portfolio, program and project management are interrelated business processes with high inherent risk. The CRISC candidate should have a comprehensive understanding of the connection between these processes and how they interrelate with risk management.

The CRISC candidate should understand how risk response activities must be included in portfolio, program and project management activities to optimize the value to be derived from the risk response plan.

Introduction

This chapter provides an overview of the portfolio, program and project management processes and:
- Explains their importance to achieving business objectives
- Introduces related key concepts
- Presents examples of common vulnerabilities and generic risk scenarios
- Lists selected key risk indicators (KRIs)
- Provides examples of common key control activities supporting the processes
- Describes the practitioner's perspective
- Offers suggested reading materials and references

Contents

This chapter contains the following topics:

Topic	Starting Page
A. Chapter Overview	227
B. Related Knowledge Statements	228
C. Key Terms and Concepts	229
D. Process Overview	230
E. Risk Management Considerations	241
F. Information Systems Control Design, Monitoring and Maintenance	244
G. The Practitioner's Perspective	250
H. Suggested Resources for Further Study	252

B. RELATED KNOWLEDGE STATEMENTS

Contents

The following table lists the applicable knowledge statements from the CRISC job practice.

No.	Knowledge Statement (KS) Knowledge of:
KS1.14	Threats and vulnerabilities related to project and program management
KS1.22	Threats and vulnerabilities associated with emerging technologies
KS4.10	Controls related to project and program management
KS5.13	Control objectives, activities and metrics related to project and program management

C. KEY TERMS AND CONCEPTS

Introduction — This section introduces terms and principles related to project and program management as well as terms that help relate the process to other key business processes.

Portfolio — A grouping of "objects of interest" (investment programs, IT services, IT projects, other IT assets or resources) managed and monitored to optimize business value

Portfolio management is different from project and program management in that the distinct objective is to create maximum value from a grouping of projects and programs.

Program — A structured grouping of interdependent projects that is both necessary and sufficient to achieve a desired business outcome and create value

> **Note:** These interdependent projects could include, but are not limited to, changes in the nature of the business, business processes, the work performed by people as well as the competencies required to carry out the work, enabling technology and organizational structure.

Project — A structured set of activities concerned with delivering a defined capability (that is necessary, but not sufficient, to achieve a required business outcome) to the enterprise based on an agreed-on schedule and budget

Scope Creep — Also called requirement creep, this refers to uncontrolled changes in a project's scope.

Exhibit 2.1: Relationship Among Portfolio, Program and Project Management

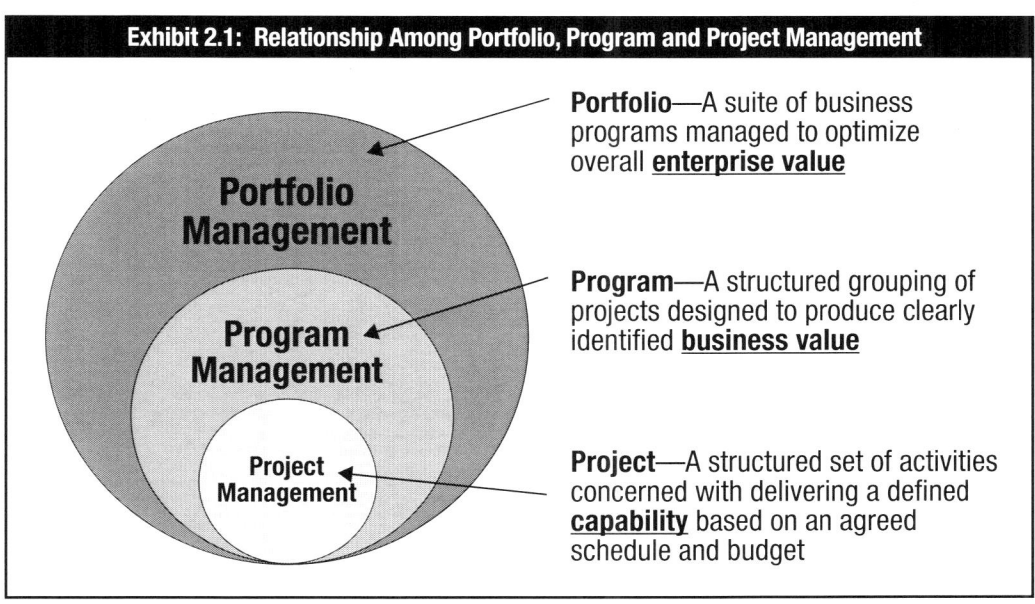

D. PROCESS OVERVIEW

Introduction

This section introduces two core processes:
1. Portfolio Management
2. Program and Project Management

For each process the section provides information about its:
- Relevance to organizational objectives and/or risk management.
- Process description
- Process goal
- Process-specific management practices
- Process-specific activities
- Process specific risk

1. PORTFOLIO MANAGEMENT

Relevance

The portfolio management process is a pervasive business process with a high inherent risk. If portfolios are not managed effectively, value optimization is virtually impossible.

The process links directly to risk response activities, which must be considered as one part of the organizations investment portfolio.

Process Description

Key objectives of the portfolio management process are to:
- Execute the strategic direction set for investments in line with the enterprise architecture vision and the desired characteristics of the investment and related services portfolios, and consider the different categories of investments and the resources and funding constraints.
- Evaluate priorities and balance programs and services, managing demand within resource and funding constraints, based on their alignment with strategic objectives, enterprise worth and risk.
- Move selected programs into the active services portfolio for execution.
- Monitor the performance of the overall portfolio of services and programs, proposing adjustments as necessary in response to program and service performance or changing enterprise priorities.

Process Goal

The goal of portfolio management is to optimize the performance of the overall portfolio of programs in response to program and service performance and changing enterprise priorities and demands.

Management Practices

Key management practices of portfolio management are:
1. Establish the target investment mix.
2. Determine the availability and sources of funds.
3. Evaluate and select programs to fund.
4. Monitor, optimize and report on investment portfolio performance.
5. Maintain portfolios.
6. Manage benefits achievement.

Part II—Risk Management and Information Systems Control in Practice
2. Portfolio, Program and Project Management
D. Process Overview

1. Establish the target investment mix.	Review and ensure clarity of the enterprise and IT strategies and current services. Define an appropriate investment mix based on cost, alignment with strategy, and financial measures such as cost and expected ROI over the full economic life cycle, degree of risk, and type of benefit for the programs in the portfolio. Adjust the enterprise and IT strategies where necessary. Related activities include: 1. Validate that IT-enabled investments and current IT services are aligned with enterprise vision, enterprise principles, strategic goals and objectives, enterprise architecture vision, and priorities. 2. Obtain a common understanding between IT and the other business functions on the potential opportunities for IT to drive and support the enterprise strategy. 3. Create an investment mix that achieves the right balance among a number of dimensions, including an appropriate balance of short- and long-term returns, financial and non-financial benefits, and high- and low-risk investments. 4. Identify the broad categories of information systems, applications, data, IT services, infrastructure, IT assets, resources, skills, practices, controls and relationships needed to support the enterprise strategy. 5. Agree on an IT strategy and goals, taking into account the inter-relationships between the enterprise strategy and the IT services, assets and other resources. Identify and leverage synergies that can be achieved.
2. Determine the availability and sources of funds.	Determine potential sources of funds, different funding options and the implications of the funding source on the investment return expectations. Related activities include: 1. Understand the current availability and commitment of funds, the current approved spending, and the actual amount spent to date. 2. Identify options for obtaining additional funds for IT-enabled investments, internally and from external sources. 3. Determine the implications of the funding source on the investment return expectations.
3. Evaluate and select programs to fund.	Based on the overall investment portfolio mix requirements, evaluate and prioritize program business cases, and decide on investment proposals. Allocate funds and initiate programs. Related activities include: 1. Recognize investment opportunities and classify them in line with the investment portfolio categories. Specify expected enterprise outcome(s), all initiatives required to achieve the expected outcomes, costs, dependencies and risk, and how all would be measured. 2. Perform detailed assessments of all program business cases, evaluating strategic alignment, enterprise benefits, risk and availability of resources. 3. Assess the impact on the overall investment portfolio of adding candidate programs, including any changes that might be required to other programs. 4. Decide which candidate programs should be moved to the active investment portfolio. Decide whether rejected programs should be held for future consideration or provided with some seed funding to determine whether the business case can be improved or discarded. 5. Determine the required milestones for each selected program's full economic life cycle. Allocate and reserve total program funding per milestone. Move the program into the active investment portfolio. 6. Establish procedures to communicate the cost, benefit and risk-related aspects of these portfolios to the budget prioritization, cost management and benefit management processes.

Part II—Risk Management and Information Systems Control in Practice
2. Portfolio, Program and Project Management
D. Process Overview

4. Monitor, optimize and report on investment portfolio performance.	On a regular basis, monitor and optimize the performance of the investment portfolio and individual programs throughout the entire investment life cycle.
	
	Related activities include:
	1. Review the portfolio on a regular basis to identify and exploit synergies, eliminate duplication between programs, and identify and mitigate risk.
	2. When changes occur, reevaluate and reprioritize the portfolio to ensure that the portfolio is aligned with the business strategy and the target mix of investments is maintained so the portfolio is optimizing overall value. This may require programs to be changed, deferred or retired, and new programs to be initiated.
	3. Adjust the enterprise targets, forecasts, budgets and, if required, the degree of monitoring to reflect the expenditures to be incurred and enterprise benefits to be realized by programs in the active investment portfolio. Incorporate program expenditures into chargeback mechanisms.
	4. Provide an accurate view of the performance of the investment portfolio to all stakeholders.
	5. Provide management reports for senior management's review of the enterprise's progress towards identified goals, stating what still needs to be spent and accomplished over what time frames.
	6. In the regular performance monitoring, include information on the extent to which planned objectives have been achieved, risk mitigated, capabilities created, deliverables obtained and performance targets met.
	7. Identify deviations for:
	• Budget control between actual and budget
	• Benefit management of:
	   – Actual vs. targets for investments for solutions, possibly expressed in terms of ROI, net present value (NPV) or internal rate of return (IRR)
	   – The actual trend of service portfolio cost for service delivery productivity improvements
	8. Develop metrics for measuring IT's contribution to the enterprise, and establish appropriate performance targets reflecting the required IT and enterprise capability targets. Use guidance from external experts and benchmark data to develop metrics.

5. Maintain portfolios.	Maintain portfolios of investment programs and projects, IT services and IT assets. Related activities include: 1. Create and maintain portfolios of IT-enabled investment programs, IT services and IT assets, which form the basis for the current IT budget and support the IT tactical and strategic plans. 2. Work with service delivery managers to maintain the service portfolios and with operations managers and architects to maintain the asset portfolios. Prioritize portfolios to support investment decisions. 3. Remove the program from the active investment portfolio when the desired enterprise benefits have been achieved or when it is clear that benefits will not be achieved within the value criteria set for the program.

Part II—Risk Management and Information Systems Control in Practice
2. Portfolio, Program and Project Management
D. Process Overview

6. Manage benefits achievement.	Monitor the benefits of providing and maintaining appropriate IT services and capabilities, based on the agreed-on and current business case. Related activities include: 1. Use the agreed-on metrics and track how benefits are achieved, how they evolve throughout the life cycle of programs and projects, how they are being delivered from IT services, and how they compare to internal and industry benchmarks. Communicate results to stakeholders. 2. Implement corrective action when achieved benefits significantly deviate from expected benefits. Update the business case for new initiatives and implement business process and service improvements as required. 3. Consider obtaining guidance from external experts, industry leaders and comparative benchmarking data to test and improve the metrics and targets.

2. PROGRAM AND PROJECT MANAGEMENT

Relevance	Program and project management are interrelated business processes with high inherent risk. Since the potential magnitude of impact is high, these processes need to be closely monitored. Risk response activities must be included in portfolio, program and project management activities to optimize the value to be derived from the risk response plan.
Process Description	Program and project management manage all programs and projects from the investment portfolio in alignment with enterprise strategy and in a coordinated way. The processes initiate, plan, control and execute programs and projects, and close with a postimplementation review.
Process Goal	The goal of program and project management is to reduce the risk of unexpected delays, costs and value erosion by improving communications to and involvement of business and end users, ensuring the value and quality of project deliverables and maximizing their contribution to the investment and services portfolio.
Management Practices	Key management practices within portfolio, program and project management are: 1. Maintain a standard approach for program and project management. 2. Initiate a program. 3. Manage stakeholder engagement. 4. Develop and maintain the program plan. 5. Launch and execute the program. 6. Monitor, control and report on the program outcomes. 7. Start up and initiate projects within a program. 8. Plan projects. 9. Manage program and project quality. 10. Manage program and project risk. 11. Monitor and control projects. 12. Manage project resources and work packages. 13. Close a project or iteration. 14. Close a program.

| 1. Maintain a standard approach for program and project management. | Maintain a standard approach for program and project management that enables governance and management review and decision making and delivery management activities focused on achieving value and goals (requirements, risk, costs, schedule and quality) for the business in a consistent manner.

Related activities include:
1. Maintain and enforce a standard approach to program and project management aligned to the enterprise's specific environment and with good practices based on a defined process and use of appropriate technology. Ensure that the approach covers the full life cycle and disciplines to be followed, including the management of scope, resources, risk, cost, quality, time, communication, stakeholder involvement, procurement, change control, integration and benefit realization.
2. Update the program and project management approach based on lessons learned from its use. |
|---|---|
| 2. Initiate a program. | Initiate a program to confirm the expected benefits and obtain authorization to proceed. This includes agreeing on program sponsorship, confirming the program mandate through approval of the conceptual business case, appointing program board or committee members, producing the program brief, reviewing and updating the business case, developing a benefits realization plan, and obtaining approval from sponsors to proceed.

Related activities include:
1. Agree on program sponsorship and appoint a program board/committee with members who have strategic interest in the program, have responsibility for the investment decision making, will be significantly impacted by the program and will be required to enable delivery of the change.
2. Confirm the program mandate with sponsors and stakeholders. Articulate the strategic objectives for the program, potential strategies for delivery, improvement and benefits that are expected to result, and how the program fits with other initiatives.
3. Develop a detailed business case for a program, if warranted. Involve all key stakeholders to develop and document a complete understanding of the expected enterprise outcomes, how they will be measured, the full scope of initiatives required, the risk involved and the impact on all aspects of the enterprise. Identify and assess alternative courses of action to achieve the desired enterprise outcomes.
4. Develop a benefits realization plan that will be managed throughout the program to ensure that planned benefits always have owners and are achieved, sustained and optimized.
5. Prepare and submit for in-principle approval the initial (conceptual) program business case, providing essential decision-making information regarding purpose, contribution to business objectives, expected value created, time frames, etc.
6. Appoint a dedicated manager for the program, with the commensurate competencies and skills to manage the program effectively and efficiently. |

Part II—Risk Management and Information Systems Control in Practice
2. Portfolio, Program and Project Management
D. Process Overview

3. Manage stakeholder engagement.	Manage stakeholder engagement to ensure an active exchange of accurate, consistent and timely information that reaches all relevant stakeholders. This includes planning, identifying and engaging stakeholders and managing their expectations. Related activities include: 1. Plan how stakeholders inside and outside the enterprise will be identified, analyzed, engaged and managed through the life cycle of the projects. 2. Identify, engage and manage stakeholders by establishing and maintaining appropriate levels of coordination, communication and liaison to ensure that they are involved in the program/project. 3. Measure the effectiveness of stakeholder engagement and take remedial actions as required. 4. Analyze stakeholder interests and requirements.
4. Develop and maintain the program plan.	Formulate a program to lay the initial groundwork and to position it for successful execution by formalizing the scope of the work to be accomplished and identifying the deliverables that will satisfy its goals and deliver value. Maintain and update the program plan and business case throughout the full economic life cycle of the program, ensuring alignment with strategic objectives and reflecting the current status and updated insights gained to date. Related activities include: 1. Define and document the program plan covering all projects, including what is needed to bring about changes to the enterprise; its image, products and services; business processes; people skills and numbers; relationships with stakeholders, customers, suppliers and others; technology needs; and organizational restructuring required to achieve the program's expected enterprise outcomes. 2. Specify required resources and skills to execute the project, including project managers and project teams as well as business resources. Specify funding, cost, schedule and interdependencies of multiple projects. Specify the basis for acquiring and assigning competent staff members and/or contractors to the projects. Define the roles and responsibilities for all team members and other interested parties. 3. Assign accountability clearly and unambiguously for each project, including achieving the benefits, controlling the costs, managing the risk and coordinating the project activities. 4. Ensure that there is effective communication of program plans and progress reports among all projects and with the overall program. Ensure that any changes made to individual plans are reflected in the other enterprise program plans. 5. Maintain the program plan to ensure that it is up to date and reflects alignment with current strategic objectives, actual progress and material changes to outcomes, benefits, costs and risk. Have the business drive the objectives and prioritize the work throughout to ensure that the program as designed will meet enterprise requirements. Review progress of individual projects and adjust the projects as necessary to meet scheduled milestones releases. 6. Throughout the program's economic life, update and maintain the business case and a benefits register to identify and define key benefits arising from undertaking the program. 7. Prepare a program budget that reflects the full economic life cycle costs and the associated financial and nonfinancial benefits.

5. Launch and execute the program.	Launch and execute the program to acquire and direct the resources needed to accomplish the goals and benefits of the program as defined in the program plan. In accordance with stage-gate or release review criteria, prepare for stage-gate, iteration or release reviews to report on the progress of the program and to be able to make the case for funding up to the following stage-gate or release review. Related activities include: 1. Plan, resource and commission the necessary projects required to achieve the program results, based on funding review and approvals at each stage-gate review. 2. Establish agreed-on stages of the development process (development checkpoints). At the end of each stage, facilitate formal discussions of approved criteria with the stakeholders. After successful completion of functionality, performance and quality reviews, and before finalizing stage activities, obtain formal approval and sign-off from all stakeholders and the sponsor/business process owner. 3. Undertake a benefits realization process throughout the program to ensure that planned benefits always have owners and are likely to be achieved, sustained and optimized. Monitor benefits delivery and report against performance targets at the stage-gate or iteration and release reviews. Perform root cause analysis for deviations from the plan and identify and address any necessary remedial actions. 4. Manage each program or project to ensure that decision making and delivery activities are focused on value by achieving benefits for the business and goals in a consistent manner, addressing risk and achieving stakeholder requirements. 5. Set up program/project management office(s) and plan audits, quality reviews, phase/stage-gate reviews and reviews of realized benefits.
6. Monitor, control and report on the program outcomes.	Monitor and control program (solution delivery) and enterprise (value/outcome) performance against plan throughout the full economic life cycle of the investment. Report this performance to the program steering committee and the sponsors. Related activities include: 1. Monitor and control the performance of the overall program and the projects within the program, including contributions of the business and IT to the projects, and report in a timely, complete and accurate fashion. Reporting may include schedule, funding, functionality, user satisfaction, internal controls and acceptance of accountabilities. 2. Monitor and control performance against enterprise and IT strategies and goals, and report to management on enterprise changes implemented, benefits realized against the benefits realization plan, and the adequacy of the benefits realization process. 3. Monitor and control IT services, assets and resources created or changed as a result of the program. Note implementation and in-service dates. Report to management on performance levels, sustained service delivery and contribution to value. 4. Manage program performance against key criteria (e.g., scope, schedule, quality, benefits realization, costs, risk, velocity), identify deviations from the plan and take timely remedial action when required. 5. Monitor individual project performance related to delivery of the expected capabilities, schedule, benefits realization, costs, risk or other metrics to identify potential impacts on program performance. Take timely remedial action when required. 6. Update operational IT portfolios reflecting changes that result from the program in the relevant IT service, asset or resource portfolios. 7. In accordance with stage-gate, release or iteration review criteria, undertake reviews to report on the progress of the program so that management can make go/no-go or adjustment decisions and approve further funding up to the following stage-gate, release or iteration.

7. Start up and initiate projects within a program.

Define and document the nature and scope of the project to confirm and develop among stakeholders a common understanding of project scope and how it relates to other projects within the overall IT-enabled investment program. The definition should be formally approved by the program and project sponsors.

Related activities include:
1. To create a common understanding of project scope among stakeholders, provide to the stakeholders a clear written statement defining the nature, scope and benefit of every project.
2. Ensure that each project has one or more sponsors with sufficient authority to manage execution of the project within the overall program.
3. Ensure that key stakeholders and sponsors within the enterprise and IT agree on and accept the requirements for the project, including definition of project success (acceptance) criteria and key performance indicators (KPIs).
4. Ensure that the project definition describes the requirements for a project communication plan that identifies internal and external project communications.
5. With the approval of stakeholders, maintain the project definition throughout the project, reflecting changing requirements.
6. To track the execution of a project, put in place mechanisms such as regular reporting and stage-gate, release or phase reviews in a timely manner with appropriate approval.

8. Plan projects.

Establish and maintain a formal, approved integrated project plan (covering business and IT resources) to guide project execution and control throughout the life of the project. The scope of projects should be clearly defined and tied to building or enhancing business capability.

Related activities include:
1. Develop a project plan that provides information to enable management to control project progress progressively. The plan should include details of project deliverables and acceptance criteria, required internal and external resources and responsibilities, clear work breakdown structures and work packages, estimates of resources required, milestones/release plan/phases, key dependencies, and identification of a critical path.
2. Maintain the project plan and any dependent plans (e.g., risk plan, quality plan, benefits realization plan) to ensure that they are up to date and reflect actual progress and approved material changes.
3. Ensure that there is effective communication of project plans and progress reports among all projects and with the overall program. Ensure that any changes made to individual plans are reflected in the other plans.
4. Determine the activities, interdependencies, and required collaboration and communication among multiple projects within a program.
5. Ensure that each milestone is accompanied by a significant deliverable requiring review and sign-off.
6. Establish a project baseline (e.g., cost, schedule, scope, quality) that is appropriately reviewed, approved and incorporated into the integrated project plan.

9. Manage program and project quality.	Prepare and execute a quality management plan, processes and practices, aligned with the quality management system (QMS) that describes the program and project quality approach and how it will be implemented. The plan should be formally reviewed and agreed on by all parties concerned and then incorporated into the integrated program and project plans. Related activities include: 1. Identify assurance tasks and practices required to support the accreditation of new or modified systems during program and project planning, and include them in the integrated plans. Ensure that the tasks provide assurance that internal controls and security solutions meet the defined requirements. 2. Provide quality assurance for the project deliverables, quality review processes, success criteria and performance metrics, and identify ownership and responsibilities, 3. Define any requirements for independent validation and verification of the quality of deliverables in the plan. 4. Perform quality assurance and control activities in accordance with the quality management plan and QMS.
10. Manage program and project risk.	Eliminate or minimize specific risk associated with programs and projects through a systematic process of planning, identifying, analyzing, responding to, and monitoring and controlling the areas or events that have the potential to cause unwanted change. Risk faced by program and project management should be established and centrally recorded. Related activities include: 1. Establish a formal project risk management approach aligned with the enterprise risk management (ERM) framework. Ensure that the approach includes identifying, analyzing, responding to, mitigating, monitoring and controlling risk. 2. Assign to appropriately skilled personnel the responsibility for executing the enterprise's project risk management process within a project and ensuring that this is incorporated into the solution development practices. Consider allocating this role to an independent team, especially if an objective viewpoint is required or a project is considered critical. 3. Perform the project risk assessment of identifying and quantifying risk continuously throughout the project. Manage and communicate risk appropriately within the project governance structure. 4. Reassess project risk periodically, including at initiation of each major project phase and as part of major change request assessments. 5. Identify owners for actions to avoid, accept or mitigate risk. 6. Maintain and review a project risk register of all potential project risk, and a risk mitigation log of all project issues and their resolution. Analyze the log periodically for trends and recurring problems to ensure that root causes are corrected.

Part II—Risk Management and Information Systems Control in Practice
2. Portfolio, Program and Project Management
D. Process Overview

11. Monitor and control projects.	Measure project performance against key project performance criteria such as schedule, quality, cost and risk. Identify any deviations from the expected. Assess the impact of deviations on the project and overall program, and report results to key stakeholders. Related activities include: 1. Establish and use a set of project criteria including, but not limited to, scope, schedule, quality, cost and level of risk. 2. Measure project performance against key project performance criteria. Analyze deviations from established key project performance criteria for cause, and assess positive and negative effects on the program and its component projects. 3. Report to identified key stakeholders project progress within the program, deviations from established key project performance criteria, and potential positive and negative effects on the program and its component projects. 4. Monitor changes to the program and review existing key project performance criteria to determine whether they still represent valid measures of progress. 5. Document and submit any necessary changes to the program's key stakeholders for their approval before adoption. Communicate revised criteria to project managers for use in future performance reports. 6. Recommend and monitor remedial action, when required, in line with the program and project governance framework. 7. Gain approval and sign-off on the deliverables produced in each iteration, release or project phase from designated managers and users in the affected business and IT functions. 8. Base the approval process on clearly defined acceptance criteria agreed on by key stakeholders prior to work commencing on the project phase or iteration deliverable. 9. Assess the project at agreed-on major stage-gates, releases or iterations and make formal go/no-go decisions based on predetermined critical success criteria. 10. Establish and operate a change control system for the project so that all changes to the project baseline (e.g., cost, schedule, scope, quality) are appropriately reviewed, approved and incorporated into the integrated project plan in line with the program and project governance framework.
12. Manage project resources and work packages.	Manage project work packages by placing formal requirements on authorizing and accepting work packages, and assigning and coordinating appropriate business and IT resources. Related activities include: 1. Identify business and IT resource needs for the project and clearly map appropriate roles and responsibilities, with escalation and decision-making authorities agreed on and understood. 2. Identify required skills and time requirements for all individuals involved in the project phases in relation to defined roles. Staff the roles based on available skills information (e.g., IT skills matrix). 3. Utilize experienced project management and team leader resources with skills appropriate to the size, complexity and risk of the project. 4. Consider and clearly define the roles and responsibilities of other involved parties, including finance, legal, procurement, HR, internal audit and compliance. 5. Clearly define and agree on the responsibility for procurement and management of third-party products and services, and manage the relationships. 6. Identify and authorize the execution of the work according to the project plan. 7. Identify project plan gaps and provide feedback to the project manager to remediate.

13. Close a project or iteration. At the end of each project, release or iteration, require the project stakeholders to ascertain whether the project, release or iteration delivered the planned results and value. Identify and communicate any outstanding activities required to achieve the planned results of the project and the benefits of the program, and identify and document lessons learned for use on future projects, releases, iterations and programs.

Related activities include:
1. Define and apply key steps for project closure, including postimplementation reviews that assess whether a project attained desired results and benefits.
2. Plan and execute postimplementation reviews to determine whether projects delivered expected benefits and to improve the project management and system development process methodology.
3. Identify, assign, communicate and track any uncompleted activities required to achieve planned program project results and benefits.
4. Regularly, and upon completion of the project, collect from the project participants the lessons learned. Review them and key activities that led to delivered benefits and value. Analyze the data and make recommendations for improving the current project as well as project management method for future projects.
5. Obtain stakeholder acceptance of project deliverables and transfer ownership.

14. Close a program. Remove the program from the active investment portfolio when there is agreement that the desired value has been achieved or when it is clear it will not be achieved within the value criteria set for the program.

Related activities include:
1. Bring the program to an orderly closure, including formal approval, disbanding of the program organization and supporting function, validation of deliverables, and communication of retirement.
2. Review and document lessons learned. Once the program is retired, remove it from the active investment portfolio.
3. Put accountability and processes in place to ensure that the enterprise continues to optimize value from the service, asset or resources. Additional investments may be required at some future time to ensure that this occurs.

E. RISK MANAGEMENT CONSIDERATIONS

Introduction

This section discusses risk management practices related to portfolio, program and project management. The following points are addressed:
- Risk factors
- Generic threats and common vulnerabilities
- Generic risk scenarios
- Key risk indicators (KRIs)

Risk Factors

Examples of factors affecting the program and project management process are:
- **Enterprise size and complexity**—Size, complexity, geographic range, etc., all affect the ease with which program and project management can be implemented and maintained.
- **Corporate culture**—Industry, traditions and management style have a high impact on risk, risk appetite, risk tolerance and risk awareness.
- **Operational model:**
  - Control environment: mature, immature, resilient
  - Applications-source: in-house, purchased off the shelf, customized
  - Application portfolio: managed, *ad hoc*
- **Risk management capabilities**—The maturity of ERM and the decisions to approve and manage projects, lack of knowledge of regulatory requirements, unclear reporting requirements
- **Program management capability**—Lack of integration and interoperability between projects
- **Project management capability**—The experience within the business to manage projects, project-specific approach instead of enterprise architecture approach
- **Project-specific factors**—Scope, time line, budget, criticality, complexity, purpose, and the level of impact on existing business and IT processes; incorrect sponsorship; failure to incorporate security requirements into the project; push to market leading to compromise of quality assurance or testing processes
- **Portfolio management capability**—The diligence with which business cases are assessed and the project portfolio is managed, nonalignment with business strategy

Generic Threats and Common Vulnerabilities

Generic threats and common vulnerabilities related to the portfolio, program and project processes are:
- Program management methodology:
  - Inappropriate, disorganized or ineffective project management
  - Different project management approaches within the enterprise
  - Lack of compliance with the enterprise's reporting structure
  - Failure to respond to project issues with optimal and approved decisions
- Stakeholder commitment:
  - Unclear responsibilities and accountabilities for ensuring cost control and project success
  - Insufficient stakeholder participation in defining requirements and reviewing deliverables
  - Reduced understanding and delivery of business benefits
- Project scope statement:
  - Misunderstanding of project objectives and requirements
  - Failure of projects to meet business and user requirements
- Project phase initiation:
  - Lack of alignment of projects to the enterprise's vision
  - Wrong prioritization of projects
  - Undetected deviations from the overall project plan
  - Poor resource allocation (e.g., inadequate skill match)
- Integrated project plan:
  - Errors in project planning and budgeting
  - Lack of project status reporting
- Project resources:
  - Gaps in skills and resources jeopardizing critical project tasks
  - Inefficient use of resources
  - Failure of outsourced suppliers to meet contract terms or time lines
- Project risk management:
  - Undetected and unforeseen project risk
  - Lack of mitigating actions for identified risk
  - Uncontrolled changes to project scope or deliverables
- Project quality plan:
  - Project deliverables failing to meet business and user requirements
  - Failure of vendor-supplied products to meet quality or performance standards
  - Gaps in expected and delivered quality within projects
  - Lack of testing and assurance
  - Implemented system or changes adversely impact existing systems and infrastructure
- Project closure:
  - Lack of feedback from users upon project completion
  - Missed opportunities from lessons learned

Generic Risk Scenarios

Generic risk scenarios associated with portfolio, program, or project management include:
- Failure to optimize value delivery from the enterprise view
- Failure to deliver desired benefits
- Exceeding costs or time allotted
- Cessation of funding
- Change in management
- Changes in business priorities/strategy
- Changes in market conditions
- Emergence of new technology/competitors
- Changes in regulation
- Malware/attacks

KRIs

Examples of KRIs are the:
- Number of projects approved outside the portfolio management process
- Number of projects not assessed for interrelationships with other projects
- Number of projects that do not satisfy the business case
- Percentage of projects that do not meet project time lines/milestones
- Number of third-party issues affecting delivery of the project
- Number of projects in which the project management office (PMO) does not provide project oversight

F. INFORMATION SYSTEMS CONTROL DESIGN, MONITORING AND MAINTENANCE

Introduction

This section provides an overview of common controls related to the portfolio, program and project process. The following points are addressed:
- Key control activities
- IS control metrics
- Monitoring practices

Key Control Activities

Key control activities of the portfolio, program and project management process are described in this section for the following aspects:
- The portfolio/program management methodology
- Stakeholder commitment/involvement
- The project scope statement
- Project phase initiation
- Project resources
- Project risk management
- The project quality plan
- Project change control
- Portfolio/project performance measurement, reporting and monitoring
- Portfolio/project closure

Key Control Activities Related to the Portfolio/Program Management Methodology

Key control activities for the portfolio/program management methodology may include the following:
- Define and document the portfolio/program, including all of the projects required to achieve the program's expected business outcomes; specify the required resources, including oversight, funding, project managers, project teams, IT resources and business resources where applicable; and gain formal approval of the document from key business and IT stakeholders.
- Assign accountability (clearly and unambiguously) for each project, including achieving the benefits, controlling the costs, managing the risk, reporting, and coordinating the project activities.
- Determine the interdependencies of multiple projects in the portfolio/program and develop a schedule for their completion that will enable the overall program schedule to be met.
- Determine the program stakeholders (inside and outside of the enterprise) and establish and maintain appropriate levels of coordination, communication and liaison with these parties.
- Establish a change control process for recording, evaluating, communicating and authorizing changes to the project scope, project requirements or system design.
- Define reporting requirements for project phases.
- Verify that the project management method covers, at a minimum, the initiating, planning, executing, controlling and closing project stages; and checkpoints and approvals.
- Verify periodically with the business that the current program as designed will meet business requirements and make adjustments as necessary; review progress of individual projects and adjust the availability of resources as necessary to meet scheduled milestones.

Part II—Risk Management and Information Systems Control in Practice
2. Portfolio, Program and Project Management
F. Information Systems Control Design, Monitoring and Maintenance

Key Control Activities Related to Stakeholder Commitment/ Involvement	Key control activities for stakeholder commitment/involvement may include the following: • Obtain the commitment and participation of key stakeholders, including management of the affected user department and key end users in the initiation, definition and authorization of a project. • During project initiation, outline ongoing key stakeholder commitment and roles and responsibilities for the duration of the project life cycle. Ongoing involvement includes, but is not limited to, project approval, project phase approval, project checkpoint reporting, project board representation, project planning, product testing, user training, user procedures documentation and project communication material development.
Key Control Activities Related to the Project Scope Statement	Key control activities for the project scope statement may include the following: • Provide an approved, clear, written statement defining the nature, scope and business benefit of every project to the stakeholders to create a common understanding of project scope. • Ensure that the project definition describes the requirements for a plan that identifies internal and external project communications. • With the approval of stakeholders, maintain the project definition throughout the project, reflecting changing requirements. • Ensure that change management processes are used to manage scope throughout the project.
Key Control Activities Related to Project Phase Initiation	Key control activities for project phase initiation may include the following: • Gain approval and sign-off from designated stakeholders for the deliverables produced in each project phase. • Assess whether the project is on schedule, within budget and aligned with the agreed-on scope; assess identified variances; and identify the impact on the project plan and realization of expected benefits. • Evaluate the project at agreed-on major review points and make formal "stop/go" decisions based on predetermined critical success criteria.
Key Control Activities Related to Project Resources	Key control activities for project resource activities may include the following: • Identify resource needs for the project and clearly map out appropriate roles and responsibilities. • Identify required skills and time requirements for all individuals involved in the project phases (e.g., IT skills matrix). • Provide specialized training as required for the project. • Utilize experienced project management resources, information security, internal audit and compliance. • Clearly define and agree on the responsibility for procurement and management of third-party products and services, and manage the relationships.
Key Control Activities Related to Project Risk Management	Key control activities for project risk management may include the following: • Establish a formal, objective project risk management framework that includes identifying, analyzing, responding to, mitigating, monitoring and controlling project risk. • Perform the project risk assessment continuously throughout the project, and manage and communicate risk appropriately. • Reassess project risk as part of major change request assessments. • Identify risk and issue owners for responses to avoid, accept or mitigate risk. • Maintain a log and review a project risk register of all potential project risk.

Key Control Activities Related to the Project Quality Plan	Key control activities for the project quality plan may include the following: • Identify ownership and responsibilities, quality review processes, and success criteria and performance metrics, all to provide quality assurance for the project deliverables • Define requirements for independent validation and verification of the quality of deliverables in the plan • Ensure that vendors and suppliers deliver products according to agreed-on requirements. • Identify requirements for test plans for the project and to ensure sufficient integration with the program.
Key Control Activities Related to Project Change Control	Key control activities for project change control may include the following: • Establish a standard change request process requiring documentation of the requested change and the expected benefits of the change. • Review change requests; estimate the potential effects on the project, including resource requirements and impact on the schedule and on other related projects. • Review the completed change request and document the approval or denial of the request. • Update and communicate the project and program plans for all approved changes and communicate approved changes.
Key Control Activities Related to Portfolio/Project Performance Measurement, Reporting and Monitoring	Key control activities for portfolio/project performance measurement, reporting and monitoring include the following: • Establish and use a set of project criteria as part of the program management framework, including, but not limited to, scope, schedule, quality, cost and level of risk. • Measure project performance against key project performance criteria and analyze deviations from established key project performance criteria. • Report the progress for the program and component projects and deviations from established key project performance criteria to identified key stakeholders. • Monitor and review existing key project performance criteria to determine whether they still represent valid measures of progress. • Recommend, implement and monitor remedial action, when required, in line with the program and project governance framework.
Key Control Activities Related to Project Closure	Key control activities for portfolio/project closure may include the following: • Define and apply key steps for project closure, including postimplementation reviews that assess whether a project attained desired results and benefits. • Plan and execute a postimplementation review to determine whether a project delivered expected benefits. • Identify and track any uncompleted activities required to achieve planned program project results and benefits. • Collect, from project participants and reviewers, the lessons learned and key activities that led to delivered benefits; analyze the data; and make recommendations for improving the project management method for future projects. **IMPORTANT:** Metrics are only as reliable as the data they are based on. The risk practitioner must ensure that a data validation process is in place.

Metrics for Process Monitoring

The following table provides examples of metrics for monitoring:
- The portfolio management process
- The program and project management process

Metrics for Monitoring the Process to Manage the Portfolio					
Attribute	**Target**	**Current Period**	**Prior Period**	**Prior Period −1**	**Prior Period −2**
Percent of IT investments that have traceability to the enterprise strategy					
Degree to which enterprise management is satisfied with IT's contribution to the enterprise strategy					
Ratio between funds allocated and funds used					
Ratio between funds available and funds allocated					
Percent of business units involved in the evaluation and prioritization process					
Level of satisfaction with the portfolio monitoring reports					
Percent of changes from the investment program reflected in the relevant IT portfolios					
Percent of investments in which realized benefits have been measured and compared to the business case					

Part II—Risk Management and Information Systems Control in Practice
2. Portfolio, Program and Project Management
F. Information Systems Control Design, Monitoring and Maintenance

Metrics for Process Monitoring *(cont.)*

Metrics for Monitoring the Process to Manage Programs and Projects					
Attribute	Target	Current Period	Prior Period	Prior Period −1	Prior Period −2
Percent of stakeholders effectively engaged					
Level of stakeholder satisfaction with involvement					
Percent of stakeholders approving enterprise need, scope, planned outcome and level of project risk					
Percent of projects undertaken without approved business cases					
Percent of activities aligned to scope and expected outcomes					
Percent of active programs undertaken without valid and updated program value maps					
Frequency of status reviews					
Percent of deviations from plan addressed					
Percent of stakeholder sign-offs for stage-gate reviews of active programs					
Number of resource issues (e.g., skills, capacity)					
Percent of expected benefits achieved					
Percent of outcomes with first-time acceptance					
Level of stakeholder satisfaction expressed at project closure review					

Example of Project Status Report

The following table provides an example of a project status report.

Quarterly Project Status Report—High-visibility Projects									
Project	Budget (US $ Million [M]/ Billion [B])	Spent (US $ Million)	Percent Complete	Business Case/ Governance	Project Management	Budget Management	Internal Control	Business Process	Third-party Management
A	200 M	100 M	60	OK	OK	OK	Alert	OK	OK
B	145 M	155 M	70	OK	Problems	Problems	OK	OK	Problems
C	1.2 B	5 M	5	OK	Alert	Alert	OK	OK	Alert
D	120 M	100 M	75	Problems	Problems	Problems	OK	OK	OK
E	75 M	50 M	50	Problems	Problems	Problems	Problems	Problems	N/A

Example of Status Report Comparison

Monitoring Consideration	Status
Business case/governance	Defined and managed
Project management	On target, milestones met
Budget management	Within budget for completion percentage
Internal control	Internal audit performed a review of the project and identified internal control issues requiring attention and potential redesign.
Business process	The business is satisfied with the deliverables.
Third-party management	Consultants are providing the agreed-on assistance.

G. THE PRACTITIONER'S PERSPECTIVE

Introduction

It is important to understand the relationship between risk and project management. Because poorly managed projects are one of the most significant risk factors to the enterprise, properly conducted portfolios, programs and projects will reduce the likelihood and impact of project-related risk. This section provides an overview of how the five CRISC domains relate to project and program management in practice:
- Domain 1—Risk Identification, Assessment and Evaluation
- Domain 2—Risk Response
- Domain 3—Risk Monitoring
- Domain 4—Information Systems Control Design and Implementation
- Domain 5—Information Systems Control Monitoring and Maintenance

Domain 1—Risk Identification, Assessment and Evaluation

The risk identification, assessment and evaluation process identifies those risk scenarios that would preclude the achievement of a portfolio's, program's, and/or project's goals and objectives, including being in scope, on time and within budget in order to optimize the overall investment.

The initial identification of risk related to project management should be integrated into the project management process and overseen by an internal senior-level governance body such as an IT/business steering committee or executive committee. The project risk assessment and evaluation can be performed as part of the project initiation phase and should be repeated for material changes affecting the risk profile during each phase or milestone review.

The practitioner should focus attention on any change that precludes successful completion of the project. These include, but are not limited to, changes in the business case; architectural changes (operating system, equipment, database design, network design); staffing; third-party participation; internal control, external control and regulatory requirements; and organizational issues.

The project risk process should require the executive sponsor's approval and a report to the steering committee responsible for development.

Domain 2—Risk Response

The risk response will be based on the risk identification, assessment and evaluation. Project management has an inherent risk due to the dynamic nature of the project environment. Most project risk is a result of the project management process being intentionally or inadvertently circumvented in the course of operations. Since these issues may be a result of scheduling and/or cost overruns, skipped tasks or hurried activities, a pre-planned response process for high-frequency and/or high-impact events should be established, documented and appropriate practice drills implemented.

Risk responses require a formal approach to issues, opportunities and events to ensure that solutions are cost-effective and in alignment with business objectives. The following should be considered:
- When preparing the risk response, identify the risk in business terms: loss of productivity, disclosure of confidential information, lost opportunity costs, etc.
- Understand the business risk appetite.
- Take an integrated approach—business, IT and project management need to provide a cohesive response.
- Risk responses requiring an investment should be supported by a well-thought-out business case that justifies the expenditure, outlines alternatives and describes the justification for the alternative selected.

Domain 3—Risk Monitoring	Risk monitoring provides the communication between the affected stakeholders. Most problems occur suddenly and are a result of human error or hardware failure. When solving problems, many times the underlying symptoms are ignored and each problem is approached as an individual incident. The function of risk monitoring is to minimize this occurrence through effective key risk indicators (KRIs). The KRIs cited in this document are generic. The identification of meaningful KRIs is essential to transparency and effective communication with stakeholders and requires attention. The KRI should describe the risk issue, provide a trend line to identify performance patterns and measure remediation processes. An issue monitoring process should be implemented to provide IT and the stakeholders with an understanding of open issues, an estimated issue closure date, a means to prioritize and track issue resolution, and status reports to interested parties.
Domain 4—Information Systems Control Design and Implementation	Risk frameworks require effectively designed controls to address inherent risk. After the application of controls, the risk framework provides a residual risk level that is within levels and costs that are acceptable to the stakeholders. Control design must be a part of the system development life cycle (SDLC) process. Subject matter experts from the business (management and operations), IT development, IT operations, internal control and information security should have a role. Requirements related to project management should be defined during the project requirements definition and should take into consideration the SDLC and architectural review processes. The cost associated with processes required to achieve the requirements should be defined and accepted by the business. This may include tollgate reviews. Automated tools should be considered for reporting project status. KRIs designed to monitor the key project-related processes should be included in the control design.
Domain 5—Information Systems Control Monitoring and Maintenance	Effective monitoring of project management should be built into an effective control monitoring system, e.g., a project management office (PMO). Timely reporting using control metrics defined previously should be distributed to stakeholders within affected IT and business units. Reviews of, and reports on high-profile projects should be prepared and delivered according to a defined schedule based on business requirements. Dashboard reporting can be effective in highlighting issues and facilitating the identification of trends. These reports should be shared with senior IT and business management. Some enterprises also share these reports with the board of director's audit committee. Annual evaluation of the maturity of the project management controls provides a barometer of controls in their current state, comparison to previous periods, and the target maturity level. Target maturity levels should be agreed on among stakeholders. Maturity assessment can be performed as a self-assessment, with oversight by an objective third party, peer review or independent assessment (internal audit or external provider).

H. SUGGESTED RESOURCES FOR FURTHER STUDY

Suggested Resources for Further Study

In addition to the resources cited throughout this manual, the following resources are suggested for further study:
- ISACA:
  - COBIT 5, 2012
  - *Enterprise Value: Governance of IT Investments, The Val IT Framework 2.0*, 2008
  - *Implementing and Continually Improving IT Governance*, 2010
  - *Systems Development and Project Management Audit/Assurance Program*, 2009
  - *The Risk IT Framework*, 2009
  - *The Risk IT Practitioner Guide*, 2009
- Office of Government Commerce (OGC):
  - *Projects in Controlled Environments 2 (PRINCE2): Directing Successful Projects With PRINCE2*, UK, 2009
  - *Projects in Controlled Environments 2 (PRINCE2): Managing Successful Projects With PRINCE2*, UK, 2009
- Project Management Institute (PMI), *A Guide to the Project Management Body of Knowledge (PMBOK)*, 4th Edition, USA, 2008

3. Change Management

A. CHAPTER OVERVIEW

Learning Objectives

Because the change management process has a high level of inherent risk, the CRISC candidate should have a comprehensive understanding of the change management process itself and how it mitigates the risk related to implementing changes.

Introduction

This chapter provides an overview of the change management process and:
- Explains its importance to achieving business objectives
- Outlines a high-level process overview
- Introduces related key concepts
- Presents examples of common risk related to change management
- Lists selected key risk indicators (KRIs)
- Provides examples of common key control activities supporting the change management process
- Offers suggested reading materials and references

Contents

This chapter contains the following topics:

Topic	Starting Page
A. Chapter Overview	253
B. Related Knowledge Statements	254
C. Key Terms and Concepts	255
D. Process Overview	256
E. Risk Management Considerations	259
F. Information Systems Control Design, Monitoring and Maintenance	261
G. The Practitioner's Perspective	254
H. Suggested Resources for Further Study	266

B. RELATED KNOWLEDGE STATEMENTS

Contents

The following table lists the applicable knowledge statements from the CRISC job practice.

No.	Knowledge Statement (KS) Knowledge of:
KS1.13	Threats and vulnerabilities related to the system development life cycle
KS1.14	Threats and vulnerabilities related to project and program management
KS1.16	Threats and vulnerabilities related to management of IT operations
KS4.12	Controls related to management of IT operations
KS5.5	Control objectives, activities and metrics related to IT operations and business processes and initiatives

C. KEY TERMS AND CONCEPTS

Introduction

This section introduces terms and principles related to change management as well as terms that help relate the process to other key business processes.

Change Management

Effective change management—within the IT service domain—helps ensure that changes to the IT infrastructure are applied using standardized methods and procedures to minimize the number and impact of any related incidents upon service. Change management comprises that aspect of the overall IT governance framework, which is enterprise sensitive, dealing with IT configuration, release and problem management issues.

> **Note:** Changes in the IT infrastructure may arise as part of project, program or service improvement initiatives, based on IT services-related problems or in response to changing legislative requirements.

Exhibit 3.1: Relationship Among Change Management, Release Management, Problem Management and Configuration Management

Exhibit 3.1 describes the relationship among change management, configuration management, release management and problem management.

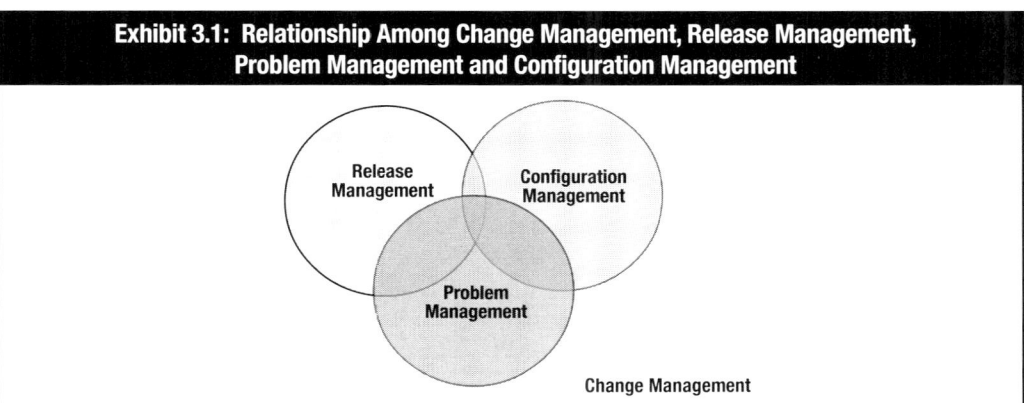

Change Management should specifically be identified as the governance framework of processes and controls within which configuration management (baseline definition and maintenance control), release management (software and hardware release controls) and problem management (issue resolution) take place.

Configuration Management

The control of changes to a set of configuration items over a system life cycle

Configuration management focuses on the processes relating to the establishment and maintenance of hardware and software baselines, hardware versions and models, software versions (commercial off-the shelf software [COTS]) and release baselines (changes to internally developed systems) and related service agreements

Release Management

Focuses on the software release life cycle processes, ranging from the initial development of a piece of software to its eventual release, and creating updated versions of the released version to help improve the software or to fix existing software bug(s). This terminology can also be applied to the hardware acquisition, deployment, upgrade and retirement processes.

Problem Management

Systematic assessment and resolution through a defined set of IT change and control activities, of technical and procedural issues arising from the use of IT systems in a production environment.

D. PROCESS OVERVIEW

Introduction
This section introduces the change management process and its importance to the achievement of business objectives. It should be remembered that the purpose of the process of change management is to allow and manage change where appropriate, and also to ensure that changes are performed correctly; only as authorized; and in accordance with a process of review, approval and feedback.

Relevance
Changes are necessary to permit upgrades to systems and processes, deploy new technologies, fix vulnerabilities and add new functionality; however, change to systems, networks, projects, controls and business processes poses a serious risk to the enterprise. As a result of the change, errors may be introduced into the system or process and system and security control functionality may be impacted. The process of change management helps mitigate the risk that changes to information systems might negatively impact the stability or integrity of the production environment or weaken the control environment.

Process Description
Change management manages all changes in a controlled manner, including standard changes and emergency maintenance relating to business processes, applications and infrastructure. This includes change standards and procedures, impact assessment, prioritization, authorization, emergency changes, tracking, reporting, closure, and documentation.

Process Goal
The goal of change management is to enable fast and reliable delivery of change to the business and mitigation of the risk of negatively impacting the stability or integrity of the changed environment.

Management Practices
Key management practices related to change management are:
1. Evaluate, prioritize and authorize change requests.
2. Manage emergency changes.
3. Track and report change status.
4. Close and document the changes.

3. Change Management
D. Process Overview

1. Evaluate, prioritize and authorize change requests.

Evaluate all requests for change to determine the impact on business processes and IT services, and to assess whether change will adversely affect the operational environment and introduce unacceptable risk. Ensure that changes are logged, prioritized, categorized, assessed, authorized, planned and scheduled.

Related activities include:
1. Use formal change requests to enable business process owners and IT to request changes to business process, infrastructure, systems or applications. Make sure that all such changes arise only through the change request management process.
2. Categorize all requested changes (e.g., business process, infrastructure, operating systems, networks, application systems, purchased/packaged application software) and relate affected configuration items.
3. Prioritize all requested changes based on the business and technical requirements; resources required; and the legal, regulatory and contractual reasons for the requested change.
4. Plan and evaluate all requests in a structured fashion. Include an impact analysis on business process, infrastructure, systems and applications, business continuity plans (BCPs), and service providers to ensure that all affected components have been identified. Assess the likelihood of adversely affecting the operational environment and the risk of implementing the change. Consider security, legal, contractual and compliance implications of the requested change. Consider also inter-dependencies among changes. Involve business process owners in the assessment process, as appropriate.
5. Formally approve each change by business process owners, service managers and IT technical stakeholders, as appropriate. Changes that are low-risk and relatively frequent should be preapproved as standard changes.
6. Plan and schedule all approved changes.
7. Consider the impact of contracted services providers (e.g., outsourced business processing, infrastructure, application development and shared services) on the change management process, including integration of organizational change management processes with change management processes of service providers and the impact on contractual terms and SLAs.

2. Manage emergency changes.

Carefully manage emergency changes to minimize further incidents and make sure the change is controlled and takes place securely. Verify that emergency changes are appropriately assessed and authorized after the change.

Related activities include:
1. Ensure that a documented procedure exists to declare, assess, give preliminary approval, authorize after the change and record an emergency change.
2. Verify that all emergency access arrangements for changes are appropriately authorized, documented and revoked after the change has been applied.
3. Monitor all emergency changes, and conduct postimplementation reviews involving all concerned parties. The review should consider and initiate corrective actions based on root causes such as problems with a business process, application system development and maintenance, development and test environments, documentation and manuals, and data integrity.
4. Define what constitutes an emergency change.

3. Track and report change status.	Maintain a tracking and reporting system to document rejected changes, communicate the status of approved and in-process changes, and completed changes. Make certain that approved changes are implemented as planned. Related activities include: 1. Categorize change requests in the tracking process (e.g., rejected, approved but not yet initiated, approved and in process, and closed). 2. Implement change status reports with performance metrics to enable management review and monitoring of both the detailed status of changes and the overall state (e.g., aged analysis of change requests). Ensure that status reports form an audit trail so changes can subsequently be tracked from inception to eventual disposition. 3. Monitor open changes to ensure that all approved changes are closed in a timely fashion, depending on priority. 4. Maintain a tracking and reporting system for all change requests.
4. Close and document the changes.	When changes are implemented, update accordingly the solution and user documentation and the procedures affected by the change. Related activities include: 1. Include changes to documentation (e.g., business and IT operational procedures, business continuity and disaster recovery documentation, configuration information, application documentation, help screens, and training materials) within the change management procedure as an integral part of the change. 2. Define an appropriate retention period for change documentation and pre- and post-change system and user documentation. 3. Subject documentation to the same level of review as the actual change.
Relationship to Problem Management	The change management process is often used to repair a vulnerability or follow up on a recommendation that is a finding from the incident and problem management processes. The CRISC candidate may want to refer to Part II, chapter 8—Problem Management for further information on this topic.

E. RISK MANAGEMENT CONSIDERATIONS

Introduction

This section discusses risk management practices related to change management. The following points are addressed:
- Risk factors
- Generic threats and common vulnerabilities
- Generic risk scenarios
- Key risk indicators (KRIs)

Risk Factors

Examples of factors affecting risk related to change management are:
- **System complexity**—Size, complexity, geographic range, etc., all affect the ease with which change management can be implemented and maintained.
- **Organizational structure**:
  - Infrastructure: insourced, facilities managed, outsourced (specific facility), outsourced (cloud)
  - Organizational structure: insourced, outsourced (consultants), hybrid
  - Control environment: mature, immature, resilient
  - Applications-source: in-house, purchased off the shelf, customized
  - Application portfolio: managed, *ad hoc* (informal)
  - Hardware/software purchasing: centralized, decentralized
  - Geographic location: single location, multiple locations, international, global
- **System criticality**—The process used to assess the effect that each change has on the core business processes of the enterprise. Effecting a change on a critical system may be difficult when the system cannot be turned off to allow the change to be implemented.
- **Change management capability**—Maturity of the change management process and the ability to monitor, document and authorize change activities
- **Rate of change**—The volatility of the systems and the rate of change to business processes, systems and procedures

Generic Threats and Common Vulnerabilities

Generic threats and common vulnerabilities related to the change management process are:
- A change control process that is not well-managed or enforced and allows changes or exceptions without proper review and validation
- Lack of a separate environment for testing and development, which causes developers and administrators to effect changes directly in production
- Unauthorized changes resulting in compromised security and unauthorized changes or access to business processes and data
- Lack of documentation of previous changes to understand the reason and effect of the change
- The inability to monitor application changes affecting business processes and data
- Emergency changes are not documented; the potential for unauthorized changes to be implemented as part of emergency changes
- Configuration documentation that does not reflect the current system configuration
- Failure to meet compliance requirements (access control, system availability, etc.)
- Changes resulting in bypassing of security controls
- Increased dependence on key individuals, lack of cross-training or knowledge of systems
- Reduced system availability and impact on performance due to changes
- Unintended processing side effects
- Adverse effects on capacity and performance of the infrastructure

Generic Risk Scenarios	Examples of generic risk scenarios related to change management are: • Change does not meet business objectives • Change causes system outages (everything worked until the change) • Data integrity or availability errors caused by revised process • Inability to implement required changes quickly enough due to inflexibility of change management process • Vendor patches not available due to outdated systems or equipment • Changes to scope of projects causing project failure
KRIs	Examples of KRIs are the: • Number of disruptions or data errors caused by an inaccurate assessment of the impact of changes • Number of incidents of application code rework caused by incorrect change specifications • Ratio of emergency fixes to total changes • Number of change requests in the backlog • Time between release of vendor patch and rollout of change • Failure to document changes via a change management system • Changes not formally tracked, reported or authorized • Number of changes promoted to production without management approval • Percentage of changes reviewed with business for feedback and customer satisfaction

F. INFORMATION SYSTEMS CONTROL DESIGN, MONITORING AND MAINTENANCE

Introduction

This section provides an overview of common controls related to the change management process. The following points are addressed:
- Key control activities
- IS control metrics
- Monitoring practices

Key Control Activities

Key control activities for the following aspects of the change management process are described in this section:
- Change standards and procedures
- Impact assessment, prioritization and authorization
- Emergency changes
- Change status tracking and reporting
- Change closure and documentation

Key Control Activities Related to Change Standards and Procedures

Set up formal change management procedures to handle, in a standardized manner, all requests (including maintenance and patches) for changes to applications, procedures, processes, networks and network configurations, system and service parameters, and the underlying platforms:
- Develop, document and promulgate a change management framework that specifies the policies and processes regarding:
  - Roles and responsibilities
  - Classification and prioritization of all changes based on business and security risk
  - Assessment of impact
  - Authorization and approval of all changes by the business process owners and IT
  - Tracking and status of changes
  - The impact on data integrity (e.g., all changes to data files being made under system and application control rather than by direct user intervention)
- Establish and maintain version control over all changes.
- Implement roles and responsibilities that involve business process owners and technical IT functions, with an appropriate segregation of duties.
- Establish record management practices and audit trails to record key steps in the change management process. Ensure timely closure of changes. Elevate and report to management changes that are not closed in a timely fashion.
- Consider the impact of contracted services providers (e.g., of infrastructure, application development and shared services) on the change management process, integration of organizational change management processes with change management processes of service providers, and the impact of the organizational change management process on contractual terms and service level agreements (SLAs).

Key Control Activities Related to Impact Assessment, Prioritization and Authorization	Assess all requests for change in a structured way to determine the impact on the operational system and its functionality. Ensure that changes are categorized, prioritized and authorized. Develop a process to allow business process owners and/or IT staff to request changes to infrastructure, systems or applications. Develop controls to ensure that all such changes are processed only through the formal change request management process: • Categorize all requested changes according to systems or areas affected (e.g., infrastructure, operating systems, networks, application systems, purchased/packaged application software). • Prioritize all requested changes so that the change management process identifies both the business as well as technical and security needs for the change. Consider legal, regulatory and contractual reasons for the requested change. • Assess the impact of all requests in a structured fashion. The assessment process should address impact analysis on infrastructure, systems and applications. Consider security, legal, contractual and compliance implications of the requested change and also interdependencies among changes. Involve business process owners in the assessment process, as appropriate. • Estimate the resources and time required to develop, test and implement the change and the impact on current workload. • Determine the acceptable window of downtime for rollout of the change • Each change should be formally approved by business process owners and IT technical stakeholders, as appropriate.
Key Control Activities Related to Emergency Changes	Establish a process for defining, escalating, testing, documenting, assessing and authorizing emergency changes that do not follow the established change process: • A documented process exists within the overall change management process to declare, assess, authorize and record an emergency change. • Emergency changes are processed in accordance with the emergency change element of the formal change management process. • Ensure that emergency access arrangements for changes are appropriately authorized, documented and revoked after the change has been applied. • Conduct a postimplementation review of all emergency changes, involving all concerned parties. The review should consider implications for aspects such as further application system maintenance and impact on development. • Ensure that all temporary files needed for the change are removed from production systems.
Key Control Activities Related to Change Status Tracking and Reporting	Establish a tracking and reporting system to document rejected changes, communicate the status of approved and in-process changes, and complete changes. Make certain that approved changes are implemented as planned: • Establish a process to allow tracking of the status of requests throughout the various stages of the change management process. • Implement change status reports with performance metrics to enable management review and monitoring of both the detailed status of changes and the overall state (e.g., aged analysis of change requests). • Monitor open changes so that all approved and completed changes are closed in a timely fashion, depending on priority.

Key Control Activities Related to Change Closure and Documentation

Whenever changes are implemented, update the associated system and user documentation and procedures accordingly:
- Ensure that documentation—including operational procedures, configuration information, application documentation, help screens and training materials—follows the same change management procedure and is considered to be an integral part of the change.
- Consider an appropriate retention period for change documentation and pre- and post-change system and user documentation.
- Update business processes for changes in hardware or software to ensure that new or improved functionality is used.
- Update business continuity and disaster recovery plans where appropriate.
- Subject documentation to the same level of testing as the actual change.

Metrics for Process Monitoring

Metrics for Monitoring the Process to Manage Changes					
Attribute	Target	Current Period	Prior Period	Prior Period − 1	Prior Period − 2
Amount of rework caused by failed changes					
Reduced time and effort required to make changes					
Number and age of backlogged change requests					
Percent of unsuccessful changes due to inadequate impact assessments					
Percent of total changes that are emergency fixes					
Number of emergency changes not authorized after the change					
Stakeholder feedback ratings on satisfaction with communications					

G. THE PRACTITIONER'S PERSPECTIVE

Introduction

Change management is one of the highest risk areas within IT operations. Business processes formerly performed manually by employees are revised and executed using programs implemented by the change management process. The introduction of unauthorized or incorrectly designed business processes can undermine the reliability and integrity of the business processes. The change management process is the primary control point for ensuring the integrity of business processes and the rollout of the change. A failure to implement the change correctly could result in data or processing integrity errors, interruption in service or financial penalties. The following section provides an overview of how the five CRISC domains (listed below) relate to change management in practice:
- Domain 1—Risk Identification, Assessment and Evaluation
- Domain 2—Risk Response
- Domain 3—Risk Monitoring
- Domain 4—Information Systems Control Design and Implementation
- Domain 5—Information Systems Control Monitoring and Maintenance

Domain 1—Risk Identification, Assessment and Evaluation

The risk identification, assessment and evaluation process is key to the determination of which changes are implemented. The risk process:
- Determines the effect the change will have on core business activities, security and the supporting infrastructure
- Evaluates alternatives to the proposed change and its effect on the core business processes
- Manages interrelationships with other business processes and potential consequences and changes to the interfacing systems
- Prioritizes proposed changes
- Provides a basis for scheduling resources and implementations

The risk practitioner should focus attention on how changes are initially identified and then entered into a change management tracking system, and on the quality of the initial risk assessment performed prior to the change project being assigned to a development team.

Domain 2—Risk Response

The risk response will be based on the risk identification, assessment and evaluation.
Risk response requires a formal approach to issues, opportunities and events. This will ensure that solutions are in alignment with the business objectives, do not negatively impact other processes, and are cost-effective. The following should be considered:
- When preparing the risk response, identify the risks in business terms: loss of productivity, disclosure of confidential information, lost opportunity costs, violation of regulatory compliance, etc.
- Understand the business risk appetite, acceptable service interruptions, confidentiality of data, compliance requirements, etc.
- Keep the business stakeholders apprised of identified risks and how IT will respond to these risks.
- Ensure that steps are taken to identify and investigate any unauthorized changes

Domain 3—Risk Monitoring

Risk monitoring has the potential to provide management with an early warning of change management issues. The key risk indicators (KRIs) described previously should identify negative trends in the change process and in the change project, which require management oversight and follow-up. An issue monitoring process should be implemented to provide IT and the stakeholders with an understanding of open issues, estimated issue closure dates, a means to prioritize and track issue resolution, and status reports to interested parties.

Domain 4—Information Systems Control Design and Implementation	The IS control objectives and activities described previously provide an excellent resource of leading practice processes to ensure an effective change management function. The risk practitioner should periodically review the control design against stated policy and control requirements. Where the design effectiveness of the control is of questionable or negative value, enhancement or replacement of the control should be considered.
Domain 5—Information Systems Control Monitoring and Maintenance	IS control monitoring and maintenance requires the following: • Timely reporting using the control metrics defined previously should be distributed to stakeholders within IT and the business units affected. • Annual evaluation of the maturity of controls provides a barometer of controls in their current state, comparison to previous periods, and the target maturity level. Target maturity levels should be agreed on by stakeholders and evaluated by senior IT and business operations management as part of the IT annual review. • Maturity assessment can be performed as a self-assessment, with oversight by an objective third party, peer review, or independent assessment (internal audit or external provider). • Change-related issue monitoring should be reported and evaluated routinely.

H. SUGGESTED RESOURCES FOR FURTHER STUDY

Suggested Resources for Further Study

In addition to the resources cited throughout this manual, the following resources are suggested for further study:

- ISACA:
  - *Change Management Audit/Assurance Program*, 2009
  - *COBIT 5*, 2012
  - *Enterprise Value: Governance of IT Investments, The Val IT Framework 2.0*, 2008
  - *Implementing and Continually Improving IT Governance*, 2010
  - *The Risk IT Framework*, 2009
  - *The Risk IT Practitioner Guide*, 2009
- ITIL Service Operations, Her Majesty's Government Cabinet Office, UK, 2011
- Kouns, Jake; Daniel Minoli; *Information Technology Risk Management in Enterprise Environments: A Review of Industry Practices and a Practical Guide to Risk Management Teams*, Wiley-Interscience, USA, 2010
- Westerman, George; Richard Hunter; *IT Risk: Turning Business Threats Into Competitive Advantage*, Harvard Business School Press, USA, 2007

4. Third-party Service Management

A. CHAPTER OVERVIEW

Learning Objectives

The CRISC candidate should have a general understanding of the third-party service management. There are many types of third-party services available, each with its own risk and benefits. The risk practitioner must carefully examine the conditions of the third-party agreement as a part of the risk management process.

Introduction

This chapter provides an overview of the third-party service management process and:
- Explains its importance to achieving business objectives
- Outlines a high-level process overview
- Introduces related key concepts
- Presents examples of common risk
- Lists selected key risk indicators (KRIs)
- Provides examples of common key control activities supporting the process
- Describes the practitioner's perspective
- Offers suggested reading materials and references

Contents

This chapter contains the following topics:

Topic	Starting Page
A. Chapter Overview	267
B. Related Knowledge Statements	268
C. Key Terms and Concepts	269
D. Process Overview	270
E. Risk Management Considerations	273
F. Information Systems Control Design, Monitoring and Maintenance	275
G. The Practitioner's Perspective	277
H. Suggested Resources for Further Study	279

B. RELATED KNOWLEDGE STATEMENTS

Contents

The following table lists the applicable knowledge statements from the CRISC job practice.

Knowledge Statement (KS)	
No.	Knowledge of:
KS1.11	Threats and vulnerabilities related to third-party management
KS4.7	Controls related to third-party management
KS5.10	Control objectives, activities and metrics related to third-party management

C. KEY TERMS AND CONCEPTS

Introduction	This section introduces terms and principles related to third-party service management as well as terms that help relate the process to other key business processes.
Infrastructure as a Service (IaaS)	Offers the capability to provision processing, storage, networks and other fundamental computing resources, enabling the customer to deploy and run arbitrary software, which can include operating systems (OSs) and applications
Nondisclosure Agreement (NDA)	A legal contract between at least two parties that outlines confidential materials that the parties wish to share with one another for certain purposes, but wish to restrict from generalized use; a contract through which the parties agree not to disclose information covered by the agreement
Operational Level Agreement (OLA)	An internal agreement covering the delivery of services that support the IT organization in its delivery of services
Platform as a Service (PaaS)	Offers the capability to deploy onto the cloud infrastructure customer-created or -acquired applications that are created using programming languages and tools supported by the provider
Service Level Agreement (SLA)	An agreement, preferably documented, between a service provider and the customer(s)/user(s) that defines minimum performance targets for a service and how they will be measured
Software as a Service (SaaS)	Offers the capability to use the provider's applications running on cloud infrastructure. The applications are accessible from various client devices through a thin client interface such as a web browser (e.g., web-based email).
Statement of Work (SOW)	A formal document that captures and defines the work activities, deliverables, and time line a vendor must execute in performance of specified work for a client. The SOW usually includes detailed requirements and pricing, with standard regulatory and governance terms and conditions

D. PROCESS OVERVIEW

Introduction

This section introduces the third-party service management process and provides information about its:
- Relevance to organizational objectives and/or risk management.
- Process description
- Process goal
- Process-specific management practices
- Process-specific activities

Relevance

As enterprises focus on their core business and look to reduce cost, outsourcing of various processes may be an appropriate alternative.

The need to ensure that services provided by third parties (suppliers, vendors and partners) meet business requirements requires an effective third-party management process. This process requires the:
- Clear definition of roles, responsibilities and expectations in third-party agreements
- Specification of process inputs and outputs
- Definition of boundaries and interfaces
- Ability to monitor and provide assurance of third-party compliance with contractual and regulatory commitments

Effective management of third-party services minimizes the business risk associated with nonperforming or noncompliant business partners.

Process Description

Managing third parties includes managing IT-related services provided by all types of suppliers to meet enterprise requirements, including the:
- Selection of suppliers
- Management of relationships
- Management of contracts
- Monitoring of supplier performance for effectiveness and compliance.

Process Goal

The goal of third-party management is to minimize the risk associated with nonperforming suppliers and ensure competitive pricing.

Management Practices

Key management practices of third-party service management are:
1. Identify and evaluate supplier relationships and contracts.
2. Select suppliers.
3. Manage supplier relationships and contracts.
4. Manage supplier risk.
5. Monitor supplier performance and compliance.

1. Identify and evaluate supplier relationships and contracts.

Identify suppliers and associated contracts and categorize them into type, significance and criticality. Establish supplier and contract evaluation criteria and evaluate the overall portfolio of existing and alternative suppliers and contracts.

Related activities include:
1. Establish and maintain criteria relating to type, significance and criticality of suppliers and supplier contracts, enabling a focus on preferred and important suppliers.
2. Establish and maintain supplier and contract evaluation criteria to enable overall review and comparison of supplier performance in a consistent way.
3. Identify, record and categorize existing suppliers and contracts according to defined criteria to maintain a detailed register of preferred suppliers that need to be managed carefully.
4. Periodically evaluate and compare the performance of existing and alternative suppliers to identify opportunities or a compelling need to reconsider current supplier contracts.

2. Select suppliers.

Select suppliers according to a fair and formal practice to ensure a viable best fit based on specified requirements. Requirements should be optimized with input from potential suppliers.

Related activities include:
1. Review all requests for information (RFIs) and requests for proposals (RFPs) to ensure that they:
   - Clearly define requirements
   - Include a procedure to clarify requirements
   - Allow vendors sufficient time to prepare their proposals
   - Clearly define award criteria and the decision process
2. Evaluate RFIs and RFPs in accordance with the approved evaluation process/criteria, and maintain documentary evidence of the evaluations. Verify the references of candidate vendors.
3. Select the supplier that best fits the RFP. Document and communicate the decision, and sign the contract.
4. In the specific case of software acquisition, include and enforce the rights and obligations of all parties in the contractual terms. These rights and obligations may include ownership and licensing of intellectual property; maintenance; warranties; arbitration procedures; upgrade terms; and fit for purpose, including security, escrow and access rights.
5. In the specific case of acquisition of development resources, include and enforce the rights and obligations of all parties in the contractual terms. These rights and obligations may include ownership and licensing of intellectual property; fit for purpose, including development methodologies; testing; quality management processes, including required performance criteria; performance reviews; basis for payment; warranties; arbitration procedures; human resource management; and compliance with the enterprise's policies.
6. Obtain legal advice on resource development acquisition agreements regarding ownership and licensing of intellectual property.
7. In the specific case of acquisition of infrastructure, facilities and related services, include and enforce the rights and obligations of all parties in the contractual terms. These rights and obligations may include service levels, maintenance procedures, access controls, security, performance review, basis for payment and arbitration procedures.

3. Manage supplier relationships and contracts.

Formalize and manage the supplier relationship for each supplier. Manage, maintain and monitor contracts and service delivery. Ensure that new or changed contracts conform to enterprise standards and legal and regulatory requirements. Deal with contractual disputes.

Related activities include:
1. Assign relationship owners for all suppliers and make them accountable for the quality of services provided.
2. Specify a formal communication and review process, including supplier interactions and schedules.
3. Agree on, manage, maintain and renew formal contracts with the supplier. Ensure that contracts conform to enterprise standards and legal and regulatory requirements.
4. Within contracts with key service suppliers, include provisions for the review of supplier site and internal practices and controls by management or independent third parties.
5. Evaluate the effectiveness of the relationship and identify necessary improvements.
6. Define, communicate and agree on ways to implement required improvements to the relationship.
7. Use established procedures to deal with contract disputes, first using, wherever possible, effective relationships and communications to overcome service problems.
8. Define and formalize roles and responsibilities for each service supplier. Where several suppliers combine to provide a service, consider allocating a lead contractor role to one of the suppliers to take responsibility for an overall contract.

4. Manage supplier risk.

Identify and manage risk relating to suppliers' ability to continually provide secure, efficient and effective service delivery.

Related activities include:
1. Identify, monitor and, where appropriate, manage risk relating to the supplier's ability to deliver service efficiently, effectively, securely, reliably and continually.
2. When defining the contract, provide for potential service risk by clearly defining service requirements, including software escrow agreements, alternative suppliers or standby agreements to mitigate possible supplier failure; security and protection of intellectual property; and any legal or regulatory requirements.

5. Monitor supplier performance and compliance.

Periodically review the overall performance of suppliers, compliance to contract requirements, and value for money, and address identified issues.

Related activities include:
1. Define and document criteria to monitor supplier performance aligned with SLAs and ensure that the supplier regularly and transparently reports on agreed-on criteria.
2. Monitor and review service delivery to ensure that the supplier is providing an acceptable quality of service, meeting requirements and adhering to contract conditions.
3. Review supplier performance and value for money to ensure that suppliers are reliable and competitive, compared with alternative suppliers and market conditions.
4. Request independent reviews of supplier internal practices and controls, if necessary.
5. Record and assess review results periodically and discuss them with the supplier to identify needs and opportunities for improvement.
6. Monitor and evaluate externally available information about the supplier.

E. RISK MANAGEMENT CONSIDERATIONS

Introduction

This section discusses risk management practices related to the third-party service management process. The following points are addressed:
- Risk factors
- Generic threats and common vulnerabilities
- Generic risk
- Key risk indicators (KRIs)

Risk Factors

Examples of factors affecting risk related to third-party service management are:
- **Availability of systems and data**—outsourcing of critical business processes may be a risk if the outsource provider is not able to maintain systems or network availability.
- **Sensitivity of data**—the outsourcing company must ensure that adequate protection is in place to prevent the disclosure of sensitive information by the vendor or the storage of sensitive data in another jurisdiction in violation of data protection regulations.
- **Criticality and uniqueness of the third-party relationship**—The reliance the business puts on a single service provider (e.g., sole, single, preferred or generic supplier)
- **Service provider attributes**—Geographic location, proximity, similarity of business and service provider organizational model and culture, financial health, language, etc.
- **Service provider portfolio management capability**—The diligence, with which service providers are identified, classified and tracked
- **Type of third-party service**—Systems development, systems infrastructure, IT operations, facilities management, applications, services
- **Level of integration**—Integration of the service provider into the enterprise
- **Existence and monitoring of SLAs**—The level and thoroughness with which service levels are established, monitored and resolved
- **Regulatory requirements**—Limitations on third-party usage and transmission or storage of data across international borders

Generic Threats and Common Vulnerabilities

Generic threats and common vulnerabilities related to the third-party service management process are:
- The inability to audit or review costs and service quality
- Limited choice of suppliers
- A supplier that is not responsive or committed to the relationship
- Unclear roles and responsibilities leading to miscommunication, poor service and increased cost
- Lack of monitoring of the vendor relationship
- Lack of accountability for errors, service disruptions, incidents or noncompliance issues

Generic Risk Scenarios

Examples of generic risk scenarios related to the third-party service management process are:
- Financial losses and reputational damage resulting from service interruption
- Inadequate service quality
- Noncompliance with regulatory and legal obligations
- Unresolved problems and issues
- Security breaches and other incidents
- An undetected service degradation
- Instances in which the provider goes bankrupt or may be bought by a competing firm

KRIs The risk practitioner can use KRIs to monitor and report on the status of risk associated with third-party service providers. Some examples of KRIs are:
- An excessive number of user complaints due to contracted services
- A purchase expenditure that is not subject to competitive procurement
- SLAs not meeting or exceeding agreed-on metrics
- An excessive number of formal disputes with suppliers
- Excessive or repeated supplier invoice disputes
- Major suppliers not subject to clearly defined requirements and service levels
- A low level of business satisfaction with the effectiveness of communication from the supplier
- A low level of supplier satisfaction with the effectiveness of communication from the business
- An excessive number of significant incidents of supplier noncompliance per time period
- Suppliers without a dedicated relationship manager

F. INFORMATION SYSTEMS CONTROL DESIGN, MONITORING AND MAINTENANCE

Introduction

This section provides an overview of common controls related to the third-party service management process. The following points are addressed:
- Key control activities
- IS control metrics
- Monitoring practices

Key Control Activities

Key control activities for the third-party service management process are:
- Identification of all supplier relationships
- Supplier relationship management
- Supplier risk management
- Supplier performance management

Key Control Activities Related to Identification of All Supplier Relationships

Identification of key control activities related to supplier relationships may include:
- Define and regularly review criteria to identify and categorize all supplier relationships according to the supplier type, significance and criticality of service. The list should include a category describing vendors as preferred, non-preferred or not recommended.
- Establish and maintain a detailed register of suppliers, including name, scope, purpose of the service, expected deliverables, service objectives and key contact details.

Key Control Activities Related to Supplier Relationship Management

Key control activities related to supplier relationship management may include:
- Define and formalize roles and responsibilities through SLAs for each service supplier. Ensure that all contracts with suppliers identify the jurisdiction for any legal disputes.
- Assign relationship owners for all suppliers and make them accountable for the quality of service(s) provided.
- Document the supplier relationship managers and communicate the information within the enterprise.
- Establish and document a formal communication process between the enterprise and the service providers.
- Ensure that contracts with key service suppliers provide for a review of supplier internal controls by management or independent third parties.
- Regularly review the reports between the enterprise and the service suppliers.
- Register incidents caused by suppliers and report them using the enterprise's internal incident management process.
- Periodically review and assess supplier performance against established and agreed-on service levels while clearly communicating suggested changes to the service supply contracts or recommendations regarding new suppliers.

Key Control Activities Related to Supplier Risk Management

Key control activities relating to supplier risk management may include:
- Identify and monitor suppliers in accordance with the enterprise's established risk management process.
- Identify and document in the contract supplier risks (and remedies) associated with the suppliers' inability to fulfill the contractual agreements.
- Consider remedies when defining the contract—including software escrow agreements, exit strategies, alternative suppliers or standby agreements—in the event of supplier failure.
- Review all contracts for legal and regulatory requirements. Ensure that contracts include which legal jurisdiction would apply in case of any dispute.

Part II—Risk Management and Information Systems Control in Practice
4. Third-party Service Management
F. Information Systems Control Design, Monitoring and Maintenance

Key Control Activities Related to Supplier Performance Management

Key control activities related to supplier performance management may include:
- Define and document criteria to monitor service suppliers' performance.
- Ensure that suppliers regularly report on agreed-on performance criteria.
- Invite users to provide feedback for assessment of supplier performance and quality of service.
- Evaluate the costs and market conditions for the service levels by benchmarking against alternative suppliers, and identify the potential for improvement.
- Define arbitration procedures to consult an arbitration committee in the event that any action must be undertaken.

Metrics for Process Monitoring

The following table provides examples of metrics for monitoring the third-party service management process.

Metrics for Monitoring the Process to Manage Third Parties					
Attribute	Target	Current Period	Prior Period	Prior Period −1	Prior Period −2
Percent of suppliers meeting agreed-on requirements					
Number of service breaches to IT-related services caused by suppliers					
Number of risk-related events leading to service incidents					
Frequency of risk management sessions with supplier					
Percent of risk-related incidents resolved acceptably (time and cost)					
Number of supplier review meetings					
Number of formal disputes with suppliers					
Percent of disputes resolved amicably in a reasonable time frame					
Number of business disruptions due to IT service incidents					
Percent of business stakeholders satisfied that IT service delivery meets agreed-on service levels					
Percent of users satisfied with the quality of IT service delivery					

G. THE PRACTITIONER'S PERSPECTIVE

Introduction

The use of third-party service providers presents a unique risk to the enterprise. Outsourcing business processes leaves the enterprise vulnerable to risks beyond its organizational borders. The third party may provide core business services that are not easily replaced if the relationship becomes contentious or service levels are unsatisfactory. For this reason, understanding and managing risk is essential. The following section provides an overview of how the five CRISC domains (listed below) relate to third-party service management in practice:
- Domain 1—Risk Identification, Assessment and Evaluation
- Domain 2—Risk Response
- Domain 3—Risk Monitoring
- Domain 4—Information Systems Control Design and Implementation
- Domain 5—Information Systems Control Monitoring and Maintenance

Domain 1— Risk Identification, Assessment and Evaluation

The decision to outsource (utilize third-party providers) and the subsequent transfer of processing responsibility must follow a systematic process, similar to a system development life cycle (SDLC) or contract awarding process.

Risk identification needs to be comprehensive, involving IT, business units, legal, audit, internal control and information security subject matter experts. All relevant issues—for example, scope of services; ability to deliver according to acceptable service levels; cultural, language and regulatory requirements; and conditions for the transfer of personally identifiable or sensitive information—need to be included in the process.

The risk assessment process should be performed several times during the outsourcing process, for example, during:
- The outsourcing decision
- Vendor selection
- Contract negotiation

The risk assessment process should be performed at the time of contract renewal, when material regulatory changes occur and business processes change, and on a scheduled, periodic basis.

Domain 2— Risk Response

The risk response will be based on the result of the risk identification, assessment and evaluation phase.

If the third-party relationship is in the development stage, risk response will focus on working with the negotiation team or the project team responsible for servicer selection to ensure that the risk is addressed in the contract and that there are provisions in place to audit and review the performance of the third-party service provider.

If the third-party relationship has been established, the risk response should include identifying the relationship manager assigned to the service provider. Using the issue monitoring system, any issues should be defined and appropriate risk mitigation processes scheduled. The remediation process should be monitored until closure.

Domain 3—Risk Monitoring

The function of risk monitoring is to continue to evaluate the risk factors that could indicate a change in risk levels and thereby ensure that the actual risk levels that are caused by the use of third-party agreements are within the acceptable risk levels of the enterprise and that the risk mitigation controls are working effectively and, as designed, to mitigate risk. The process of continuous or periodic monitoring will minimize the development of unacceptable levels of risk through the use of effective key risk indicators (KRIs). The identification of meaningful KRIs is essential to transparency and effective communication with stakeholders and requires attention to the needs of the business and the risk culture of the enterprise.

KRIs should describe the risk, provide a trend line to identify performance patterns, and measure remediation processes.

The issue monitoring process described in risk response should be the primary vehicle to ensure that risks are being adequately mitigated.

Domain 4—Information Systems Control Design and Implementation

Requirements related to third-party management should be defined during the requirements definition phase of the SDLC. As the risk associated with the use of a third-party service provider is identified, the appropriate control requirements should be documented. As the controls are designed, the costs associated with the controls (or the level of residual risk to be accepted) required to achieve the risk management requirements should be defined and accepted by the enterprise. This may include designing audit and detection systems or services, additional monitoring and problem and change management processes, etc.

SLA and relationship management controls need definition and implementation. These are the primary oversight processes required for third-party management.

Automated tools should be designed to report when problems related to third parties are cited and should also report on the corresponding remediation process performed.

KRIs designed to monitor the key third-party processes should be included in the control design.

Domain 5—Information Systems Control Monitoring and Maintenance

The control metrics described previously are effective monitoring tools. Timely reporting using KRIs defined previously should be distributed to stakeholders within IT and the business units affected.

The relationship manager needs to be integrated with business and IT management because many early warning signs are recognizable to the relationship manager before the trend lines of the monitoring system generate an alert.

Annual evaluation of the maturity of controls provides a barometer of controls in their current state, comparison to previous periods, and the target maturity level. Target maturity levels should be agreed on by stakeholders and evaluated by senior management as part of the IT annual review.

Maturity assessment can be performed as a self-assessment, with oversight by an objective third-party, peer review, or independent assessment (internal audit or external provider).

H. SUGGESTED RESOURCES FOR FURTHER STUDY

Suggested Resources for Further Study

In addition to the resources cited throughout this manual, the following resources are suggested for further study:
- ISACA:
  - COBIT 5, 2012
  - *Enterprise Value: Governance of IT Investments, The Val IT Framework 2.0*, 2008
  - *Implementing and Continually Improving IT Governance*, 2010
  - *The Risk IT Framework*, 2009
  - *The Risk IT Practitioner Guide*, 2009
- International Organization for Standardization (ISO):
  - ISO 27001, *Information security management—Specification with guidance for use*, Switzerland, 2005. This is the replacement for BS7799-2. It is intended to provide the foundation for third-party audit and is harmonized with other management standards, such as ISO/IEC 9001 and 14001.
  - ISO/IEC 27002:2005, *Information technology—Security techniques—Code of practice for information security management,* Appendix 1, Switzerland, 2005
- National Institute of Standards and Technology (NIST), *Security Controls in External Environments*, Special Publication (SP) 800-53, Revision 3, Section 2.4, Appendix 2, USA, 2009

Part II—Risk Management and Information Systems Control in Practice
5. Continuity Management
A. Chapter Overview

5. Continuity Management

A. CHAPTER OVERVIEW

Learning Objectives

Risk is an ever-changing element of business and the environment in which the enterprise operates. The risk practitioner should have a general understanding of the continuity management process and how continuity management interrelates with risk management.

Introduction

This chapter provides an overview of the continuity management process and:
- Explains its importance to achieving business objectives
- Outlines a high-level process overview
- Introduces related key concepts
- Presents examples of common risk
- Lists selected key risk indicators (KRIs)
- Provides examples of common key control activities supporting the process
- Describes the practitioner's perspective
- Offers suggested reading materials and references

Contents

This chapter contains the following topics:

Topic	Starting Page
A. Chapter Overview	281
B. Related Knowledge Statements	282
C. Key Terms and Concepts	283
D. Process Overview	284
E. Risk Management Considerations	289
F. Information Systems Control Design, Monitoring and Maintenance	291
G. The Practitioner's Perspective	295
H. Suggested Resources for Further Study	297

B. RELATED KNOWLEDGE STATEMENTS

Contents

The following table lists the applicable knowledge statements from the CRISC job practice.

No.	Knowledge Statement (KS) Knowledge of:
KS1.15	Threats and vulnerabilities related to business continuity and disaster recovery management
KS4.11	Controls related to business continuity and disaster recovery management
KS5.15	Control objectives, activities and metrics related to business continuity and disaster recovery management

C. KEY TERMS AND CONCEPTS

Introduction	This section introduces terms and principles related to continuity management as well as terms that help relate the process to other key business processes.
Availability	Ensuring timely and reliable access to and use of information
Recovery Point Objective (RPO)	The RPO is determined based on the acceptable data loss in case of a disruption of operations. It indicates the earliest point in time that is acceptable to recover the data. The RPO effectively quantifies the permissible amount of data loss in case of interruption. (See also Recovery Time Objective [RTO].)
Recovery Strategy	An approach by an enterprise that will ensure its recovery and continuity in the face of a disaster or other major outage
Recovery Testing	A test to check the system's ability to recover after a software or hardware failure.
Recovery Time Objective (RTO)	The amount of time allowed for the recovery of a business function or resource after a disaster occurs. (See also Recovery Point Objective [RPO].)
Resilience	The ability of a system or network to resist failure or to recover quickly from any disruption, usually with minimal recognizable effect

D. PROCESS OVERVIEW

Introduction

This section introduces the continuity management process and provides information about its:
- Relevance to organizational objectives and/or risk management.
- Process description
- Process goal
- Process-specific management practices
- Process-specific activities

Relevance

The need for providing continuous business and IT services requires developing, maintaining and testing of both business and IT continuity plans, utilizing offsite backup storage and providing periodic continuity plan training.

Effective continuity management minimizes the likelihood and impact of a major interruption on key business functions and processes.

Continuity management is critical to the ongoing operation of an enterprise after an incident has occurred—after an incident has interrupted the process by impacting one or more of the process dependencies. Continuity management provides a framework to protect the interests of the enterprise and its key stakeholders by identifying critical business processes and their dependencies, developing recovery strategies, documenting recovery plan tests, testing the effectiveness of those plans, and ensuring that the plans are kept up to date.

Process Description

Continuity management establishes and maintains a plan to enable the business and IT to respond to incidents and disruptions in order to continue operation of critical business processes and required IT services and maintain availability of information at a level acceptable to the enterprise.

Process Goal

The goal of continuity management is to continue critical business operations and maintain availability of information at a level acceptable to the enterprise in the event of a significant disruption.

Management Practices

Key management practices of providing continuity management are:
1. Define the business continuity policy, objectives and scope.
2. Maintain a continuity strategy.
3. Develop and implement a business continuity response.
4. Exercise, test and review the business continuity plan (BCP).
5. Review, maintain and improve the continuity plan.
6. Conduct continuity plan training.
7. Manage backup arrangements.
8. Conduct postresumption review.

1. Define the business continuity policy, objectives and scope.	Define the business continuity policy and scope aligned with enterprise and stakeholder objectives. Related activities include: 1. Identify internal and outsourced business processes and service activities that are critical to enterprise operations or necessary to meet legal and/or contractual obligations. 2. Identify key stakeholders and roles and responsibilities for defining and agreeing on continuity policy and scope. 3. Define and document the agreed-on minimum policy objectives and scope for business continuity, and embed the need for continuity planning in the enterprise culture. 4. Identify essential supporting business processes and related IT services.
2. Maintain a continuity strategy.	Evaluate business continuity management options and choose a cost-effective and viable continuity strategy that will ensure enterprise recovery and continuity in the face of a disaster or other major incident or disruption. Related activities include: 1. Identify potential scenarios likely to give rise to events that could cause significant disruptive incidents. 2. Conduct a business impact analysis (BIA) to evaluate the impact over time of a disruption to critical business functions and the effect that a disruption would have on them. 3. Establish the minimum time required to recover a business process and supporting IT based on an acceptable length of business interruption and maximum tolerable outage. 4. Assess the likelihood of threats that could cause loss of business continuity and identify measures that will reduce the likelihood and impact through improved prevention and increased resilience. 5. Analyze continuity requirements to identify possible strategic business and technical options. 6. Identify potential scenarios likely to give rise to events that could cause significant disruptive incidents. 7. Determine the conditions and owners of key decisions that will cause the continuity plans to be invoked. 8. Identify resource requirements and costs for each strategic technical option and make strategic recommendations. 9. Obtain executive business approval for selected strategic options.

3. Develop and implement a business continuity response.	Develop a BCP based on the strategy that documents the procedures and information in readiness for use in an incident to enable the enterprise to continue its critical activities. Related activities include: 1. Define the incident response actions and communications to be taken in the event of disruption. Define related roles and responsibilities, including accountability for policy and implementation. 2. Develop and maintain operational BCPs containing the procedures to be followed to enable continued operation of critical business processes and/or temporary processing arrangements, including links to plans of outsourced service providers. 3. Ensure that key suppliers and outsource partners have effective continuity plans in place. Obtain audited evidence as required. 4. Define the conditions and recovery procedures that would enable resumption of business processing, including updating and reconciliation of information databases to preserve information integrity. 5. Define and document the resources required to support the continuity and recovery procedures, considering people, facilities and IT infrastructure. 6. Define and document the information backup requirements required to support the plans, including plans and paper documents as well as data files, and consider the need for security and offsite storage. 7. Determine required skills for individuals involved in executing the plan and procedures. 8. Distribute the plans and supporting documentation securely to appropriately authorized interested parties and make sure they are accessible in all disaster scenarios.
4. Exercise, test and review the BCP.	Test the continuity arrangements on a regular basis to exercise the recovery plans against predetermined outcomes and to allow innovative solutions to be developed, and help to verify over time that the plan will work as anticipated. Related activities include: 1. Define objectives for exercising and testing the business, technical, logistical, administrative, procedural and operational systems of the plan to verify completeness of the BCP in meeting business risk. 2. Define and agree on with stakeholders exercises that are realistic, validate continuity procedures, and include roles and responsibilities and data retention arrangements that cause minimum disruption to business processes. 3. Assign roles and responsibilities for performing continuity plan exercises and tests. 4. Schedule exercises and test activities as defined in the continuity plan. 5. Conduct a postexercise debriefing and analysis to consider the achievement. 6. Develop recommendations for improving the current continuity plan based on the results of the review.

5. Review, maintain and improve the continuity plan.

Conduct a management review of the continuity capability at regular intervals to ensure its continued suitability, adequacy and effectiveness. Manage changes to the plan in accordance with the change control process to ensure that the continuity plan is kept up to date and continually reflects actual business requirements.

Related activities include:
1. Review the continuity plan and capability on a regular basis against any assumptions made and current business operational and strategic objectives.
2. Consider whether a revised business impact assessment may be required, depending on the nature of the change.
3. Recommend and communicate changes in policy, plans, procedures, infrastructure, and roles and responsibilities for management approval and processing via the change management process.
4. Review the continuity plan on a regular basis to consider the impact of new or major changes to enterprise organization, business processes, outsourcing arrangements, technologies, infrastructure, operating systems and application systems.

6. Conduct continuity plan training.

Provide all concerned internal and external parties with regular training sessions regarding the procedures and their roles and responsibilities in case of disruption.

Related activities include:
1. Define and maintain training requirements and plans for those performing continuity planning, impact assessments, risk assessments, media communication and incident response. Ensure that the training plans consider frequency of training and training delivery mechanisms.
2. Develop competencies based on practical training including participation in exercises and tests.
3. Monitor skills and competencies based on the exercise and test results.

7. Manage backup arrangements.

Maintain availability of business-critical information.

Related activities include:
1. Back up systems, applications, data and documentation according to a defined schedule, considering:
   - Frequency (monthly, weekly, daily, etc.)
   - Mode of backup (e.g., disk mirroring for real-time backups vs. DVD-ROM for long-term retention)
   - Type of backup (e.g., full vs. incremental)
   - Type of media
   - Automated online backups
   - Data types (e.g., voice, optical)
   - Creation of logs
   - Critical end-user computing data (e.g., spreadsheets)
   - Physical and logical location of data sources
   - Security and access rights
   - Encryption
2. Ensure that systems, applications, data and documentation maintained or processed by third parties are adequately backed up or otherwise secured. Consider requiring return of backups from third parties. Consider escrow or deposit arrangements.
3. Define requirements for onsite and offsite storage of backup data that meet the business requirements. Consider the accessibility required to back up data.
4. Roll out BCP awareness and training.
5. Periodically test and refresh archived and backup data.

| 8. Conduct postresumption review. | Assess the adequacy of the BCP following the successful resumption of business processes and services after a disruption.

Related activities include:
1. Assess adherence to the documented BCP.
2. Determine the effectiveness of the plan, continuity capabilities, roles and responsibilities, skills and competencies, resilience to the incident, technical infrastructure, and organizational structures and relationships.
3. Identify weaknesses or omissions in the plan and capabilities and make recommendations for improvement.
4. Obtain management approval for any changes to the plan and apply via the enterprise change control process. |
|---|---|

E. RISK MANAGEMENT CONSIDERATIONS

Introduction

This section discusses risk management practices related to the continuity management process. The following points are addressed:
- Risk factors
- Generic threats and common vulnerabilities
- Generic risk scenarios
- Key risk indicators (KRIs)

Risk Factors

Examples of factors affecting the risk related to the continuity management process are:
- **Enterprise size and complexity**—Size, complexity, geographic location, etc., all affect the ease with which continuity management can be implemented and maintained.
- **Recovery strategies**—In-house, outsourced (vendor-supplied), hot sites, warm sites, cold sites
- **Geographic location**—The hazards inherent to the location
- **Process complexity**—The complexity of the business processes and the amount of resources (time, location, people and technology) required for continuity
- **Internal controls and incident response capability**—A poorly controlled process is more likely to be subject to failure and difficulty for recovery
- **Use of third-party services**—Outsourcing of IT services places reliance on the service provider's ability to provide continuous services.
- **Time sensitivity of business processes**—Can be impacted by many factors such as regulatory, contractual, financial, work flow peaks, and legal time constraints on completing process activities
- **Controls**—Loss of controls during business continuity and recovery procedures can affect continuity management

> **Note:** Many enterprises do not know their business priorities and interdependencies well enough to create adequate recovery plans.

Generic Threats and Common Vulnerabilities

Generic threats and common vulnerabilities related to the continuity management process are:
- Continuity practices are not sufficient to support business processes. Recovery plans are not aligned with business goals or the BIA.
- A lack of documentation and/or training results in an increased dependency on key individuals.
- Unrealistic time lines for the recovery of critical products or services
- Lack of required recovery resources (hardware, software and personnel)
- Plans are not up to date and fail to reflect changes to business needs and technology.
- Inadequate recovery steps and processes exist.
- Backups of data and systems are incomplete.

Generic Risk Scenarios

Examples of generic risk scenarios related to continuity management are:
- Business is unable to restore critical functions or services in a timely manner.
- Confidential information in the plans is compromised.
- The business is unable to respond effectively to a crisis.
- There is a loss of market share or customer confidence.
- There is a loss of revenue.
- Violations of regulatory obligations exist.

KRIs   Examples of KRIs related to continuity management are the:
- Periodic review of mission-critical business process
- Number of tests performed
- Percentage of outstanding issues resolved from previous incidents, tests or audits
- Tests of backups to ensure integrity and completeness
- Amount of training of staff for roles related to business continuity planning

F. INFORMATION SYSTEMS CONTROL DESIGN, MONITORING AND MAINTENANCE

Introduction

This section provides an overview of common controls related to continuity management. The following points are addressed:
- Key control activities
- IS control metrics
- Monitoring practices

Key Control Activities

Key control activities for continuity management are:
- A business and IT continuity management framework
- Documented and tested business and IT continuity planning
- Identification of critical business and IT resources
- Maintenance of the business and IT continuity plan
- Testing the business and IT continuity plan
- Business and IT continuity plan training
- Distribution of the business and IT continuity plan
- Business and IT services recovery and resumption
- Offsite backup storage
- Postresumption review (application of lessons learned)

Key Control Activities Related to the IT Continuity Management Framework

Key control activities may include assigning responsibility for and establishing an enterprisewide business continuity management process that includes:
- A completed BIA
- A prioritized recovery strategy
- Established policies
- Tools
- Roles and responsibilities
- Testing
- Maintenance

Key Control Activities Related to IT Continuity Planning	Key control activities may include:
	• Create an IT continuity plan that includes the:
	  – Conditions and responsibilities for activating and/or escalating the plan
	  – Prioritized recovery strategy, including the necessary sequence of activities
	  – Minimum recovery requirements to maintain adequate business operations and service levels with diminished resources
	  – Emergency procedures
	  – IT processing resumption procedures
	  – Maintenance and test schedule
	  – Awareness, education and training activities
	  – Critical assets and resources and up-to-date personnel contact information
	  – Alternative processing facilities
	  – Alternative suppliers for critical resources
	• Define the underlying assumptions (e.g., level and scope of outage covered by the plan) in the IT continuity plan and which systems (i.e., computer systems, network components and other IT infrastructure) and sites are to be included. Describe alternative processing options for each site.
	• Ensure that the IT continuity plan includes a defined checklist of recovery events and a form for event logging.
	• Establish and maintain detailed information for every recovery site, including assigned staff and logistics (e.g., transport of media to the recovery site). This information should include the:
	  – Processing requirements for each site
	  – Length of time a recovery site may be available
	  – Location
	  – Resources (e.g., systems, staff, support) available at each location
	  – Utility companies on which the site depends
	• Define the response and recovery team structures, including the reporting requirements.
	• Define and prioritize communication processes and define responsibility for communication (e.g., public, press, government).
	• Create emergency procedures to ensure the safety of all affected parties, including coverage of occupational health and safety requirements (e.g., counseling services) and coordination with public authorities.

Key Control Activities Related to Critical IT Resources	Key control activities may include:
	• Define priorities for all applications, systems and sites that are in line with business objectives (usually through the execution of a BIA).
	• Include these priorities in the continuity plan. When defining priorities, consider the:
	  – Business risk and IT operational risk
	  – Interdependencies
	  – Data classification framework
	  – SLAs and operational level agreements (OLAs)
	  – Costs

Key Control Activities Related to Maintenance of the IT Continuity Plan	Key control activities may include: • Maintain a change history of the IT continuity plan. Ensure proper version management of the plan, e.g., through the use of document management systems, and that all distributed copies are the same version. • Involve the business continuity and IT continuity manager(s) in the change management process to ensure awareness of important changes that would require updates to the IT continuity plan. • Update the IT continuity plan as described by the IT continuity framework. Triggering events for the update of the plan include: – Important architecture changes – Important business changes – Key staff changes or organizational changes – Incidents/disasters and the lessons learned – Results from continuity plan tests
Key Control Activities Related to Testing the IT Continuity Plan	Key control activities may include: • Schedule IT continuity tests on a regular basis or after major changes in the IT infrastructure or to the business and related applications. • Ensure that test scenarios are realistic. • Ensure that all outstanding issues related to continuity planning are analyzed and resolved in an appropriate time frame. • Measure and report the success or failure of the test and, therefore, the continuity and contingency ability for services to the risk management process.
Key Control Activities Related to IT Continuity Plan Training	Key control activities may include: • On a regular basis (at least annually) or upon plan changes, provide training to the required staff members with respect to their roles and responsibilities. • Assess all needs for training periodically and update all schedules appropriately. • Measure and document training attendance, training results and coverage.
Key Control Activities Related to Distribution of the IT Continuity Plan	Key control activities may include: • Define a proper distribution list for the IT continuity plan and keep this list up to date. Include people and locations in the list on a need-to-know basis.
Key Control Activities Related to IT Services Recovery and Resumption	Key control activities may include: • Activate the IT continuity plan when conditions require it. • Maintain an activity and problem log during recovery activities to be used during postresumption review. • Ensure that vendors can live up to the terms of support agreements.

Part II—Risk Management and Information Systems Control in Practice
5. Continuity Management
F. Information Systems Control Design, Monitoring and Maintenance

Key Control Activities Related to Offsite Backup Storage	Key control activities may include:
	• Provide protection for the data commensurate with the value and security classification
	• Ensure that the backup facilities are not subject to the same risks (e.g., geography, weather, key service provider, utilities, human factors) as the primary site.
	• Perform regular testing of:
	  – The quality of the backups and media
	  – The ability to meet the committed recovery time frame
	• Ensure that the backups contain all data, programs and associated resources needed for recovery according to the plan.
	• Provide sufficient recovery instructions and adequate labeling of backup media.
	• Maintain an inventory of all backups and backup media.

Key Control Activities Related to Postresumption Review	Key control activities may include: • Using the problem and activity log of recovery activities • Identifying the shortcomings of the plan after reestablishing normal processing • Identifying opportunities for improvement to include in the next update of the IT continuity plan.
Metrics for Process Monitoring	The following tables provide examples of metrics for monitoring the continuity management process.

Metrics for Monitoring the Process to Manage Continuity					
Attribute	Target	Current Period	Prior Period	Prior Period −1	Prior Period −2
Percent of IT services meeting uptime requirements					
Percent of successful and timely restoration from backup or alternate media copies					
Percent of backup media transferred and stored securely					
Number of critical business systems not covered by the plan					
Number of exercises and tests that have achieved recovery objectives					
Frequency of tests					
Percent of agreed-on improvements to the plan that have been reflected in the plan					
Percent of issues identified that have been subsequently addressed in the plan					
Percent of internal and external stakeholders that have received training					
Percent of issues identified that have been subsequently addressed in the training materials					

G. THE PRACTITIONER'S PERSPECTIVE

Introduction

Continuity is not just a technology issue. Some of most significant areas of business risk in the continuity management domain come from inadequate understanding of, and involvement by, the business. Historically, continuity processes have been "owned" by technology departments. In some cases, continuity plans were limited to the resumption and recovery of technology resources. The following section provides an overview of how the five CRISC domains (listed below) relate to the continuity management process in practice:
- Domain 1—Risk Identification, Assessment and Evaluation
- Domain 2—Risk Response
- Domain 3—Risk Monitoring
- Domain 4—Information Systems Control Design and Implementation
- Domain 5—Information Systems Control Monitoring and Maintenance

Domain 1— Risk Identification, Assessment and Evaluation

The risk practitioner must identify any risk that the enterprise is not prepared for in the case of an adverse event that could impact its ability to meet business requirements. This would include reviewing whether the enterprise has plans in place, whether those plans would be adequate to recover operational capability, whether the leadership is in place to manage a disaster should it happen and whether the staff and plan are up to date with training and according to the current business environment.

Continuity management begins with the BIA. A BIA identifies the risk associated with the business processes and the supporting IT functions that are required to ensure continuous business processing. Once the BIA identifies the required processing functions, plans can be developed that address these concerns within the risk tolerances and costs the enterprise can accept.

Outsourced environments add an additional complexity. The service provider is responsible for the BCP, but the customer is responsible for the BIA, establishing requirements (depending on the outsourcing agreement) and the customer processes.

Domain 2— Risk Response

BCPs and disaster recovery plans (DRPs) are the risk response based on the priorities identified in the BIA. How the enterprise responds to known and unforeseen risk will determine the effectiveness of the continuity management. Different plans should be in place depending on the nature of the incident and the priority of the services and systems affected.

Practice drills should be organized.

During the drills, potential situations should be included in the simulation, e.g., individuals who are not available, software that is missing, delays.

Domain 3— Risk Monitoring

The KRIs described previously provide a monitoring process for key issues that would normally be experienced by a business.

The risk practitioner should ensure that the plans are tested on a regular basis, staff has been provided adequate training, and the plan is being updated in alignment with changes to the enterprise.

Domain 4—Information Systems Control Design and Implementation	The controls identified in the IS controls and objectives provide a resource of best practices for an IT operations unit.

The risk practitioner should ensure that the recovery alternative (hot site, warm site, cold site) chosen will meet business expectations.

The offsite storage requirement incorporates data management processes affecting data restoration. |
| **Domain 5—Information Systems Control Monitoring and Maintenance** | Monthly and annual continuity management monitoring provides a monitoring benchmark. This report should be provided to IT management and business stakeholders on a monthly basis and summarized on an annual basis.

Timely reporting using the control metrics defined previously should be distributed to stakeholders within IT and the business units affected.

Annual evaluation of the maturity of controls provides a barometer of controls in their current state, comparison to previous periods, and the target maturity level. Target maturity levels should be agreed on by stakeholders and evaluated by senior management as part of the annual review.

Maturity assessment can be performed as a self-assessment, with oversight by an objective third-party, peer review, or independent assessment (internal audit or external provider).

Issue monitoring related to continuity management should be reported and evaluated routinely. |

H. SUGGESTED RESOURCES FOR FURTHER STUDY

Suggested Resources for Further Study

In addition to the resources cited throughout this manual, the following resources are suggested for further study:
- ISACA:
  - COBIT 5, 2012
  - *Enterprise Value: Governance of IT Investments, The Val IT Framework 2.0*, 2008
  - *Implementing and Continually Improving IT Governance*, 2010
  - *The Risk IT Framework*, 2009
  - *The Risk IT Practitioner Guide*, 2009
- British Standards Institution (BSI), BS 25999, *A Code of Practice for Business Continuity Management*, UK, 2007
- Business Continuity Institute (BCI), *Good Practice Guidelines 2010*, UK, 2010
- International Organization for Standardization (ISO), ISO/IEC 22301:2012, *Societal security—Business continuity management systems—Requirements*, www.iso.org/iso/catalogue_detail?csnumber=50038

Page intentionally left blank

Part II—Risk Management and Information Systems Control in Practice
6. Information Security Management
A. Chapter Overview

6. Information Security Management

A. CHAPTER OVERVIEW

Learning Objectives

Information security management is a major risk element. The CRISC candidate should have a comprehensive understanding of information security management processes and how information security management interrelates with risk management.

Introduction

This chapter provides an overview of the information security management process and:
- Explains its importance to achieving business objectives
- Outlines a high-level process overview
- Introduces related key concepts
- Presents examples of common risk
- Lists selected key risk indicators (KRIs)
- Provides examples of common key control activities supporting the process
- Describes the practitioner's perspective
- Offers suggested reading materials and references

Contents

This chapter contains the following topics:

Topic	Starting Page
A. Chapter Overview	299
B. Related Knowledge Statements	300
C. Key Terms and Concepts	301
D. Process Overview	302
E. Risk Management Considerations	306
F. Information Systems Control Design, Monitoring and Maintenance	309
G. The Practitioner's Perspective	316
H. Suggested Resources for Further Study	318

B. RELATED KNOWLEDGE STATEMENTS

Contents

The following table lists the applicable knowledge statements from the CRISC job practice.

No.	Knowledge Statement (KS) Knowledge of:
KS1.10	Information security concepts
KS1.22	Threats and vulnerabilities associated with emerging technologies
KS4.6	Controls related to information security
KS5.9	Control objectives, activities and metrics related to information security

C. KEY TERMS AND CONCEPTS

Introduction	This section introduces terms and principles related to information security management as well as terms that help relate the process to other key business processes.
Access Controls	The processes, rules and deployment mechanisms that control access to information systems, resources and physical access to premises
Accountability	The ability to map a given activity or event back to the responsible party
Authentication	The act of verifying identity (i.e., user, system)
Availability	Ensuring timely and reliable access to and use of information
Data Classification	The assignment of a level of sensitivity to data (or information) that results in the specification of controls for each level of classification. Levels of sensitivity of data are assigned according to predefined categories as data are created, amended, enhanced, stored or transmitted. The classification level is an indication of the value or importance of the data to the enterprise.
Confidentiality	Preserving authorized restrictions on access and disclosure, including means for protecting privacy and proprietary information
Integrity	Guarding against improper information modification or destruction, and includes ensuring information nonrepudiation and authenticity
Nonrepudiation	The assurance that a party cannot later deny originating data; provision of proof of the integrity and origin of the data and that can be verified by a third party

D. PROCESS OVERVIEW

Introduction
This section introduces the information security management process and its importance to the achievement of business objectives. The risk practitioner is expected to influence and be active in the design and monitoring of information systems controls.

Relevance
The need to maintain the integrity, confidentiality and availability of information and the protection of IT assets requires a security management process. An information security management process is almost always based on risk management and the need to implement risk-based, cost-effective controls that will protect information systems from adverse events. The information security management process includes establishing and maintaining information security roles and responsibilities, policies, standards and procedures. Security management also includes performing security monitoring, periodic testing and implementing corrective actions for identified security weaknesses or incidents. Effective security management protects all IT assets by reducing the likelihood of an adverse event and/or by minimizing the business impact of security incidents.

Process Objectives
Key objectives of the information security management process are to:
- Define and maintain an information security plan.
- Define and integrate an enterprisewide security architecture model
- Define, establish and operate an identity (account) management process.
- Monitor potential and actual security incidents.
- Periodically review and validate user access rights and privileges.
- Establish and maintain procedures to safeguard cryptographic keys.
- Implement and maintain technical and procedural controls to protect information in storage, process and transmission as it flows between systems and across networks.
- Conduct regular vulnerability assessments.

Process Description
Information security management protects enterprise information to maintain the level of information security risk acceptable to the enterprise in accordance with the security policy. It establishes and maintains information security roles and access privileges and perform security monitoring.

Process Goal
The goal of managing security services is to minimize the business impact of operational information security vulnerabilities and incidents.

Management Practices
Key management practices of providing information security management are:
1. Protect against malware.
2. Manage network and connectivity security.
3. Manage endpoint security.
4. Manage user identity and logical access.
5. Manage physical access to IT assets.
6. Manage sensitive documents and output devices.
7. Monitor the infrastructure for security-related events.

Part II—Risk Management and Information Systems Control in Practice
6. Information Security Management
D. Process Overview

1. Protect against malware.	Implement and maintain preventive, detective and corrective measures in place (especially up-to-date security patches and virus control) across the enterprise to protect information systems and technology from malware (e.g., viruses, worms, spyware, spam). Related activities include: 1. Communicate malicious software awareness and enforce prevention procedures and responsibilities. 2. Install and activate malicious software protection tools on all processing facilities, with malicious software definition files that are updated as required (automatically or semiautomatically). 3. Distribute all protection software centrally (version and patch-level) using centralized configuration and change management. 4. Regularly review and evaluate information on new potential threats (e.g., reviewing vendors' products and services security advisories). 5. Filter incoming traffic, such as email and downloads, to protect against unsolicited information (e.g., spyware, phishing emails). 6. Conduct periodic training about malware in email and Internet usage. Train users to not install shared or unapproved software.
2. Manage network and connectivity security.	Use security measures and related management procedures to protect information over all methods of connectivity. Related activities include: 1. Based on risk assessments and business requirements, establish and maintain a policy for security of connectivity. 2. Allow only authorized devices to have access to corporate information and the enterprise network. Configure these devices to force password entry. 3. Implement network filtering mechanisms, such as firewalls and intrusion detection software, with appropriate policies to control inbound and outbound traffic. 4. Encrypt information in transit according to its classification. 5. Apply approved security protocols to network connectivity. 6. Configure network equipment in a secure manner. 7. Establish trusted mechanisms to support the secure transmission and receipt of information. 8. Carry out periodic penetration testing to determine adequacy of network protection. 9. Carry out periodic testing of system security to determine adequacy of system protection.
3. Manage endpoint security.	Ensure that endpoints (e.g., laptop, desktop, server, and other mobile and network devices or software) are secured at a level that is equal to or greater than the defined security requirements of the information processed, stored or transmitted. Related activities include: 1. Configure operating systems in a secure manner. 2. Implement device lockdown mechanisms. 3. Encrypt information in storage according to its classification. 4. Manage remote access and control. 5. Manage network configuration in a secure manner. 6. Implement network traffic filtering on endpoint devices. 7. Protect system integrity. 8. Provide physical protection of endpoint devices. 9. Dispose of endpoint devices securely.

| 4. Manage user identity and logical access. | Ensure that all users have information access rights in accordance with their business requirements and coordinate with business units that manage their own access rights within business processes.

Related activities include:
1. Maintain user access rights in accordance with business function and process requirements. Align the management of identities and access rights to the defined roles and responsibilities, based on least-privilege, need-to-have and need-to-know principles.
2. Uniquely identify all information processing activities by functional roles, coordinating with business units to ensure that all roles are consistently defined, including roles that are defined by the business itself within business process applications.
3. Authenticate all access to information assets based on their security classification, coordinating with business units that manage authentication within applications used in business processes to ensure that authentication controls have been properly administered.
4. Administer all changes to access rights (creation, modifications and deletions) to take effect at the appropriate time based only on approved and documented transactions authorized by designated management individuals.
5. Segregate and manage privileged user accounts.
6. Perform regular management review of all accounts and related privileges.
7. Ensure that all users (internal, external and temporary) and their activity on IT systems (business application, IT infrastructure, system operations, development and maintenance) are uniquely identifiable. Uniquely identify all information processing activities by user.
8. Maintain an audit trail of access to information classified as highly sensitive. |
|---|---|
| 5. Manage physical access to IT assets. | Define and implement procedures to grant, limit and revoke access to premises, buildings and areas according to business needs, including emergencies. Access to premises, buildings and areas should be justified, authorized, logged and monitored. This should apply to all persons entering the premises, including staff, temporary staff, clients, vendors, visitors or any other third party.

Related activities include:
1. Manage the requesting and granting of access to the computing facilities. Formal access requests are to be completed and authorized by management of the IT site, and the request records retained. The forms should specifically identify the areas to which the individual is granted access.
2. Ensure that access profiles remain current. Base access to IT sites (server rooms, buildings, areas or zones) on job function and responsibilities.
3. Log and monitor all entry points to IT sites. Register all visitors, including contractors and vendors, to the site.
4. Instruct all personnel to display visible identification at all times. Prevent the issuance of identity cards or badges without proper authorization.
5. Require visitors to be escorted at all times while onsite. If an unaccompanied, unfamiliar individual who is not wearing staff identification is identified, alert security personnel.
6. Restrict access to sensitive IT sites by establishing perimeter restrictions, such as fences, walls, and security devices on interior and exterior doors. Ensure that the devices record entry and trigger an alarm in the event of unauthorized access. Examples of such devices include badges or key cards, keypads, closed-circuit television and biometric scanners.
7. Conduct regular physical security awareness training. |

6. Manage sensitive documents and output devices.	Establish appropriate physical safeguards, accounting practices and inventory management over sensitive IT assets, such as special forms, negotiable instruments, special-purpose printers or security tokens. Related activities include: 1. Establish procedures to govern the receipt, use, removal and disposal of special forms and output devices into, within and out of the enterprise. 2. Assign access privileges to sensitive documents and output devices based on the least-privilege principle, balancing risk and business requirements. 3. Establish an inventory of sensitive documents and output devices, and conduct regular reconciliations. 4. Establish appropriate physical safeguards over special forms and sensitive devices. 5. Destroy sensitive information and protect output devices (e.g., degaussing of electronic media, physical destruction of memory devices, making shredders or locked paper baskets available to destroy special forms and other confidential papers).
7. Monitor the infrastructure for security-related events.	Using intrusion detection tools, monitor the infrastructure for unauthorized access and ensure that any events are integrated with general event monitoring and incident management. Related activities include: 1. Log security-related events reported by infrastructure security monitoring tools, identifying the level of information to be recorded based on a consideration of risk. Retain them for an appropriate period to assist in future investigations. 2. Define and communicate the nature and characteristics of potential security-related incidents so they can be easily recognized and their impacts understood to enable a commensurate response. 3. Regularly review the event logs for potential incidents. 4. Maintain a procedure for evidence collection in line with local forensic evidence rules and ensure that all staff are made aware of the requirements. 5. Ensure that security incident tickets are created in a timely manner when monitoring identified potential security incidents.

E. RISK MANAGEMENT CONSIDERATIONS

Introduction This section discusses risk management practices related to information security management. The following points are addressed:
- Risk factors
- Generic threats and common vulnerabilities
- Generic risk scenarios
- Key risk indicators (KRIs)

Relevance An information security management process can only be effective and efficient if it is properly managed and based on a valid risk assessment.
Otherwise, the likelihood is very high that security investments are made in the wrong areas, inappropriate controls are implemented or that chosen solutions do not meet regulatory requirements. The risk practitioner must consider the following when assessing the risk factors related to information security management:
- Emerging risk factors
- New technology
- Changes in business strategy or culture
- Incident handling
- Response to previously identified risk
- Accountability and ownership for security management and risk

Risk Factors Examples of factors affecting information security management are:
- **Enterprise size and complexity**—Size, complexity, geographic range, number of users to be managed, types of devices to be secured, etc., all affect the ease with which information security management can be implemented and maintained.
- **Operational model:**
    - Infrastructure: insourced, facilities managed, outsourced—specific facility, outsourced—cloud
    - Organizational structure: in-sourced, outsourced (consultants), hybrid
    - Control environment: mature, immature, resilient
    - Applications source: in-house, purchased off the shelf, customized
    - Use or lack of use of an identity management system (i.e., single sign-on technology)
- **Risk management capabilities**—The enterprise's risk culture and integration of information security functions with risk management

Part II—Risk Management and Information Systems Control in Practice
6. Information Security Management
E. Risk Management Considerations

Generic Threats and Common Vulnerabilities

Generic threats and common vulnerabilities related to information security management are:
- Management of information security:
  - There is a lack of information security governance.
  - IT, security, and business objectives are misaligned.
  - Data and information assets are unprotected.
  - There is a lack of an integrated enterprisewide approach to security.
- Information security plan:
  - The information security plan is incomplete or not cost-effective.
  - Gaps exist between known risk and planned and implemented information security measures.
  - Users are not aware of, or do not comply with, the information security plan.
  - Unauthorized changes are made to user identities/permissions.
  - There is a lack of protection for business-critical systems.
  - Security requirements are not specified for all systems.
  - Segregation-of-duty violations exist.
  - Incidents are not solved in a timely manner or according to a standard handling procedure.
- Security, testing, surveillance and monitoring:
  - Undetected security breaches exist.
  - Security logs are unreliable or are not being monitored.
  - Unauthorized access is available to cryptographic keys.
- Malicious software prevention, detection and correction:
  - Systems and data are prone to virus attacks.
  - Antivirus signature files are out of date.
  - Ineffective countermeasures exist.
- Network security:
  - Firewall rules do not reflect the enterprise security policy.
  - Undetected unauthorized modifications to firewall rules are made.
  - The overall security architecture is compromised.
  - Security breaches are not detected in a timely manner.
- Exchange of sensitive data:
  - Sensitive information is unprotected.
  - Inadequate physical security measures exist.
  - Unauthorized external connections to remote sites exist.

Generic Risk Scenarios

Examples of generic risk scenarios for information security management are:
- IT program selection (unknown breaches or noncompliance due to poorly governed IT security program)
- New technologies
- Technology selection
- IT investment decision making
- Accountability over IT
- Integration of IT with business processes
- State of infrastructure technology
- Aging of application software
- Architecture agility and flexibility

KRIs

Examples of KRIs include:
- Numbers of security incidents
- Number of vacant skilled positions in the IT department
- A high turnover ratio or unfilled positions among information security staff
- Excessive security projects that are delayed or cancelled
- Actual information security expenditure significantly over or under budget
- Excessive user violations
- Excessive time to investigate and close security violations
- Number of security vulnerabilities identified by scans
- Number of malware attacks affecting operations
- Percentage of user accounts that are not configured correctly

F. INFORMATION SYSTEMS CONTROL DESIGN, MONITORING AND MAINTENANCE

Introduction	This section provides an overview of common controls related to information security management. The following points are addressed: • Key control activities • IS control metrics • Monitoring practices
Key Control Activities	Key control activities are provided for the following aspects of information security management: • Management of information security • The information security plan • Personnel management • Physical security of IT assets • Identity management/account management • Security testing, surveillance and monitoring • Security incident handling • Protection of security technology • Cryptographic key management • Prevention, detection and correction of malware • Network security • Exchange of sensitive data • Disaster recovery plans

Key Control Activities Related to Management of Information Security	Key control activities may include:
	• Define a charter/policy for information security:
	  – Define the security management function:
	  – Scope and objectives
	  – Responsibilities
	  – Drivers (e.g., compliance, risk, performance)
	• Ensure that the information security policy reflects the requirements of the business and is signed off on by management.
	• Set up an adequate organizational structure and reporting line for information security. Define the interaction with business functions such as risk management, compliance, physical security, business continuity management and audit.
	• Implement an information security management reporting mechanism, reporting on the status of information security so that appropriate management actions can be taken:
	  – Ensure proper oversight of security through a steering committee or oversight body.
	  – Prioritize proposed security initiatives, including required levels of policies, standards and procedures.
	  – Review policy to verify that it refers to the organizational risk appetite relative to information security, and that the charter clearly includes:
	    · Scope and objectives of the security management function
	    · Responsibilities of the security management function
	    · Compliance and risk drivers

Key Control Activities Related to Management of Information Security *(cont.)*	• Ensure security roles and responsibilities are assigned.
	• Confirm that detailed security policy, standards and procedures exist. Examples of policies, standards and procedures may include:
	  – A security compliance policy
	  – Management risk acceptance
	  – An external communications security policy
	  – A firewall policy
	  – An email security policy
	  – An agreement to comply with IS policies
	  – A laptop/desktop computer security policy
	  – An Internet usage policy
	  – A mobile device (i.e., universal serial bus [USB], external hard drives) policy
	• Confirm that a plan for managing security incidents is in place.
	• Ensure that the information security department supports compliance and audit functions.
	• Ensure that the information technology projects are aligned with business needs.
	• Ensure that information security is considered in all business processes and IT initiatives.

Key Control Activities Related to the Information Security Plan	Key control activities may include:
	• Define and maintain an overall information security plan that includes:
	  – A complete set of security policies and standards in line with the established information security policy framework
	  – Procedures to implement, monitor and enforce the policies and standards
	  – Ongoing risk evaluation
	  – Roles and responsibilities
	  – Staffing requirements
	  – Security awareness and training
	  – Enforcement practices
	  – Investments in required security resources
	• Collect information security requirements from IT tactical operations and projects for integration into the overall information security plan.
	• Translate the overall information security plan into enterprise information security baselines for all major platforms and integrate it into the configuration baseline.
	• Provide information security requirements and implementation advice to other processes, including the development of automated solution requirements for service level agreements (SLAs) and operating level agreements (OLAs), and the development of application software and IT infrastructure components.
	• Ensure that there is an up-to-date and complete asset inventory of all IT systems, equipment, applications, configurations and documentation.
	• Determine whether a process exists to periodically update the information security plan.
	• Determine whether enterprise information security baselines for all major platforms are commensurate with the overall information security plan.

Key Control Activities Related to Personnel Management	Key control activities may include: • Ensure that all employees, contractors and anyone else accessing the systems of the enterprise are informed of security policies. • Ensure that all staff attends periodic security awareness programs. • Ensure that staff is trained in security according to job responsibilities. • Ensure that incidents of misuse are investigated and reported to management for appropriate action. • Ensure that hiring and termination practices include security practices such as nondisclosure agreements (NDAs), noncompete agreements and tracking of all equipment provided to the employee.

Key Control Activities Related to Physical Security of IT Assets	Key control activities may include: • Ensure that access is restricted to all data processing centers and areas containing sensitive information. • Have visitor logs and controls over access to buildings and nonpublic areas. • Have access controls, locks and tracing software on all equipment.

Key Control Activities Related to Identity Management/Account Management	Key control activities may include:
	• Establish and communicate policies and procedures to uniquely identify, authenticate and authorize access mechanisms and access rights for all users on a need-to-know/need-to-have/least privilege basis, based on predetermined and preapproved roles.
	• Clearly state the accountability of any user for any action on any of the systems and/or applications involved.
	• Ensure that roles and access authorization criteria for assigning user access rights take into account:
	  – Sensitivity of information and applications involved (data classification)
	  – Policies for information protection and dissemination (legal, regulatory, internal policies and contractual requirements)
	  – Roles and responsibilities as defined within the enterprise
	  – The need-to-have access rights associated with the function
	  – Standard, but individual, user access profiles for common job roles in the enterprise
	  – Requirements to implement appropriate segregation of duties
	• Establish a method for authenticating and authorizing users to establish responsibility and enforce access rights in line with sensitivity of information and functional application requirements and infrastructure components, and in compliance with applicable laws, regulations, internal policies and contractual agreements.
	• Define and implement a procedure for identifying new users and recording, approving and maintaining access rights.
	• Ensure that a timely information flow is in place that reports changes in jobs (i.e., people in, people out, job responsibility changes). Grant, revoke and adapt user access rights in coordination with HR and departments for users who are new, who have left the enterprise, or who have changed roles or jobs.

Key Control Activities Related to Security Testing, Surveillance and Monitoring	Key control activities may include:
	• Implement monitoring, testing, review and other controls to:
	  – Promptly prevent/detect errors in the results of processing
	  – Promptly identify attempted, successful and unsuccessful security breaches and incidents
	  – Determine whether the actions taken to resolve a breach of security are effective
	• Conduct effective and efficient security testing procedures at regular intervals to:
	  – Verify that identity management procedures are effective
	  – Validate that security-relevant system parameter settings are defined correctly and are in compliance with the information security baseline
	  – Validate that network security controls/settings are configured properly and are in compliance with the information security baseline
	  – Validate that security monitoring procedures are working properly
	  – Consider, where necessary, obtaining expert reviews of the security perimeter

Key Control Activities Related to Security Incident Handling	Key control activities may include: • Describe what a security incident entails. Communicate and distribute information related to security incidents, or relevant parts, to previously identified people who need to be notified. • Establish security incident response processes and incident response teams. Ensure that all team members have been training appropriately. • Ensure that security incidents and appropriate follow-up actions, including root cause analysis, follow the existing incident and problem management processes. • Define measures to protect confidentiality of information related to security incidents.
Key Control Activities Related to Protection of Security Technology	Protection of security technology activities may include: • Ensure that all hardware, software and facilities related to the security function and controls, e.g., security tokens and encryptions, are tamperproof. • Secure security documentation and specifications to prevent unauthorized access. • Make the security design of dedicated security technology (e.g., encryption algorithms) strong enough to resist exposure, even if the security design is made available to unauthorized individuals. • Evaluate the protection mechanisms on a regular basis (at least annually) and perform updates to the protection of the security technology, if necessary.

Key Control Activities Related to Cryptographic Key Management	Key control activities may include:
	• Ensure that there are appropriate procedures and practices in place for the generation, storage and renewal of the root key, including dual custody and observation by witnesses.
	• Make sure that procedures are in place to determine when any cryptographic key (especially the root key) renewal is required (e.g., the root key is compromised or expired).
	• Create and maintain a written certification practice statement that describes the practices that have been implemented in the certification authority, registration authority and directory when using a public-key-based encryption system.
	• Create cryptographic keys in a secure manner. When possible, enable only individuals not involved with the operational use of the keys to create the keys. Verify the credentials of key requestors (e.g., registration authority).
	• Ensure that cryptographic keys are distributed in a secure manner (e.g., offline mechanisms) and stored securely:
	  – In an encrypted form, regardless of the storage media used (e.g., write-once disk with encryption)
	  – With adequate physical protection (e.g., sealed, dual custody vault), if stored on paper
	• Create a process that identifies and revokes compromised keys. Notify all stakeholders as soon as possible of the compromised key.
	• Verify the authenticity of the counterparty before establishing a trusted path.

Key Control Activities Related to Prevention, Detection and Correction of Malware	Key control activities may include: • Establish, document, communicate and enforce a malicious software prevention policy in the enterprise. Ensure that people in the enterprise are aware of the need for protection against malicious software, and their responsibilities. • Install and activate malicious software protection tools on all processing facilities, with malicious software definition files that are updated as required (automatically or semi-automatically). • Use both network-based and host-based malware tools. • Distribute all protection software centrally (version and patch-level) using centralized configuration and change management. • Regularly review and evaluate information on new potential threats. • Filter incoming traffic, such as email and downloads, to protect against unsolicited information (e.g., spyware, phishing emails). • Ensure that suspicious activity is restricted to a sandbox or other quarantine area.
Key Control Activities Related to Network Security	Key control activities may include: • Establish, maintain, communicate and enforce a network security policy (e.g., provided services, allowed traffic, types of connections permitted) that is reviewed and updated on a regular basis (at least annually). • Establish and regularly update the standards and procedures for administering all networking components (e.g., core routers, demilitarized zone [DMZ], virtual private network [VPN] switches, wireless). • Properly secure network devices with special mechanisms and tools (e.g., authentication for device management, secure communications, strong authentication mechanisms). Implement active monitoring and pattern recognition to protect devices from attack. • Configure operating systems with minimal features enabled (e.g., features that are necessary for functionality and are hardened for security applications). Remove all unnecessary services, functionalities and interfaces (e.g., graphical user interface [GUI]). Apply all relevant security patches and major updates to the system in a controlled and timely manner. • Plan the network security architecture (e.g., DMZ architectures, internal and external network segmentation, intrusion prevention and detection systems [IPS/IDS] placement, and wireless equipment) to address processing and security requirements. Ensure that documentation contains information on how traffic is exchanged through systems and how the structure of the enterprise's internal network is hidden from the outside world (i.e., network address translation [NAT]). • Subject devices to reviews by experts who are independent of the implementation or maintenance of the devices.
Key Control Activities Related to Exchange of Sensitive Data	Key control activities may include: • Determine, by using the established information classification scheme, how the data should be protected when exchanged. • Apply appropriate application controls to protect the data exchange. • Apply appropriate infrastructure controls, based on information classification and technology in use, to protect the data exchange.
Key Control Activities Related to Disaster Recovery Plans	Key control activities may include: • Ensure that disaster recovery plans have been created for all critical systems and networks. • Review training, maintenance and distribution procedures for all disaster recovery plans to ensure that plans are up-to-date and will work effectively in a crisis. • Ensure that backups of all data, configurations, documentation and applications are being conducted and tested for completeness.

Metrics for Process Monitoring

The following tables provide examples of metrics for monitoring the following processes:
- Information security program management
- Information security services management

Metrics for Monitoring the Process to Manage Security (Program)					
Attribute	Target	Current Period	Prior Period	Prior Period −1	Prior Period −2
Number of key security roles clearly defined					
Number of security related incidents					
Level of stakeholder satisfaction with the security plan throughout the enterprise					
Number of security solutions deviating from the plan					
Number of security solutions deviating from the enterprise architecture					
Number of services with confirmed alignment to the security plan					
Number of security incidents caused by nonadherence to the security plan					
Number of solutions developed with confirmed alignment to the security plan					

Metrics for Process Monitoring *(cont.)*

Metrics for Monitoring the Process to Manage Security (Services)					
Attribute	Target	Current Period	Prior Period	Prior Period − 1	Prior Period − 2
Number of vulnerabilities discovered					
Number of firewall breaches					
Percent of individuals receiving awareness training relating to use of endpoint devices					
Number of incidents involving endpoint devices					
Number of unauthorized devices detected on the network or in the end-user environment					
Average time between change and update of accounts					
Number of accounts (vs. number of authorized users/staff)					
Percent of periodic tests of environmental security devices					
Average rating for physical security assessments					
Number of physical security-related incidents					
Number of incidents relating to unauthorized access to information					

G. THE PRACTITIONER'S PERSPECTIVE

Introduction

The following section provides an overview of how the five CRISC domains (listed below) relate to information security management in practice:
- Domain 1—Risk Identification, Assessment and Evaluation
- Domain 2—Risk Response
- Domain 3—Risk Monitoring
- Domain 4—Information Systems Control Design and Implementation
- Domain 5—Information Systems Control Monitoring and Maintenance

Domain 1—Risk Identification, Assessment and Evaluation

Identification is the process of transforming uncertainties and issues related to how well an enterprise's assets are being protected into distinct (tangible) areas of risk. The objective of this activity is to anticipate risk before it becomes a problem and to incorporate this information into the enterprise's information security risk management process.

Risk identification, assessment and evaluation are ongoing processes. Initially, risk identification establishes the baseline of information security requirements, prioritization of activities, and an execution plan. Information security management is all about risk. Risk changes as technology changes and new vulnerabilities are identified. A structured approach to risk identification is required. This process should include IT operations, the business risk management and other stakeholders.

Risk assessment should be initiated as conditions change; if no change is identified, risk assessment should be reviewed according to a defined schedule based on business requirements.

Domain 2—Risk Response

The risk response will be dependent on the risk identified.

The enterprise should develop risk mitigation strategies—including the development of managerial controls (such as policy), technical controls (such as access control) and physical controls (such as fire prevention) where appropriate in order to reduce the level of risk to an acceptable level.

An emergency response team should be established to address security incidents and report them through the problem/incident management system. This system should have a specific category for information security issues.

Anticipated issues should have pre-established responses, with practice drills, documentation of the event, and review of the process after the event.

Domain 3—Risk Monitoring

The key risk indicators (KRIs) described previously are an example of metrics that can identify trends and identify new risk affecting the enterprise. KRIs should be analyzed and reported to management as part of an ongoing monitoring process. In addition, where significant risks are identified, the risk assessment process should be initiated.

The risk practitioner should regularly validate that security controls are working correctly and producing the desired result.

Part II—Risk Management and Information Systems Control in Practice
6. Information Security Management
G. The Practitioner's Perspective

Domain 4—Information Systems Control Design and Implementation	The suggested controls described in the IS control objectives and activities provides a best practice resource for designing and implementing effective systems security controls. Information security controls must support and align with business requirements and not prohibit the business from operating effectively.
Domain 5—Information Systems Control Monitoring and Maintenance	Control monitoring should be integrated into the information security management and the incident reporting systems. The risk practitioner must ensure that IS controls are interoperable, consistent and comprehensive. This includes the review of the overall enterprise systems architecture to ensure that there are no gaps in the security control framework and that the controls are arranged in a layered defense model. Timely reporting using control metrics defined previously should be distributed to stakeholders within IT and the business units affected. Annual evaluation of the maturity of controls provides a barometer of controls in their current state, comparison to previous periods, and the target maturity level. Target maturity levels should be agreed on by stakeholders and evaluated by senior IT and business operations management as part of the IT annual review.

H. SUGGESTED RESOURCES FOR FURTHER STUDY

Suggested Resources for Further Study

In addition to the resources cited throughout this manual, the following resources are suggested for further study:
- ISACA:
  - COBIT 5, 2012
  - *Enterprise Value: Governance of IT Investments, The Val IT Framework 2.0*, 2008
  - *Implementing and Continually Improving IT Governance*, 2010
  - *The Risk IT Framework*, 2009
  - *The Risk IT Practitioner Guide*, 2009
- Alberts, Christopher; Audrey Dorofee; *Managing Information Security Risks: The OCTAVE^{SM} Approach*, Addison-Wesley Professional, USA, 2002
- International Organization for Standardization (ISO), ISO/IEC 27002:2005, *Information technology—Security techniques—Code of practice for information security management*, Switzerland, 2005

7. Configuration Management

A. CHAPTER OVERVIEW

Learning Objectives

The CRISC candidate should have a general understanding of configuration management processes and how configuration management interrelates with risk management.

Introduction

This chapter provides an overview of configuration management and:
- Explains its importance to achieving business objectives
- Outlines a high-level process overview
- Introduces related key concepts
- Provides examples of common risk
- Lists selected key risk indicators (KRIs)
- Provides examples of common key control activities supporting the process
- Describes the practitioner's perspective
- Offers suggested reading materials and references

Contents

This chapter contains the following topics:

Topic	Starting Page
A. Chapter Overview	320
B. Related Knowledge Statements	321
C. Key Terms and Concepts	322
D. Process Overview	323
E. Risk Management Considerations	325
F. Information Systems Control Design, Monitoring and Maintenance	327
G. The Practitioner's Perspective	329
H. Suggested Resources for Further Study	332

B. RELATED KNOWLEDGE STATEMENTS

Contents

The following table lists the applicable knowledge statements from the CRISC job practice.

No.	Knowledge Statement (KS) Knowledge of:
KS1.9	Information systems architecture (e.g., platforms, networks, applications, databases and operating systems)
KS4.5	Information systems architecture (e.g., platforms, networks, applications, databases and operating systems)
KS5.8	Control objectives, activities and metrics related to information systems architecture (e.g., platforms, networks, applications, databases and operating systems)

C. KEY TERMS AND CONCEPTS

Introduction	This section introduces terms and principles related to configuration management as well as terms that help relate the process to other key business processes.
Asset	Something of either tangible or intangible value that is worth protecting, including people, information, infrastructure, finances and reputation
Asset Management	Accounts for all IT assets and optimizes the value provided by these assets
Change Management	Effective change management—within the IT service domain—helps ensure that changes to the IT infrastructure are applied using standardized methods and procedures to minimize the number and impact of any related incidents upon service. Change management comprises that aspect of the overall IT governance framework which is enterprise sensitive and deals with IT configuration, release and problem management issues. **Note:** Changes in the IT infrastructure may arise as part of project, program or service improvement initiatives, based on problems related to IT services or in response to changing legislative requirements.
Configuration Item (CI)	Component of an infrastructure—or an item, such as a request for change, associated with an infrastructure—which is (or is to be) under the control of configuration management **Note:** CIs may vary widely in complexity, size and type, from an entire system (including all hardware, software and documentation) to a single module or a minor hardware component.
Configuration Management	The control of changes to a set of configuration items over a system life cycle Configuration management focuses on the processes relating to the establishment and maintenance of hardware and software baselines, hardware versions and models, software versions (commercial off-the shelf software [COTS]), and release baselines (changes to internally developed systems) and related service agreements.
Release Management	Release management focuses on the software release life cycle processes, ranging from the initial development of a piece of software to its eventual release, and creating updated versions of the released version to help improve the software or to fix existing software bug(s). This terminology can also be applied to the hardware acquisition, deployment, upgrade and retirement processes.
Risk Culture	The set of shared values and beliefs that governs attitudes toward risk-taking, care and integrity, and determines how openly risk and losses are reported and discussed
Problem Management	Systematic assessment and resolution through a defined set of IT change and control activities, of technical and procedural issues arising from the use of IT systems in a production environment

Exhibit 7.1: Relationship Among Change Management, Configuration Management, Release, Management and Problem Management

Exhibit 7.1 describes the relationship among change management, configuration management, release management and problem management.

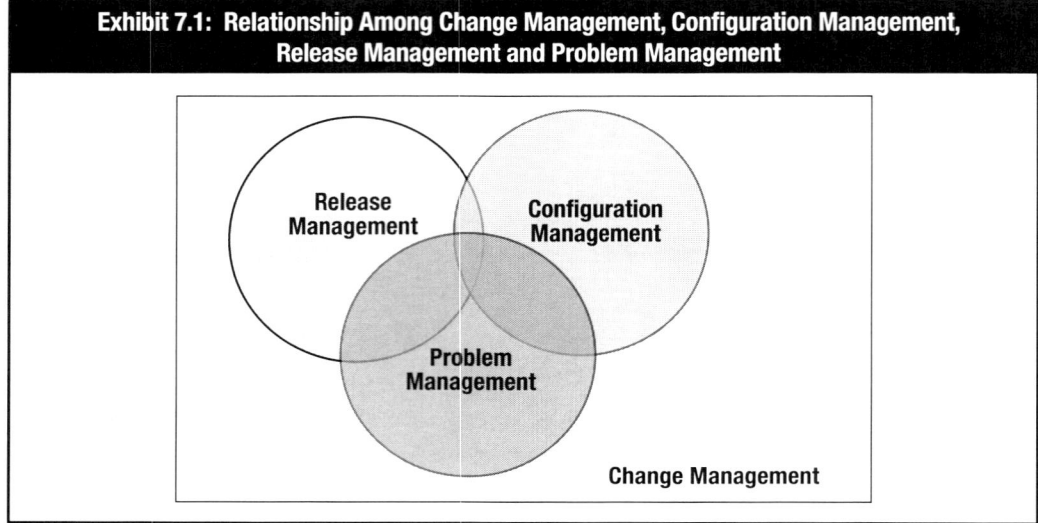

Change management should be identified as the governance framework of processes and controls within which configuration management (baseline definition and maintenance control), release management (software and hardware release controls) and problem management (issue resolution) take place.

D. PROCESS OVERVIEW

Introduction
This section introduces the configuration management process and its importance to the achievement of business objectives.

Relevance
Ensuring the integrity of hardware and software configurations requires the establishment and maintenance of an accurate and complete inventory of all information technology assets (i.e., software, databases, routers, switches, firewalls, operating systems, applications and communications devices) and a configuration repository (configuration management database [CMDB]). This process includes collecting initial configuration information, establishing baselines, verifying and auditing configuration information, and updating the configuration repository as needed. Effective configuration management facilitates greater system availability, minimizes production issues and resolves issues more quickly.

Uncontrolled changes to configurations may result in unauthorized access to systems, networks or data, errors in processing, unavailability of systems and loss of data integrity.

For many enterprises, configuration management also includes patch and vulnerability management and change management processes.

Process Description
Configuration management defines and maintains descriptions and relationships between key resources and capabilities required to deliver IT-enabled services, including collecting configuration information, establishing baselines, verifying and auditing configuration information, and updating the configuration repository.

Process Goal
The goal of configuration management is to provide sufficient information about service assets to enable the service to be effectively managed, assess the impact of changes and deal with service incidents.

Management Practices
Key management practices of configuration management are:
1. Establish and maintain a configuration model.
2. Establish and maintain a configuration repository and baseline.
3. Maintain and control configuration items.
4. Produce status and configuration reports.
5. Verify and review integrity of the configuration repository.

1. Establish and maintain a configuration model.
Establish and maintain a logical model of the services, assets and infrastructure, and how to record CIs and the relationships among them. Include the CIs considered necessary to manage services effectively and to provide a single reliable description of the assets in a service.

Related activities include:
1. Define and agree on the scope and level of detail for configuration management (i.e., which services, assets and infrastructure configurable items to include).
2. Establish and maintain a logical model for configuration management, including information on configuration item types, configuration item attributes, relationship types, relationship attributes and status codes.

2. Establish and maintain a configuration repository and baseline.	Establish and maintain a configuration management repository and create controlled configuration baselines. Related activities include: 1. Identify and classify configuration items and populate the repository. 2. Create, review and formally agree on configuration baselines of a service, application or infrastructure.
3. Maintain and control configuration items.	Maintain an up-to-date repository of configuration items by populating with changes. Related activities include: 1. Regularly identify all changes to configuration items. 2. Review proposed changes to configuration items against the baseline to ensure completeness and accuracy. 3. Update configuration details for approved changes to configuration items. 4. Create, review and formally agree on changes to configuration baselines whenever needed.
4. Produce status and configuration reports.	Define and produce configuration reports on status changes of configuration items. Related activities include: 1. Identify status changes of configuration items and report against the baseline. 2. Match all configuration changes with approved requests for change to identify any unauthorized changes. Report unauthorized changes to change management. 3. Identify reporting requirements from all stakeholders, including content, frequency and media. Produce reports according to the identified requirements.
5. Verify and review integrity of the configuration repository.	Periodically review the configuration repository and verify completeness and correctness against the desired target. Related activities include: 1. Periodically verify live configuration items against the configuration repository by comparing physical and logical configurations and using appropriate discovery tools, as required. 2. Report and review all deviations for approved corrections or action to remove any unauthorized assets. 3. Periodically verify that all physical configuration items, as defined in the repository, physically exist. Report any deviations to management. 4. Set and periodically review the target for completeness of the configuration repository based on business need. 5. Periodically compare the degree of completeness and accuracy against targets and take remedial action, as necessary, to improve the quality of the repository data.

E. RISK MANAGEMENT CONSIDERATIONS

Introduction

This section discusses risk management practices related to the configuration management process. The following points will be addressed:
- Risk factors
- Generic threats and common vulnerabilities
- Generic risk scenarios
- Key risk indicators (KRIs)

Risk Factors

Examples of factors affecting the configuration management process are:
- **Enterprise size and complexity**—Size, complexity, geographic range, international borders, industry, regulatory requirements, etc., all affect the ease with which configuration management can be implemented and maintained.
- **Operational model:**
  - Infrastructure: insourced, facilities managed, outsourced—specific facility, outsourced—cloud
  - Organizational structure: insourced, outsourced (consultants), hybrid
  - Infrastructure design: mainframe, distributed, virtualized
  - Information sensitivity and criticality
  - User community
  - Control environment: mature, immature, resilient
  - Application source: in-house, purchased off the shelf, customized
  - Application portfolio: managed, *ad hoc* (unmanaged)
  - Hardware/software acquisition: centralized, decentralized
  - Budgetary responsibility: centralized, decentralized
- **Risk management capabilities:**
  - The enterprise risk management (ERM) function extends to the IT function or the IT function operates in its own silo
  - The absence of ERM, the maturity of IT risk management processes
  - Risk tolerance of IT and business
  - IT risk levels aligned with business risks
- **Portfolio management capability**—The alignment of the application portfolio with the hardware and systems software architecture
- **Staff expertise**—The skill and knowledge levels of staff responsible for administration of devices

Generic Threats and Common Vulnerabilities

Generic threats and common vulnerabilities related to the configuration management process are:
- Uncontrolled change management of configuration, causing business disruptions
- Inability to roll out changes to all systems and affected components
- Lack of asset inventory; inability to accurately account for assets
- Failure of changes to comply with the overall technology architecture
- Unauthorized changes to hardware and software that are not discovered, which could result in security breaches
- Assets that are not properly protected
- Vendors not supporting older equipment
- Documentation that fails to reflect the current architecture
- Inability to fall back to the baseline configuration
- Inability to assess the impact of a change because of inaccurate information
- Failure to identify business-critical components
- Misused assets
- Increased costs for problem solving

Generic Risk Scenarios	Examples of generic risk scenarios related to configuration management are: • Business is unable to restore critical functions or services in a timely manner • Security breaches due to unpatched vulnerabilities in systems • Violation of regulatory obligations • Failure to deploy patches in a timely manner and meet regulatory requirements • Audit findings regarding unauthorized configuration changes
KRIs	Examples of KRIs are the: • Number of business compliance issues caused by improper configuration of assets • Number of deviations identified between the configuration repository and actual asset configurations • Percent of licenses purchased and not accounted for in the repository • Average time period (lag) between identifying a discrepancy and rectifying it • Number of discrepancies relating to incomplete or missing configuration information • Percent of configuration items in line with service levels for performance, security and availability • Excessive number of unique workstation images • Number of configuration deviations or policy overrides to approved architecture • Time between the release of vendor patches and deployment across the enterprise • Number of staff that have administrator level access to devices

F. INFORMATION SYSTEMS CONTROL DESIGN, MONITORING AND MAINTENANCE

Introduction

This section provides an overview of common controls related to configuration management. The following points are addressed:
- Key control activities
- IS control metrics
- Monitoring practices

Key Control Activities

Key control activities for configuration management are:

Policies and Procedures:
- Define and implement a policy requiring that all configuration items and their attributes and versions be identified and maintained.
- Establish a configuration policy and baseline for each operating platform and operating function (database server, file server, web server, etc.). Define a policy that integrates incident, change and problem management procedures with the maintenance of the configuration repository.

Asset Inventory:
- Record all assets—including new hardware and software, procured or internally developed—within the configuration management data repository.
- Tag physical assets according to a defined policy. Consider using an automated mechanism, such as barcodes.
- Define a process to identify critical configuration items in relationship to business functions (component failure impact analysis).
- Provide a unique identifier to a configuration item so the item can be easily tracked and related to physical asset tags and financial records.
- Implement a configuration repository to capture and maintain configuration management items.
- Implement a tool to enable the effective logging of configuration management information within a repository.

Configuration Baseline:
- Define and document configuration baselines for components across development, test and production environments, to enable identification of system configuration at specific points in time (past, present and planned).
- Establish an approved configuration baseline for each operating platform and operating function.
- Establish a process to revert to the baseline configuration in the event of problems, if determined appropriate after an initial investigation.

Change Monitoring Management:
- Install mechanisms to monitor changes against the defined repository and baseline. Provide management reports for exceptions, reconciliation and decision making.
- Define a process to record new, modified and deleted configuration items and their relative attributes and versions. Identify and maintain the relationships between configuration items in the configuration repository.
- Establish a process to maintain an audit trail for all changes to configuration items.
- To validate the integrity of configuration data, implement a process to ensure that configuration items are monitored. Compare recorded data against actual physical existence, and ensure that errors and deviations are reported and corrected.
- Using automated discovery tools where appropriate, reconcile actual installed software and hardware periodically against the configuration database, license records and physical tags.

Part II—Risk Management and Information Systems Control in Practice
7. Configuration Management
F. Information Systems Control Design, Monitoring and Maintenance

IS Control Objectives and Activities *(cont.)*	Compliance: • Define and implement a process to ensure that valid licenses are in place to prevent the inclusion of unauthorized software. • Periodically review (against the policy for software usage) the existence of any software in violation or in excess of current policies and license agreements. Report deviations for correction.
Metrics for Process Monitoring	The following tables provide examples of metrics for monitoring the configuration management process.

Metrics for Monitoring the Process to Manage the Configuration					
Attribute	Target	Current Period	Prior Period	Prior Period −1	Prior Period −2
Number of deviations between the configuration repository and live configuration					
Number of discrepancies relating to incomplete or missing configuration information					

G. THE PRACTITIONER'S PERSPECTIVE

Introduction

The following section provides an overview of how the five CRISC domains (listed below) relate to configuration management in practice:
- Domain 1—Risk Identification, Assessment and Evaluation
- Domain 2—Risk Response
- Domain 3—Risk Monitoring
- Domain 4—Information Systems Control Design and Implementation
- Domain 5—Information Systems Control Monitoring and Maintenance

Domain 1—Risk Identification, Assessment and Evaluation

Configuration management focuses on:
- Achieving processing service level agreements (SLAs)
- Secure computing
- Achieving compliance
- Responding to events to minimize operational disruption
- Tracking assets

The risk identification, assessment and evaluation process addresses those issues that would preclude the achievement of the above focus areas.

The initial identification of risk should be integrated into the systems architecture, hardware/software acquisition, and systems development processes. Assessment and evaluation can be performed as part of the initial process; however, configuration risk is also associated with subsequent reconfiguration of the assets, requiring risk monitoring (Domain 3) as well as reassessment of Domain 1. The risk identification process should include an annual review of previous assessments and evaluations for material changes affecting the risk profile. Any changes that have been made to a system without proper documentation and approval should be thoroughly investigated and reported to management.

The practitioner should focus attention on new operating systems introduced into the environment, new operating platforms (i.e., Windows servers dedicated to email, web services, database management), virtualization, network design and modification (firewalls, intrusion detection, etc.), and system software patches, updates or upgrades.

Domain 2— Risk Response	The risk response will be based on the risk identification, assessment and evaluation. Configuration management has an inherent risk in the dynamic nature of the controls. Many configuration-based controls can be circumvented intentionally or accidentally in the course of normal maintenance. Since these issues may be a result of human error or software "bugs," a pre-planned response process for high-risk and/or high-probability issues should be established, documented and appropriate practice drills implemented.

Systems should be hardened against threats by disabling unneeded services. An error in the configuration of a system or device may present an opportunity for a successful attack. Therefore, all changes must be reviewed and gaps in the configuration management process remedied as quickly as possible.

As vendors become aware of threats, they will release patches to close any vulnerability that the threat could exploit. For this reason, it is critically important for an enterprise to test and deploy patches through an approved process as quickly as possible.

Risk responses require a formal approach to issues, opportunities and events to ensure that solutions are in alignment with the business objectives and are cost effective. The following should be considered:
• When preparing the risk response, identify the risks in business terms: loss of productivity, disclosure of confidential information, lost opportunity costs, etc.
• Understand the business risk appetite, acceptable service interruptions, confidentiality of data, compliance requirements, etc.
• Keep the business stakeholders apprised of identified risks and how IT will respond to these risks. In defining the response, be specific and describe it in nontechnical terms. Transparency is key.
• The risk response requiring major changes to system configuration should be documented to describe the justification for the alternative selected, and how the changes support the business objective.
• The impact of the configuration change on the business environment and the need to have a maintenance window to allow for the deployment of the new configuration
• When building the response, consider how the response will be measured (leading to Domain 3: Risk Monitoring). |
| **Domain 3— Risk Monitoring** | Risk monitoring provides the communication between the affected stakeholders. Most configuration incidents do not suddenly occur. There are symptoms, sometimes ignored. The function of risk monitoring is to minimize this occurrence through effective key risk indicators (KRIs). The KRIs cited in this document are generic. The identification of meaningful KRIs is essential to transparency and effective communication with stakeholders and requires attention:
• The KRI should describe the risk issue and provide a trend line to identify performance patterns and measure remediation processes.
• An issue monitoring process should be implemented to provide IT and the stakeholders with an understanding of open issues, estimated issue closure dates, a means to prioritize and track issue resolution, and to provide status reports to interested parties. |

Domain 4—Information Systems Control Design and Implementation	Risk frameworks require effectively designed controls that reduce inherent risk to a residual risk level that is within levels and costs acceptable to the stakeholders: • Control design must be a part of the systems development, acquisition and maintenance processes. • Configuration requirements should be defined during the system development life cycle (SDLC) requirements definition. The costs associated with processes required to achieve the requirements should be defined and accepted by the business. This may include detection systems or services, additional monitoring and change management processes, etc. • Automated tools should be designed to report when configuration changes are implemented and these changes should be matched to configuration change requests. • The user acceptance tests of new or modified systems should include configuration tests to ensure that the settings and approved processes (services, started tasks and other utilities) are according to design. • KRIs designed to monitor the key configuration processes should be included in the control design.
Domain 5—Information Systems Control Monitoring and Maintenance	Effective monitoring of configuration management, operations and changes should be built into an effective monitoring system: • Timely reporting using control metrics defined previously should be distributed to stakeholders within IT and the business units affected. • Annual evaluation of the maturity of controls provides a barometer of controls in their current state, comparison to previous periods, and the target maturity level. Target maturity levels should be agreed on by stakeholders and evaluated by senior management as part of the IT annual review. • Maturity assessment can be performed as a self-assessment, with oversight by an objective third party, peer review or independent assessment (internal audit or external provider). • Configuration-related issue monitoring should be reported and evaluated routinely.

H. SUGGESTED RESOURCES FOR FURTHER STUDY

Suggested Resources for Further Study

In addition to the resources cited throughout this manual, the following resources are suggested for further study:
- ISACA:
  - COBIT 5, 2012
  - *Enterprise Value: Governance of IT Investments, The Val IT Framework 2.0*, 2008
  - *Implementing and Continually Improving IT Governance*, 2010
  - *The Risk IT Framework*, 2009
  - *The Risk IT Practitioner Guide*, 2009
- National Institute of Standards and Technology (NIST), *Guide for Security-Focused Configuration Management of Information Systems*, Special Publication (SP) 800-128, USA, 2011, *http://csrc.nist.gov/publications/nistpubs/800-128/sp800-128.pdf*

8. Problem Management

A. CHAPTER OVERVIEW

Learning Objectives	The CRISC candidate should have a general understanding of the problem management process and how problem management interrelates with risk management.
Introduction	This chapter provides an overview of problem management and: • Explains its importance to achieving business objectives • Outlines a high-level process overview • Introduces related key concepts • Presents examples of common risk • Lists selected key risk indicators (KRIs) • Provides examples of common key control activities supporting the process • Describes the practitioner's perspective • Offers suggested reading materials and references
Contents	This chapter contains the following topics:

Topic	Starting Page
A. Chapter Overview	333
B. Related Knowledge Statements	334
C. Key Terms and Concepts	335
D. Process Overview	337
E. Risk Management Considerations	340
F. Information Systems Control Design, Monitoring and Maintenance	342
G. The Practitioner's Perspective	345
H. Suggested Resources for Further Study	347

B. RELATED KNOWLEDGE STATEMENTS

Contents

The following table lists the applicable knowledge statements from the CRISC job practice.

	Knowledge Statement (KS)
No.	Knowledge of:
KS5.6	Control objectives, activities and metrics related to incident and problem management
KS1.22	Threats and vulnerabilities associated with emerging technologies

C. KEY TERMS AND CONCEPTS

Introduction	This section introduces terms and principles related to problem management as well as terms that help relate the process to other key business processes.
Event	Something that happens at a specific place or time. i.e., any measureable occurrence
Incident	Any event that is not part of the standard operation of a service and that causes, or may cause, an interruption to, or a reduction in, the quality of that service
Problem	In IT, the unknown underlying cause of one or more incidents
Problem Escalation Procedure	The process of escalating a problem up from junior to senior support staff, and ultimately to higher levels of management
Problem Management	Problem management includes the activities required to diagnose the root cause of incidents and to determine the resolution to underlying problems.

The COBIT 5 process DSS03 *Manage Problems* describes the problem management process as follows: Identify and classify problems and their root causes and provide timely resolution to prevent recurring incidents. Provide recommendations for improvements.

> **Note:** It is also responsible for ensuring that the resolution is implemented through the appropriate control procedures, especially change management and release management.

Relationship Between Events, Incidents and Problems

Exhibit 8.1 shows the relationship between events, incidents, and problems.

Logs, reports, audits and alarms all provide notice of events. Most events are harmless—e.g., a person logs in or enters a building—but some events have the potential to interrupt or disrupt normal business processes. Such events are incidents.

The incident management process is used to prepare for, detect, respond to and mitigate the effects of incidents.

Some (but not all) incidents and problems may indicate the need to change or amend business processes. This would prompt the use of the change management processes.

A feedback loop ensures that lessons learned from any incident are used to improve the affected business process, the event monitoring process or the incident management process.

Some incidents may be the result of an underlying or systemic problem in the business process; application; or managerial, technical and physical controls. This would trigger the problem management process to address the underlying issue or root cause of the problem.

The problem management process would also feed back into the event and incident management processes.

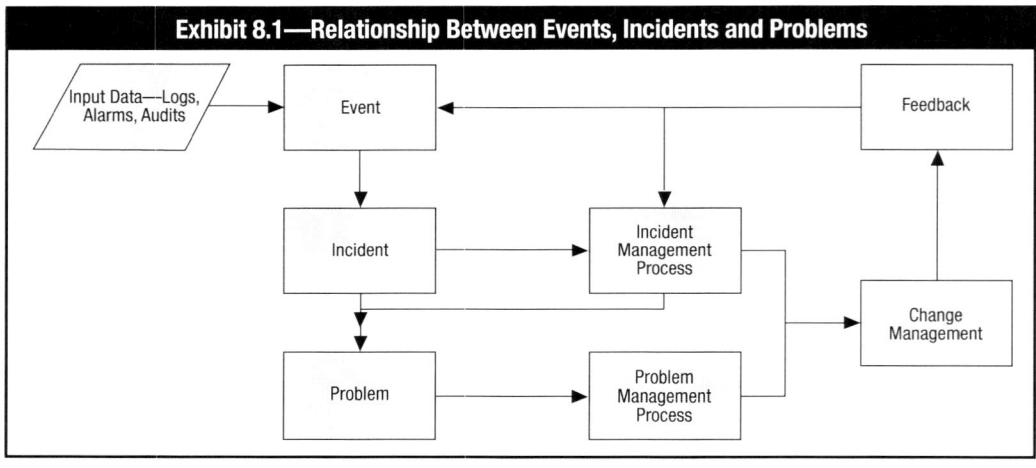

Relevance to Change Management

Problem management is closely aligned with change management. The incident and problem management processes will frequently require making changes to systems, business processes, equipment or network configurations, applications or applications development projects, or monitoring and reporting structures.

The CRISC candidate should refer back to Part II chapter 3, Change Management as they review this chapter on problem management.

D. PROCESS OVERVIEW

Introduction

This section introduces the problem management process and provides information about its:
- Relevance to organizational objectives and/or risk management.
- Process description
- Process goal
- Process-specific management practices
- Process-specific activities.

> **NOTE**: Problem management is frequently interrelated with change management, because incidents and problems often identify where changes are needed. Further details on change management can be found in Part II chapter 3 of this manual.

Relevance

Effective problem management requires:
- Identification and classification of problems
- Root cause analysis
- Determining an approved solution
- Assignment of resolution procedures through case management
- Monitoring of the resolution process
- Escalation to appropriate functions, where necessary
- Closure of the case
- Analysis of cases for process improvement

An effective problem management process maximizes system availability, improves service levels, reduces costs, and improves customer convenience and satisfaction.

Process Description

Problem management identifies and classifies problems and their root causes and provides timely resolution to prevent recurring incidents and provide recommendations for improvements.

Process Goal

The goal of problem management is to increase availability, improve service levels, reduce costs, and improve customer convenience and satisfaction by reducing the number of operational problems.

Management Practices

Key management practices of problem management are:
1. Identify and classify problems.
2. Investigate and diagnose problems.
3. Raise known errors.
4. Resolve and close problems.
5. Perform proactive problem management.

| 1. **Identify and classify problems.** | Define and implement criteria and procedures to report problems identified, including problem classification, categorization and prioritization.

Related activities include:
1. Identify problems through the correlation of incident reports, error logs and other problem identification resources. Determine priority levels and categorization to address problems in a timely manner based on business risk and service definition.
2. Handle all problems formally with access to all relevant data, including information from the change management system and IT configuration/asset and incident details.
3. Define appropriate support groups to assist with problem identification, root cause analysis and solution determination to support problem management. Determine support groups based on predefined categories, such as hardware, network, software, applications and support software.
4. Define priority levels through consultation with the business to ensure that problem identification and root cause analysis are handled in a timely manner according to the agreed-on service level agreements (SLAs). Base priority levels on business impact and urgency.
5. Report the status of identified problems to the service desk so customers and IT management can be kept informed.
6. Maintain a single problem management catalogue to register and report problems identified and to establish audit trails of the problem management processes, including the status of each problem (i.e., open, reopen, in progress or closed). |
|---|---|
| 2. **Investigate and diagnose problems.** | Investigate and diagnose problems using relevant subject management experts to assess and analyze root causes.

Related activities include:
1. Identify problems that may be known errors by comparing incident data with the database of known and suspected errors (e.g., those communicated by external vendors) and classify problems as a known error.
2. Associate the affected configuration items to the established/known error.
3. Produce reports to communicate the progress in resolving problems and to monitor the continuing impact of problems not solved. Monitor the status of the problem-handling process throughout its life cycle, including input from change and configuration management. |
| 3. **Raise known errors.** | As soon as the root causes of problems are identified, create known-error records and an appropriate workaround, and identify potential solutions.

Related activities include:
1. As soon as the root causes of problems are identified, create known-error records and develop a suitable workaround.
2. Identify, evaluate, prioritize and process (via change management) solutions to known errors based on a cost-benefit business case and business impact and urgency. |

| 4. Resolve and close problems. | Identify and initiate sustainable solutions addressing the root cause, raising change requests via the established change management process if required to resolve errors. Ensure that the personnel affected are aware of the actions taken and the plans developed to prevent future incidents from occurring.

Related activities include:
1. Close problem records either after confirmation of successful elimination of the known error or after agreement with the business on how to alternatively handle the problem.
2. Inform the service desk of the problem closure schedule, e.g., the schedule for fixing the known errors, the possible workaround or the fact that the problem will remain until the change is implemented, and the consequences of the approach taken. Keep affected users and customers informed as appropriate.
3. Throughout the resolution process, obtain regular reports from change management on progress in resolving problems and errors.
4. Monitor the continuing impact of problems and known errors on services.
5. Review and confirm the success of resolutions of major problems.
6. Make sure the knowledge learned from the review is incorporated into a service review meeting with the business customer. |
|---|---|
| 5. Perform proactive problem management. | Collect and analyze operational data (especially incident and change records) to identify emerging trends that may indicate problems. Log problem records to enable assessment.

Related activities include:
1. Capture problem information related to IT changes and incidents and communicate it to key stakeholders. This communication could take the form of reports to and periodic meetings among incident, problem, change and configuration management process owners to consider recent problems and potential corrective actions.
2. Ensure that process owners and managers from incident, problem, change and configuration management meet regularly to discuss known problems and future planned changes.
3. To enable the enterprise to monitor the total costs of problems, capture change efforts resulting from problem management process activities (e.g., fixes to problems and known errors) and report on them.
4. Produce reports to monitor the problem resolution against the business requirements and SLAs. Ensure the proper escalation of problems, e.g., escalation to a higher management level according to agreed-on criteria, contacting external vendors, or referring to the change advisory board to increase the priority of an urgent request for change (RFC) to implement a temporary workaround.
5. To optimize the use of resources and reduce workarounds, track problem trends.
6. Identify and initiate sustainable solutions (permanent fix) addressing the root cause, and raise change requests via the established change management processes. |

E. RISK MANAGEMENT CONSIDERATIONS

Introduction

This section discusses risk management practices related to the problem management process. The following points are addressed:
- Risk factors
- Generic threats and common vulnerabilities
- Generic IT risk scenarios
- Key risk indicators (KRIs)

Relevance to Risk Management

Every enterprise will encounter problems and incidents that affect normal business operations and the ability of the enterprise to meet its goals and objectives. Problem management ensures that the enterprise is able to identify the root cause of incidents, prepared to address problems through an incident management process, and then also able to detect and react to any incident in an effective manner. This should work to contain the incident and prevent the degree of damage from expanding to a point where the very existence of the enterprise may be in jeopardy.

The controls chosen, monitored, and reported on, by the risk practitioner, including the problem management process, are based on the likelihood and potential impact that a problem could have on the enterprise.

Risk Factors

Examples of factors affecting the problem management process are:
- **Enterprise size and complexity**—Size, complexity, geographic range, etc., all affect the ease with which problem management can be implemented and maintained.
- **Operational model:**
  - Infrastructure: insourced, facilities managed, outsourced—specific facility, outsourced—cloud
  - Organizational structure: insourced, outsourced (consultants), hybrid
  - Infrastructure design: mainframe, distributed, virtualized
  - Control environment: mature, immature, resilient
  - Applications source: in-house, purchased off the shelf, customized
  - Application portfolio: managed, *ad hoc*
  - Hardware/software purchasing: centralized, decentralized
  - Budgetary responsibility: centralized, decentralized
  - Hardware/software acquisition: centralized, decentralized
  - External risk factors such as vendor issues (e.g., support, bankruptcy, etc.)
- **Risk management capabilities:**
  - The enterprise risk management (ERM) function extends to the IT function or the IT function operates in its own silo.
  - In the absence of ERM, the maturity of IT risk management processes
  - Risk tolerance of IT and the business
  - IT risk levels aligned with business risks
- **Portfolio management capability**—The diligence with which business cases are assessed
- **Risk culture**—The risk-related patterns of behaviors, assumptions and beliefs that exist throughout an enterprise, together with the geographic location, political and economic climate, etc.

Generic Threats and Common Vulnerabilities	Generic threats and common vulnerabilities related to the problem management process are: • Poor employee morale (labor disruptions, layoffs, management issues) • Unreliable IT systems and business services • Loss of information • Repeated system failure • Problems and incidents not solved in a timely manner or not satisfying SLAs or regulatory requirements • Lack of audit trails of problems, incidents and their solutions for proactive problem and incident management • Increasing number of problems • Unplanned expenses • Zero day attacks • Availability of vendor patches/solutions
Generic Risk Scenarios	Generic risk and potential root causes related to the problem management process are: • Financial loss • Impact on employee and customer confidence • Recurring disruptions to IT and business services • Compromise of information • Increased likelihood of, and actual, problem recurrence • Failure to comply with SLAs or regulatory requirements • Successful malware and penetration attacks
KRIs	Examples of KRIs are the: • Percent of major incidents for which problems were logged • Percent of workarounds defined for open problems • Percent of problems logged as part of the proactive problem management activity • Number of problems for which a satisfactory resolution that addressed root causes were found • Excessive time to detect and react to a problem • High number of recurring problems categorized as critical with an impact on the business (customers, core functions) • High number of problems originating from a single functional area (applications development, scheduling, end user, technical support, information security, third party) • High number of problems not resolved according to the SLA • High number of problems requiring escalation • High number of emergency changes that were initiated to fix a problem introduced directly after a change

F. INFORMATION SYSTEMS CONTROL DESIGN, MONITORING AND MAINTENANCE

Introduction

This section provides an overview of common controls related to problem management. The following points are addressed:
- Key control activities
- IS control metrics
- Monitoring practices

Key Control Activities

Key control activities for problem management are:
- Policy
- Process
- Monitoring

Key Control Activities Related to Policy

Key control activities include:
- Establish and maintain a single problem management system to register and report problems identified and to establish audit trails of the problem management processes, including the status of each problem (i.e., open, reopened, in progress, closed). The system should register each problem, including the:
  - Needed information to understand the problem
  - Relevant documentation of the problem
  - Contact persons
  - Time the problem was identified
  - Method by which the problem was identified/detected
  - Known consequences/impact
  - Actual problem owner (department)
  - Workaround performed, if any
  - How and when of the solutions implemented
  - Identification of the root cause
  - Lessons learned
  - Status of problem
- Identify and initiate sustainable solutions (permanent fix) addressing the root cause, and initiate change requests via the established change management process.
- Define priority levels through consultation with the business to ensure that problem identification and root cause analysis are handled in a timely manner according to the agreed-on service level agreements (SLAs). Base priority levels on business impact and urgency.
- Define appropriate support groups to assist with problem identification, root cause analysis and solution determination to support problem management. Determine the support groups based on predefined categories such as hardware, network, software, applications and support software.

Key Control Activities Related to Process

Key control activities include:
- Define and implement a problem-handling process that has access to all relevant data, including information from the change management system and IT configuration/asset and incident details, to effectively address the root cause(s). Identify problems through the correlation of incident reports, error logs and other problem identification resources. Determine priority levels and categorization to address problems in a timely manner.
- Define problem ownership and the team responsible for problem management and reporting. Team members may include:
  – A senior management representative
  – Information technology experts
  – Representatives of the business units
  – Communications and public relations
  – Human resources (HR)
  – Legal
  – Internal audit
  – Health and safety
  – Outside experts

> **Note:** Not all team members will be contacted for each incident, but the people required are identified and can be deployed to the problem location as quickly as possible. This requires a communication system that can reach each team member as needed. Team members should be chosen based on the skills, knowledge or authority that they will provide the team. All team members should be aware of the need to maintain confidentiality of the incident.

- Ensure that process owners and managers from problem, change and configuration management meet regularly to discuss future planned changes.
- Inform the service desk (so users and customers can be informed) of the schedule of problem closure—e.g., the schedule for fixing the known errors, the possible workaround or the fact that the problem will remain until the change is implemented—and the consequences of the approach taken.
- Develop and implement a process to capture problem information related to IT changes and communicate it to key stakeholders. This communication could take the form of reports to and periodic meetings among problem, change and configuration management process owners to consider recent problems and potential corrective actions.
- Define and implement a process to close problem records either after confirmation of successful elimination of the known error or after agreement with the business on how to alternatively handle the problem.
- To maximize resources and reduce turnaround, define and implement problem management procedures for the tracking of problem trends.
- Produce reports to monitor the problem resolution against the business and customer SLAs. Ensure the proper escalation of problems—e.g., escalation to a higher management level according to agreed-on criteria, contacting external vendors or referring to the change advisory board—to increase the priority of an urgent request for change to implement a temporary workaround.
- Identify and implement a process for problems to be assigned and analyzed in a timely manner to determine the root cause. Identify problems by comparing incident data with the database of known and suspected errors (e.g., those communicated by external vendors). Upon successful root cause identification, classify problems as known errors. Associate the affected configuration items to the established/known error.

Key Control Activities Related to Monitoring	Key control activities include: • To determine the overall improvement of availability of IT services, monitor changes resulting from the problem management process. Monitor how the results of the problem management process decrease repeat incidents and reactive support requirements. • Produce reports to communicate the progress in resolving problems and to monitor the continuing impact of problems not solved. Monitor the status of the problem handling process throughout its life cycle, including input from change and configuration management. • To enable the enterprise to monitor the total costs of problems, develop and implement a process to capture change efforts resulting from problem management process activities (e.g., fixes to problems and known errors) and report on them. • Report the status of identified problems to the service desk so customers and IT management can be kept informed. • Review industry listings of vulnerabilities. (Common Weakness Enumeration [CWE] *cwe.mitre.org*; Common Vulnerabilities and Exposure [CVE] *cve.mitre.org*; SANS Twenty Critical Security Controls *www.sans.org/critical-security-controls*)
IS Control Metrics for Measuring Performance	The following IS control metrics measure the performance of problem management: • Number of recurring problems categorized as critical with an impact on the business (customers, core functions) • Number of problems by functional responsibility (applications, scheduling, end user, technical support, information security, third party) • Number of open/new/closed problems, by severity • Percent of problems that recur (within a time period), by severity • Percent of problems resolved according to the SLA • Percent of problems not resolved according to the SLA • Percent of problems requiring escalation • Aging of open items, problems not yet closed off • Average and standard deviation of time lag between problem identification and resolution • Average and standard deviation of time lag between problem resolution and closure • Average duration between the logging of a problem and the identification of the root cause • Number of emergency changes that were invoked to fix a problem • Number of changes required due to previous incidents that are not yet completed
Metrics for Process Monitoring	The following table provides examples of metrics for monitoring problem management.

Metrics for Monitoring the Process to Manage Problems					
Attribute	**Target**	**Current Period**	**Prior Period**	**Prior Period −1**	**Prior Period −2**
Decrease in number of recurring incidents caused by unresolved problems					
Percent of major incidents for which problems were logged					
Percent of workarounds defined for open problems					
Percent of problems logged as part of the proactive problem management activity					
Number of problems for which a satisfactory resolution that addressed root causes were found					

G. THE PRACTITIONER'S PERSPECTIVE

Introduction

Problem management is a key process in the mitigation of risk. The problem management process identifies, categorizes and follows an issue from its origination to closure. Because problems may originate from several sources within the enterprise, a central clearinghouse of all incidents is critical to the assurance that all issues have been remediated in a timely manner.

Outsourcing adds a level of complexity. Communication between the customer and servicer requires documentation, monitoring and management oversight. In addition, escalation processes need to be established where normal processes are not effective. The following section provides an overview of how the five CRISC domains (listed below) relate to problem management in practice:
- Domain 1—Risk Identification, Assessment and Evaluation
- Domain 2—Risk Response
- Domain 3—Risk Monitoring
- Domain 4—Information Systems Control Design and Implementation
- Domain 5—Information Systems Control Monitoring and Maintenance

Domain 1—Risk Identification, Assessment and Evaluation

Problem management focuses on:
- Identifying issues that could affect the ability of the enterprise to perform operational processes
- Prioritizing problems and determining remediation actions
- Ensuring the monitoring of the status of remediation and closure activities

Once a problem has been identified, the risk identification, assessment and evaluation function is the determinant used to establish the prioritization of the issue, the solution to be implemented and the process to be initiated for remediation. Risk identification, assessment and evaluation are the key elements of problem management.

Domain 2—Risk Response

The risk response will be based on the risk identification, assessment and evaluation. Problem management has an inherent risk in the dynamic nature of the controls. Most problems are a result of inadequate testing, ineffective internal program quality controls, poor internal systems design, software "bugs" introduced by outside parties, human error, and other unexpected incidents.

Risk responses require a formal approach to issues, opportunities and events to ensure that solutions are in alignment with the business objectives and are cost effective. The following should be considered:
- When preparing the risk response, identify the risks in business terms: loss of productivity, disclosure of confidential information, lost opportunity costs, etc.
- Understand the business risk appetite, acceptable extent of service interruptions, confidentiality of data, compliance requirements, etc.
- Keep the business stakeholders apprised of identified risks and how IT will respond to these risks. In defining the response, be specific and describe in nontechnical terms. Transparency is essential.
- Risk response requiring an investment in assets or resources should be supported by a well-thought-out business case that justifies the expenditure, outlines alternatives, describes the justification for the alternative selected, and supports the business objective.
- When building the response, consider how the response will be measured (leading to Domain 3: Risk Monitoring).

Domain 3—Risk Monitoring	Risk monitoring provides the communication between the affected stakeholders. Most problems occur with little warning. The key risk indicators (KRIs) are a tool to identify trends and initiate action based on the trends before the issue becomes a defined "problem." The KRIs cited in this document are generic, but provide a good basis for monitoring most problems. The problem management process should utilize the KRI-identified trends to initiate a problem management assessment and remediation process.

Domain 4—Information Systems Control Design and Implementation	Risk frameworks require effectively designed controls that reduce inherent risk to a residual risk level that is within levels and costs acceptable to the stakeholders.
	Problem management policies should be established, defining:
	• A "problem"
	• A system which records the:
	  – Problem
	  – Risk rating
	  – Remediation priority
	  – Assignment of remediation responsibility
	  – Cost estimate to remediate
	  – Target completion date
	  – Escalation details
	  – Closure date
	  – Actual remediation costs
	• Target time from problem identification to closure by risk category
	• Escalation policy
	• How to detect the problem earlier (new controls, closer monitoring)
	
	Automated tools should be designed to report when problems are identified, track all data elements in the policy, and maintain a status of the problem as it progresses through the remediation cycle.

Domain 5—Information Systems Control Monitoring and Maintenance	Effective monitoring of problem management, configuration management, operations, and changes should be built into an effective monitoring system: • Control metrics defined previously should be distributed to stakeholders within IT and the business units affected on a monthly basis, describing current and past period activity. • Annual evaluation of the maturity of controls provides a barometer of controls in their current state, comparison to previous periods, and the target maturity level. Target maturity levels should be agreed on by stakeholders and evaluated by senior management as part of the IT annual review. • Maturity assessment can be performed as a self-assessment, with oversight by an objective third party, peer review, or independent assessment (internal audit or external provider). Problems requiring a preset investment, affecting critical processes, and/or exceeding defined resolution duration should be reported and evaluated according to a defined policy.

H. SUGGESTED RESOURCES FOR FURTHER STUDY

Suggested Resources for Further Study

In addition to the resources cited throughout this manual, the following resources are suggested for further study:
- ISACA:
  - *COBIT 5*, 2012
  - *Enterprise Value: Governance of IT Investments, The Val IT Framework 2.0*, 2008
  - *Implementing and Continually Improving IT Governance*, 2010
  - *Information Security Governance: Guidance for Boards of Directors and Executive Management, 2nd Edition*, 2006
  - *The Risk IT Framework*, 2009
  - *The Risk IT Practitioner Guide*, 2009
- Blokdijk, Gerard; Claire Engle; Jackie Brewster; *IT Risk Management Guide—Risk Management Implementation Guide: Presentations, Blueprints, Templates; Complete Risk Management Toolkit Guide for Information Technology Processes and Systems*, Emereo Publishing, Australia, 2008
- Kouns, Jake; Daniel Minoli; *Information Technology Risk Management in Enterprise Environments: A Review of Industry Practices and a Practical Guide to Risk Management Teams*, Wiley-Interscience, USA, 2010
- Westerman, George; Richard Hunter; *IT Risk: Turning Business Threats Into Competitive Advantage*, Harvard Business School Press, USA, 2007

9. Knowledge Management

A. CHAPTER OVERVIEW

Learning Objectives

The CRISC candidate should have a general understanding of the knowledge management processes and how knowledge management interrelates with risk management.

Introduction

This chapter provides an overview of the knowledge management process and:
- Explains its importance to achieving business objectives
- Outlines a high-level process overview
- Introduces related key concepts
- Presents examples of common risk
- Lists selected key risk indicators
- Provides examples of common key control activities supporting the process
- Describes the practitioner's perspective
- Offers suggested reading materials and references

Contents

This chapter contains the following topics:

Topic	Starting Page
A. Chapter Overview	349
B. Related Knowledge Statements	350
C. Key Terms and Concepts	351
D. Process Overview	352
E. Risk Management Considerations	355
F. Information Systems Control Design, Monitoring and Maintenance	357
G. The Practitioner's Perspective	360
H. Suggested Resources for Further Study	367

B. RELATED KNOWLEDGE STATEMENTS

Contents

The following table lists the applicable knowledge statements from the CRISC job practice.

No.	Knowledge Statement (KS) Knowledge of:
KS1.12	Threats and vulnerabilities related to data management
KS1.22	Threats and vulnerabilities associated with emerging technologies
KS4.8	Controls related to data management
KS5.11	Control objectives, activities and metrics related to data management

C. KEY TERMS AND CONCEPTS

Introduction	This section introduces terms and principles related to knowledge management as well as terms that may help relate the process to other key business processes.
Availability	Ensuring timely and reliable access to and use of information
Confidentiality	Preserving authorized restrictions on access and disclosure, including means for protecting privacy and proprietary information
Data Classification	The assignment of a level of sensitivity to data (or information) that results in the specification of controls for each level of classification. Levels of sensitivity of data are assigned according to predefined categories as data are created, amended, enhanced, stored or transmitted. The classification level is an indication of the value or importance of the data to the enterprise. **Note:** Data classification is an indication of the value of the data to the enterprise and the level of protection required to safeguard the data.
Data Classification Scheme	An enterprise scheme for classifying data by factors such as criticality, sensitivity and ownership
Data Custodian	The individual(s) and department(s) responsible for the storage and safeguarding of computerized data
Data Owner	The individual(s)—normally a manager or director—who has responsibility for the integrity, accurate reporting and use of computerized data
Database Management System (DBMS)	A software system that controls the organization, storage and retrieval of data in a database
Integrity	Guarding against improper information modification or destruction, and includes ensuring information nonrepudiation and authenticity

D. PROCESS OVERVIEW

Introduction

This section introduces the knowledge management process and provides information about its:
- Relevance to organizational objectives and/or risk management.
- Process description
- Process goal
- Process-specific management practices
- Process-specific activities

Relevance

Knowledge is one of the most valuable assets possessed by an enterprise. Each enterprise needs to manage its knowledge throughout the information life cycle—from the initial receipt/creation of new data, throughout the phases of data entry and processing, storage, transmission, reporting, output, and deletion. Effective knowledge management identifies and maintains data protection and handling requirements, including data ownership; the controls needed to maintain the confidentiality, integrity and availability of the data; and comply with the regulations that may affect how the enterprise protects, uses, shares, retains and deletes data.

Effective knowledge management helps ensure the quality, timeliness and availability of business data.

Process Description

Knowledge management maintains the availability of relevant, current, validated and reliable knowledge to support all process activities and to facilitate decision making. It plans for the identification, gathering, organization, maintenance, use and retirement of knowledge.

Process Goal

Provide the knowledge required to support all staff in their work activities and for informed decision making and enhanced productivity.

Management Practices

Key management practices of knowledge management are:
1. Devise and implement a scheme to nurture and facilitate a knowledge-sharing culture.
2. Identify, validate and classify diverse sources of internal and external information required to enable effective use and operation of business processes and IT services.
3. Organize information based on classification criteria. Identify and create meaningful relationships between information elements and enable use of information. Identify owners and define and implement levels of access to knowledge resources.
4. Propagate available knowledge resources to relevant stakeholders and communicate how these resources can be used to address different needs (e.g., problem solving, learning, strategic planning and decision making).
5. Measure the use and evaluate the currency and relevance of information. Retire obsolete information.

Part II—Risk Management and Information Systems Control in Practice
9. Knowledge Management
D. Process Overview

1. Nurture and facilitate a knowledge-sharing culture.	Devise and implement a scheme to nurture and facilitate a knowledge-sharing culture. Related activities include: 1. Proactively communicate the value of knowledge to encourage knowledge creation, use, reuse and sharing. 2. Encourage the sharing and transfer of knowledge by identifying and leveraging motivational factors. 3. Create an environment, tools and artefacts that support the sharing and transfer of knowledge. 4. Embed knowledge management practices into other IT processes. 5. Set management expectations and demonstrate appropriate attitude regarding the usefulness of knowledge and the need to share enterprise knowledge.
2. Identify and classify sources of information.	Identify, validate and classify diverse sources of internal and external information required to enable effective use and operation of business processes and IT services. Related activities include: 1. Identify potential knowledge users, including owners of information who may need to contribute and approve knowledge. Obtain knowledge requirements and sources of information from identified users. 2. Consider content types (procedures, processes, structures, concepts, policies, rules, facts, classifications), artefacts (documents, records, video, voice), and structured and unstructured information (experts, social media, email, voice mail, RSS feeds). 3. Classify sources of information based on a content classification scheme (e.g., information architecture model). Map sources of information to the classification scheme. 4. Collect, collate and validate information sources based on information validation criteria (e.g., understandability, relevance, importance, integrity, accuracy, consistency, confidentiality, currency and reliability).
3. Organize and contextualize information into knowledge.	Organize information based on classification criteria. Identify and create meaningful relationships between information elements and enable use of information. Identify owners and define and implement levels of access to knowledge resources. Related activities include: 1. Identify shared attributes and match sources of information, creating relationships between information sets (information tagging). 2. Create views to related data sets, considering stakeholder and organizational requirements. 3. Devise and implement a scheme to manage unstructured knowledge not available through formal sources (e.g., expert knowledge). 4. Publish and make knowledge accessible to relevant stakeholders based on roles and access mechanisms.

4. Use and share knowledge.	Propagate available knowledge resources to relevant stakeholders and communicate how these resources can be used to address different needs (e.g., problem solving, learning, strategic planning and decision making). Related activities include: 1. Identify potential knowledge users by knowledge classification. 2. Transfer knowledge to knowledge users based on a needs gap analysis and effective learning techniques and access tools. 3. Educate and train users on available knowledge, access to knowledge and use of knowledge access tools.
5. Evaluate and retire information.	Measure the use and evaluate the currency and relevance of information. Retire obsolete information. Related activities include: 1. Measure the use and evaluate the usefulness, relevance and value of knowledge elements. Identify related information that is no longer relevant to enterprise knowledge requirements. 2. Define the rules for knowledge retirement and retire knowledge accordingly.

E. RISK MANAGEMENT CONSIDERATIONS

Introduction

This section discusses risk management practices related to the knowledge management process. The following points will be addressed:
- Risk factors
- Generic threats and common vulnerabilities
- Generic risk scenarios
- Key risk indicators (KRIs)

Risk Factors

Examples of factors affecting the knowledge management process are:
- **Enterprise size and complexity**—Size, complexity, geographic range, international borders, industry, regulatory requirements, etc., all affect the ease with which knowledge management can be implemented and maintained.
- **Operational model:**
  - Infrastructure: insourced, facilities managed, outsourced—specific facility, outsourced—cloud
  - Organizational structure: insourced, outsourced (consultants), hybrid
  - Infrastructure design: mainframe, distributed, virtualized
  - Control environment: mature, immature, resilient
  - Database management models, data storage models and data warehousing
- **Risk management capabilities**—The enterprise risk management (ERM) function extends to and is practiced by the IT function. In the absence of an ERM, the maturity of IT risk management processes would determine the level of risk. IT risk levels should be aligned with business risk.

Generic Threats and Common Vulnerabilities

Generic threats and common vulnerabilities related to the knowledge management process are:
- Knowledge management fails to support business requirements—data are not available, accurate (integrity) or kept confidential according to business requirements.
- Third-party contracts do not address data security requirements.
- Business, legal and regulatory data storage and retention requirements are not met.
- Cross-border transmission or storage of data may violate privacy laws.
- Media integrity is compromised. There is use of outdated or unreliable media.
- Backup media are unavailable when needed.
- Unauthorized personnel have access to sensitive data or can discover sensitive data through inference or aggregation.
- Sensitive data are not encrypted.
- Database views are not correctly defined.
- Data files are deleted in error.
- Data are not properly classified.
- Insufficient input validation of data leads to data corruption (allowable data range values).
- Data are not converted correctly between systems or during upgrades.
- Data format and metadata is defined incorrectly.
- Reports containing sensitive data are available to unauthorized personnel.
- Data are stored on personnel devices or portable media.
- There is data remembrance after deletion.

Generic Risk Scenarios	Generic risk scenarios related to the knowledge management process are: • Security breaches exist. Data is disclosed to unauthorized personnel. • Data retention and disposal laws are violated. • Processing errors are made due to inaccurate data. • Processing is affected by data backups or recovery processes. • Personal or portable media devices are lost or stolen. • Data extracts are incorrect. • Media and data sources are not securely disposed of.
KRIs	Examples of KRIs are: • Scheduled routine review of knowledge management process to ensure alignment with business requirements • Number of security breaches related to knowledge management • Media is not inventoried on a routine basis • Age of backup media (tapes, hard drives) • Number of incidents of media being reported lost or stolen • Number of times data critical to the business process cannot be recovered • Job failures caused by insufficient storage capacity • Audits of third-party agreements to ensure compliance with security controls

F. INFORMATION SYSTEMS CONTROL DESIGN, MONITORING AND MAINTENANCE

Introduction	This section provides an overview of common controls related to knowledge management. The following points are addressed: • Key control activities • IS control metrics • Monitoring practices
Key Control Activities	Key control activities for knowledge management focus on: • Business alignment • Data security • Media management • Data retention • Data disposal
Key Control Activities Related to Business Alignment	Key control activities may include: • Define the business information requirements for the management of data handling. • Periodically reassess the controls and usage of critical data and systems that affect business operations, in alignment with risk management and business requirements • Support business continuity planning.

Key Control Activities Related to Data Security	Key control activities may include:
	• Work with data owners to clearly identify sensitive data and the need for confidentiality of the data, and applicable laws and regulations.
	• Implement a policy to protect sensitive data from unauthorized access and insecure transmission and transport, including controls such as:
	  – Encryption
	  – Message authentication codes
	  – Hash totals
	  – Bonded couriers
	  – Tamper-resistant packaging for physical transport
	• Create and maintain awareness programs regarding secure handling and processing of sensitive data.
	• Secure facilities that handle sensitive information processing facilities. This may include:
	  – Perimeter security
	  – Fire protection
	  – Access controls
	  – Power conditioning
	  – Air quality controls
	• Report any security breaches to management.
	• Ensure that data handled by third parties is protected according to contractual and regulatory requirements.

Key Control Activities Related to Data Handling	Key control activities may include: • Assign responsibilities for media library management. • Monitor and review the physical storage environment of stored and archived data. Consider: — Access controls — Media age and usage — Inventory of media — Disposal of media — Compatibility of old media with new equipment **Note:** Media storage technology presents risk to the ability to recover data when needed. Media used should be routinely tested for viability (degradation of media). Additionally, periodic reviews should take place to ensure there is appropriate media reading technology to read media used to archive data or that data retained on older technologies is transferred to newer media so that the data can be retrieved when needed.
Key Control Activities Related to Data Retention	Key control activities related to backup, restoration and retention may include: • Agree on the time frame required for restoration of data (e.g., would it be fast enough to recover from tape?). • Define requirements for onsite and offsite storage of backup data. Consider the accessibility required to access backup data. • Prioritize data recovery based on business requirements and IT service continuity procedures. • Establish storage and retention procedures that address the enterprise security policy and change management procedures, including encryption and authentication. Back up the keys and certificates used for encryption and authentication. • Institute policies and procedures for retention of data and their subsequent destruction according to the sensitivity of the data. • Establish storage and retention arrangements to satisfy legal, regulatory and business requirements for documents, data, archives, programs, reports and messages (incoming and outgoing). • Define policies and procedures for the backup of systems, applications, data and documentation that considers factors including: — Type of backup (e.g., full vs. incremental) — Type of backup media — Security and access rights — Encryption — Frequency of backup (e.g., disk mirroring for real-time backups vs. DVD-ROM for long-term retention) • Assign responsibilities for taking and monitoring backups. • Ensure that systems, applications, data and documentation maintained or processed by third parties are adequately backed up or otherwise secured. Consider requiring the return of backups from third parties. Consider escrow or deposit arrangements for vendor-supplied source code. • Schedule, perform and log backups in accordance with established policies and procedures. • Periodically perform sufficient restoration tests to ensure that all components of backups can be effectively restored.

Key Control Activities Related to Data Disposal	Key control activities may include:
	• Clearly define responsibility for the development and communication of policies for the disposal of data and equipment.
	  – Sanitize equipment and media containing sensitive information prior to reuse or disposal. Media containing highly sensitive data should be physically destroyed.
	  – Ensure that data and equipment is destroyed according to policy.
	• Transport and store nonsanitized equipment and media in a secure way throughout the disposal process.
	• Remove destroyed data and equipment from inventory and continuity plans.
	• Require disposal contractors to have the necessary physical security and procedures to store and handle the equipment and media before and during disposal.

IS Control Metrics for Monitoring	*See table below*

Metrics for Monitoring the Process to Manage Knowledge					
Attribute	Target	Current Period	Prior Period	Prior Period −1	Prior Period −2
Percent of information categories covered					
Volume of information classified					
Percent of categorized information validated					
Percent of available knowledge actually used					
Number of users trained in using and sharing knowledge					
Level of satisfaction of users					
Percent of knowledge repository used					
Frequency of update					

G. THE PRACTITIONER'S PERSPECTIVE

Introduction

The practitioner's perspective for knowledge management is organized first by the following topics:
- Data classification and business alignment
- Media management
- Data disposal
- Data security

Then an overview is provided of how the five CRISC domains (listed below) relate to knowledge management in practice, if applicable:
- Domain 1—Risk Identification, Assessment and Evaluation
- Domain 2—Risk Response
- Domain 3—Risk Monitoring
- Domain 4—Information Systems Control Design and Implementation
- Domain 5—Information Systems Control Monitoring and Maintenance

DATA CLASSIFICATION AND BUSINESS ALIGNMENT

Domain 1—Risk Identification, Assessment and Evaluation

Risk identification, assessment, and evaluation processes for data classification address those issues that would affect the integrity, availability and security of corporate data.

The initial identification of risk should be integrated into the systems architecture definition, database management selection, software acquisition, and systems development processes. The assessment and evaluation of the risk associated with the knowledge management and data classification processes requires routine monitoring. The risk identification process should include a periodic assessment of the controls over, and handling of, critical data that could affect business operations or be subject to legal regulations.

The practitioner should focus attention on changes to the knowledge management environment, including changes in data processing, geographic presence and regulatory requirements. Special focus should be directed to changing business relationships and an expanded (global) customer base.

Domain 2—Risk Response	The risk response should address new knowledge management requirements resulting from an enterprise risk assessment, strategic changes in the business requirements, systems development activities and other management initiatives.

The challenge from the perspective of risk response is to balance the risk of unauthorized access to data with the need for the business to operate efficiently and allow authorized access according to business requirements.

Data protection starts with the identification of what data are sensitive and/or critical, and the determination of the ownership of the data. The data access controls must do exactly that—control access—which means that access is granted to authorized individuals or processes, but denied to unauthorized personnel. Even the granting of access must have a response to the risk of improper disclosure, modification or destruction of data by enforcing principles such as separation of duties, least privilege, mutual exclusivity, dual control and need to know.

Sensitive data must be protected at all times and in all forms—whether the data are paper, magnetic, optical or video. This is especially challenging in a world of personal devices and portable media. Policies and procedures should govern the conditions under which data are stored or transmitted and ensure that encryption is used to protect against the compromise of data even if a device that contains data is lost or stolen. |
| **Domain 3—Risk Monitoring** | Knowledge management and classification are management functions. The participation of IT in the development of data management policies and practices can provide management with sound advice and a good source of direction. The knowledge management program cannot be trusted unless it can be monitored. Monitoring tools (logs, cameras, data loss prevention tools, etc.) need to be deployed to ensure that data are only used appropriately and that any breaches can be detected.

Risk monitoring for personally identifiable information and sensitive data protection focuses on the logging and reporting of all access requests—whether permitted or denied—to data and providing those reports to management or regulatory agencies as required to provide for proper governance of data.

The risk practitioner will have to ensure that the management and use of data is compliant with organizational policy and external regulations. This may require implementation of monitoring controls that capture a record of activity related to data access. The risk of mishandling data—whether accidentally or through an intentional breach—may be severe and result in significant financial, reputational and legal impact on the enterprise. |

MEDIA MANAGEMENT AND DATA HANDLING

Domain 1—Risk Identification, Assessment and Evaluation

Media management and data handling concerns the maintenance and handling of the media used to store or transport data. IT management should focus on the following:
- Where media are stored—temperature, humidity, sunlight, etc.
- How media are maintained and protected—age, number of uses, labeling
- Controlled access to media libraries and encryption of sensitive data on storage forms such as:
  - Desktop and laptop local disk drives
  - Universal serial bus (USB) memory sticks (thumb drives)
  - Smartphones
- User awareness of how to handle data, including:
  - Sharing data with coworkers or outside parties
  - Disposal of confidential reports
  - Modification of data
  - Accessing data if not required for job function

Recognize the limitations of media storage technology and ensure that data continue to be readable (degradation of media).

The risk identification, assessment and evaluation process should be routinely performed as business storage requirements change, new technologies are introduced or limitations to existing technology are identified.

The cost of media management should be assessed against the risk to determine whether these costs are appropriate or can be reduced through efficiencies or newer technology.

IT management should address the risk of media management through accurate inventory procedures and robust reporting on events involving lost or degraded media, including investigation and resolution of those investigations.

Domain 2—Risk Response

Risk responses require a formal approach to issues, opportunities and events that were identified in the risk identification, assessment and evaluation phase. The risk responses may include the need to track media usage and storage, provide user training, label all sensitive data, install shredders for secure disposal, encrypt sensitive data, mask data that do not need to be displayed to a user (e.g., a credit card number), use database views or constrained user interfaces (menu or limited buttons on an ATM), and log all data access requests. The risk practitioner must ensure that risk response solutions are in alignment with the business objectives and regulatory requirements, and are cost effective.

Media management is perceived as an operational issue. A robust inventory tracking, testing and incident reporting and resolution system will demonstrate IT management focus and promote transparency with the business community.

The risk practitioner should develop procedures to prevent and detect data loss or theft through logging all data access requests, reviewing user access permissions on a regular basis, monitoring social media and web sites for data leakage and checking that sensitive data are discarded properly. Some of the practical steps used to protect data from loss is a clear screen and clean desk policy that requires a person to ensure that no sensitive data are left on their desk or computer screen when they are not present at their desk. Screen filters can also be used to prevent shoulder surfing.

Domain 3—Risk Monitoring	The risk practitioner should develop procedures to prevent and detect data loss or theft through logging all data access requests, reviewing user access permissions on a regular basis, monitoring social media and web sites for data leakage and checking that sensitive data is discarded properly. Metrics that will assist in the monitoring of media management issues are the: • Number of media inventories performed during the period • Number of incidents of unauthorized access to media libraries • Number of missing media identified during periodic inventory • Number of incidents in which media were lost • Number of user IDs set up with incorrect permissions • Number of cases of sensitive data left unattended in a insecure area
Domain 4—Information Systems Control Design and Implementation	Media management controls address: • Discarding media that is at end of life • Media inventory control • Media access control • Viability of media storage as technology changes • The establishment of data access controls—labeling, encryption • Hardware architecture design • Data center design— secure storage areas IS Data Handling controls include: • IT operations • Data loss prevention tools • Firewall configurations • User training • Encryption, masking • Replication and backup
Domain 5—Information Systems Control Monitoring and Maintenance	Effective monitoring of media operations, incident reports and issue monitoring should be part of IT operations and included in reports to management. Timely reporting using control metrics should be distributed to stakeholders within IT and the business units affected. Annual evaluation of the maturity of controls provides a barometer of controls in their current state, comparison to previous periods, and the target maturity level. Target maturity levels should be agreed on by stakeholders and evaluated by senior management as part of the IT annual review. Maturity assessment can be performed as a self-assessment, with oversight by an objective third party, peer review or independent assessment (internal audit or external provider). Logs should be reviewed on a regular basis to detect trends or instances of unauthorized data access activity.

DATA DISPOSAL

Domain 1—Risk Identification, Assessment and Evaluation

The risk of improper disposal of data is a serious matter for any enterprise that handles sensitive data. Secure disposal of equipment that has contained sensitive information requires clearly written and strongly enforced procedures.

The risk identification, assessment and evaluation process should be routinely performed since the methods used for disposal of media need to be adapted for new technologies.

Policies and procedures for secure data destruction or equipment disposal require the logging of the destruction, testing of the discarded equipment to ensure thorough data deletion and inspecting garbage containers to ensure that sensitive reports or papers have not been discarded insecurely.

Domain 2—Risk Response

Disposal of data is considered an operational issue and should be addressed in the IT operations function.

The method of disposal will depend on the level of risk associated with the data. Media that have contained very sensitive (classified) data may need to be destroyed in a multistep process including overwriting, degaussing and even physical destruction. Other media that have contained less sensitive data may only need to be overwritten or degaussed.

The proliferation of data on small portable devices is a serious risk. Smartphones, USB drives, DVDs and portable hard drives can contain large amounts of sensitive data and may be shared with unauthorized users, may not be adequately tracked, and may be lost or disposed of inappropriately. The risk practitioner should encourage the development of policy, procedures, encryption, labeling and other controls to mitigate this risk.

Domain 3—Risk Monitoring

Metrics that will assist in the monitoring of data disposal issues are the:
- Records of secure disposal of media
- Number of completed scheduled reviews of disposal process to ensure compliance with policy and procedures
- Number of cases in which sensitive data was able to be retrieved after media were disposed
- Number of reviews of third-party contracts to ensure secure data handling and disposal
- Number of incidents in which media containing sensitive data were reported lost

Domain 4—Information Systems Control Design and Implementation	Data disposal controls address: • Disposal policies • Management oversight to enforce policies • Identification of data loss • Equipment available for secure disposal (i.e., shredders, degaussing equipment, incinerators) Controls need to be established regarding: • IT operations for data within the custody of the data center • Data maintained in user departments • Recycle bins for data, followed by data sanitization • Awareness programs • Strong data disposal security awareness for end users to address smartphone, memory stick and disk drive concerns • Contractual arrangements and oversight with third-party vendors maintaining and destroying company data
Domain 5—Information Systems Control Monitoring and Maintenance	Effective monitoring of data disposal processes includes: • Reports on data destruction from vendors • Incident reporting and issue management processes to identify operational critical risk issues • Filtering refuse for media containing sensitive data • Timely reporting of previously defined control metrics to stakeholders within IT and affected business units • Annual evaluation of the control maturity provides a barometer of controls in their current state, comparison to previous periods, and the target maturity level. Target maturity levels should be agreed on by stakeholders and evaluated by senior management as part of the IT annual review. • Maturity assessment can be performed as a self-assessment, with oversight by an objective third party, peer review or independent assessment (internal audit or external provider).

DATA SECURITY

Domain 1—Risk Identification, Assessment and Evaluation	The primary issues within data security are: • Balancing the risk of data disclosure or unauthorized access with the ability to access data to accomplish the operational tasks • Identifying and classifying data risk • Determining data ownership • Maintaining the risk classification of data as applications, access methods (network and user hardware) and infrastructure change • Conducting awareness programs for all staff regarding data protection requirements and procedures
Domain 2—Risk Response	Risk response focuses on the actions taken after a data security incident has been identified. The response includes triaging incidents: • Identify the incident. • Determine the potential for loss, the effect on the enterprise, and its stakeholders, compliance, assets and reputation. • Determine pre-established actions based on known risk. • Perform drills based on predefined incident scenarios.

Domain 3—Risk Monitoring	Metrics include the: • Number of attempts to gain unauthorized access to data • Number of security breaches related to knowledge management per month • Number of data breaches by severity • Average period to resolve security breaches related to knowledge management • Asset value lost due to security breaches related to knowledge management during the period
Domain 4—Information Systems Control Design and Implementation	IS controls for data security include the following: • Building security controls into all systems development or business process reengineering projects • Identifying risk within the initial design phases of systems development • Enforcement of logical access controls to sensitive data or applications • Incorporate data security into systems and processes to maximize efficiency
Domain 5—Information Systems Control Monitoring and Maintenance	The monitoring of IS data security controls includes: • Monitoring processes and controls should be integrated into the information security management and incident reporting systems. • Timely reporting using the control metrics defined previously should be distributed to stakeholders within IT and the business units affected. • Annual evaluation of the maturity of controls provides a barometer of controls in their current state, comparison to previous periods, and the target maturity level. Target maturity levels should be agreed on by stakeholders and evaluated by senior management as part of the IT annual review. • Maturity assessment can be performed as a self-assessment, with oversight by an objective third party, peer review or independent assessment (internal audit or external provider).

H. SUGGESTED RESOURCES FOR FURTHER STUDY

Suggested Resources for Further Study

In addition to the resources cited throughout this manual, the following resources are suggested for further study:
- ISACA:
  - COBIT 5, 2012
  - *Enterprise Value: Governance of IT Investments, The Val IT Framework 2.0*, 2008
  - *Implementing and Continually Improving IT Governance*, 2010
  - *The Risk IT Framework*, 2009
  - *The Risk IT Practitioner Guide*, 2009
- European Commission, "European Union (EU) Data Protection Directive," *http://ec.europa.eu/justice/data-protection/index_en.htm*
- National Institute of Standards and Technology (NIST), *Guide for Mapping Types of Information and Information Systems to Security Categories*, Special Publication (SP) 800-60 USA, 2008
- Parliament of Australia, "Privacy Amendment (Enhancing Privacy Protection) Bill 2012," *www.aph-gov.au/Parliamentary_Business/Bills_Legislation*

Page intentionally left blank

10. IT Operations Management

A. CHAPTER OVERVIEW

Learning Objectives	The CRISC candidate should have a general understanding of IT operations management processes and how IT operations management interrelates with risk management.
Introduction	This chapter provides an overview of IT operations management and: • Explains its importance to achieving business objectives • Outlines a high-level process overview • Introduces related key concepts • Presents examples of common risk • Lists selected key risk indicators (KRIs) • Provides examples of common key control activities supporting the process • Describes the practitioner's perspective • Offers suggested reading materials and references
Contents	This chapter contains the following topics:

Topic	Starting Page
A. Chapter Overview	369
B. Related Knowledge Statements	370
C. Key Terms and Concepts	371
D. Process Overview	372
E. Risk Management Considerations	376
F. Information Systems Control Design, Monitoring and Maintenance	378
G. The Practitioner's Perspective	381
H. Suggested Resources for Further Study	382

B. RELATED KNOWLEDGE STATEMENTS

Contents

The following table lists the applicable knowledge statements from the CRISC job practice.

No.	Knowledge Statement (KS) — Knowledge of:
KS1.16	Threats and vulnerabilities related to management of IT operations
KS1.22	Threats and vulnerabilities associated with emerging technologies
KS4.12	Controls related to management of IT operations
KS5.5	Control objectives, activities and metrics related to IT operations and business processes and initiatives

C. KEY TERMS AND CONCEPTS

Introduction	This section introduces terms and principles related to IT operations management as well as terms that may help relate the process to other key business processes.
Operational Level Agreement (OLA)	An internal agreement covering the delivery of services that support the IT organization in its delivery of services
Privilege	The level of trust with which a system object is imbued
Problem Escalation Procedure	The process of escalating a problem up from junior to senior support staff, and ultimately to higher levels of management
Service Level Agreement (SLA)	An agreement, preferably documented, between a service provider and the customer(s)/user(s) that defines minimum performance targets for a service and how they will be measured

D. PROCESS OVERVIEW

Introduction	This section introduces the risk associated with the IT operations management process and its importance to the achievement of business objectives.
Relevance	Complete and accurate processing of data requires effective management of data processing procedures, job scheduling and monitoring, and diligent maintenance of hardware and software. This process includes defining operating policies and procedures for effective management of scheduled processing, protecting sensitive output, monitoring infrastructure and application performance and ensuring preventive maintenance of hardware and software. Effective IT operations management helps maintain data integrity and reduces business delays and IT operating costs.
Process Description	IT operations management coordinates and executes the activities and operational procedures required to deliver internal and outsourced IT services, including the execution of predefined standard operating procedures and the required monitoring activities.
Process Goal	The goal of IT operations management is to deliver IT operational service outcomes as planned
Management Practices	Key management practices of the IT operations management are: 1. Perform operational procedures. 2. Manage outsourced IT services. 3. Monitor IT infrastructure. 4. Manage the environment. 5. Manage facilities.
1. Perform operational procedures.	Maintain and perform operational procedures and operational tasks reliably and consistently. Related activities include: 1. Develop and maintain operational procedures and related activities to support all delivered services. 2. Maintain a schedule of operational activities, perform the activities, and manage the performance and throughput of the scheduled activities. 3. Verify that all data expected for processing are received and processed completely, accurately and in a timely manner. Deliver output in accordance with enterprise requirements. Support restart and reprocessing needs. Ensure that users are receiving the right outputs in a secure and timely manner. 4. Ensure that applicable security standards are met for the receipt, processing, storage and output of data in a way that meets enterprise objectives, the enterprise's security policy and regulatory requirements. 5. Schedule, take and log backups in accordance with established policies and procedures.

2. Manage outsourced IT services.

Manage the operation of outsourced IT services to maintain the protection of enterprise information and reliability of service delivery.

Related activities include:
1. Ensure that the enterprise's requirements for security of information processes are adhered to in accordance with contracts and SLAs with third parties hosting or providing services.
2. Ensure that the enterprise's operational business and IT processing requirements and priorities for service delivery are adhered to in accordance with contracts and SLAs with third parties hosting or providing services.
3. Integrate critical internal IT management processes with those of outsourced service providers, covering, for example, performance and capacity planning, change management, configuration management, service request and incident management, problem management, security management, business continuity, and the monitoring of process performance and reporting.
4. Plan for independent audit and assurance of the operational environments of outsourced providers to confirm that agreed-on requirements are being adequately addressed.

> **NOTE:** Managing third parties is addressed in detail in Part II, chapter 4.

3. Monitor IT infrastructure

Monitor the IT infrastructure and related events. Store sufficient chronological information in operations logs to enable the reconstruction, review and examination of the time sequences of operations and the other activities surrounding or supporting operations.

Related activities include:
1. Log events, identifying the level of information to be recorded based on a consideration of risk and performance.
2. Identify and maintain a list of infrastructure assets that need to be monitored based on service criticality and the relationship between configuration items and services that depend on them.
3. Define and implement rules that identify and record threshold breaches and event conditions. Find a balance between generating spurious minor events and significant events so event logs are not overloaded with unnecessary information.
4. Produce event logs and retain them for an appropriate period to assist in future investigations.
5. Establish procedures for monitoring event logs and conduct regular reviews.
6. Ensure that incident tickets are created in a timely manner when monitoring identifies deviations from defined thresholds.

> **NOTE:** General concepts related to monitoring are addressed in Part I, chapter 5.

4. Manage the environment.

Maintain measures for protection against environmental factors. Install specialized equipment and devices to monitor and control the environment.

Related activities include:
1. Identify natural and man-made disasters that might occur in the area within which the IT facilities are located. Assess the potential effect on the IT facilities.
2. Identify how IT equipment, including mobile and offsite equipment, is protected against environmental threats. Ensure that the policy limits or excludes eating, drinking and smoking in sensitive areas, and prohibits storage of stationery and other supplies posing a fire hazard within computer rooms.
3. Situate and construct IT facilities to minimize and mitigate susceptibility to environmental threats.
4. Regularly monitor and maintain devices that proactively detect environmental threats (e.g., fire, water, smoke, humidity).
5. Respond to environmental alarms and other notifications. Document and test procedures, which should include prioritization of alarms and contact with local emergency response authorities, and train personnel in these procedures.
6. Compare measures and contingency plans against insurance policy requirements and report results. Address points of noncompliance in a timely manner.
7. Ensure that IT sites are built and designed to minimize the impact of environmental risk (e.g., theft, air, fire, smoke, water, vibration, terror, vandalism, chemicals, explosives). Consider specific security zones and/or fireproof cells (e.g., locating production and development environments/servers away from each other).
8. Keep the IT sites and server rooms clean and in a safe condition at all times (i.e., no mess, no paper or cardboard boxes, no filled trash cans, no flammable chemicals or materials).

5. Manage facilities. Manage facilities, including power and communications equipment, in line with laws and regulations, technical and business requirements, vendor specifications, and health and safety guidelines.

Related activities include:
1. Examine the IT facilities' requirement for protection against power fluctuations and outages, in conjunction with other business continuity planning requirements. Procure suitable uninterruptible supply equipment (e.g., batteries, generators) to support business continuity planning.
2. Regularly test the uninterruptible power supply's mechanisms, and ensure that power can be switched to the supply without any significant effect on business operations.
3. Ensure that the facilities housing the IT systems have more than one source for dependent utilities (e.g., power, telecommunications, water, gas). Separate the physical entrance of each utility.
4. Confirm that cabling external to the IT site is located underground or has suitable alternative protection. Determine that cabling within the IT site is contained within secured conduits, and wiring cabinets have access restricted to authorized personnel. Properly protect cabling against damage caused by fire, smoke, water, interception and interference.
5. Ensure that cabling and physical patching (data and phone) are structured and organized. Cabling and conduit structures should be documented (e.g., blueprint building plan and wiring diagrams).
6. Analyze the facilities housing's high-availability systems for redundancy and fail-over cabling requirements (external and internal).
7. Ensure that IT sites and facilities are in ongoing compliance with relevant health and safety laws, regulations, guidelines, and vendor specifications.
8. Educate personnel on a regular basis on health and safety laws, regulations, and relevant guidelines. Educate personnel on fire and rescue drills to ensure knowledge and actions taken in case of fire or similar incidents.
9. Record, monitor, manage and resolve facilities incidents in line with the IT incident management process. Make available reports on facilities incidents where disclosure is required in terms of laws and regulations.
10. Ensure that IT sites and equipment are maintained according to the supplier's recommended service intervals and specifications. The maintenance must be carried out only by authorized personnel.
11. Analyze physical alterations to IT sites or premises to reassess the environmental risk (e.g., fire or water damage). Report results of this analysis to business continuity and facilities management.

E. RISK MANAGEMENT CONSIDERATIONS

Introduction

This section discusses risk management practices related to the IT operations management process. The following points are addressed:
- Risk factors
- Generic threats and common vulnerabilities
- Generic risk scenarios
- Key risk indicators (KRIs)

Risk Factors

Examples of factors affecting the operations management process are:
- **Enterprise size and complexity**—Size, complexity, geographic range, etc., all affect the ease with which operations management can be implemented and maintained.
- **Operational model**—Insourced, outsourced (consultants), mainframe, distributed computing, etc.
- **Maturity of operations and change control procedures**—poorly defined, not consistently followed, lack of oversight/enforcement

Generic Threats and Common Vulnerabilities

Generic threats and common vulnerabilities related to the IT operations management process are:
- Operations procedures:
  - Errors and rework due to misunderstanding of procedures
  - Inefficiencies due to unclear and/or nonstandard procedures
  - The inability to deal quickly with operational problems, new staff and operational changes
  - Misuse of elevated privileges by operators and administrators
  - Use of equipment or software for personal use
  - Illegal copying of software
  - Unauthorized installation of equipment or programs
- Job scheduling:
  - Resource utilization peaks
  - Problems with scheduling of ad hoc jobs
  - Reruns or restarts of jobs
- IT infrastructure monitoring:
  - Undetected infrastructure problems and occurrence of incidents
  - Infrastructure problems causing greater operational and business impact than if they had been prevented or detected earlier
  - Poorly utilized and deployed infrastructure resources
- Sensitive documents and output devices:
  - Misuse of sensitive IT assets, leading to financial losses and other business impacts
- Preventive maintenance for hardware:
  - The inability to account for all sensitive IT assets
  - Infrastructure problems that could have been avoided or prevented
  - Warranties violated due to noncompliance with maintenance requirements

Generic Risk Scenarios

Examples of generic risk scenarios for operations management are:
- System failures
- Technology selection
- IT investment decision making
- Accountability over IT
- Challenge of deploying new technologies
- Architectural agility and flexibility
- State of infrastructure technology
- Aging of application software
- Regulatory compliance
- Malware
- Database integrity errors
- Software and system performance
- Contractual compliance

KRIs

Examples of KRIs are a/an:
- Excessive number of operational incidents
- Excessive number of hours of unplanned downtime resulting from operational incidents
- Quantity of unauthorized equipment or software detected in scans
- Number of instances of service level agreements (SLAs) exceeding thresholds
- High average downtime due to operational incidents
- High number of hours lost due to late completion of scheduled work
- Excessive average time to research and remediate operations incidents

F. INFORMATION SYSTEMS CONTROL DESIGN, MONITORING AND MAINTENANCE

Introduction

This section provides an overview of common controls related to IT operations management. The following points are addressed:
- Key control activities
- IS control metrics
- Monitoring practices

Key Control Activities

Key control activities for IT operations management are:
- Operations processes and instructions
- Job scheduling
- IT infrastructure monitoring
- Sensitive documents and output devices
- Preventive maintenance for hardware

Key Control Activities Related to Operations Processes and Instructions

Key control activities may include:
- Develop, implement and maintain standard IT operational procedures covering the definition of roles and responsibilities, including those of external service providers.
- Train support personnel in operational procedures and related tasks for which they are responsible. Ensure cross-training to avoid dependency on just one person.
- Define procedures and responsibilities for formal handover of duties (e.g., for shift change, planned or unplanned absence).
- Define procedures for exception handling in line with the incident management and change management procedures and to address security aspects.
- Ensure that segregation of duties is in line with the associated risk, security and audit requirements.

Key Control Activities Related to Job Scheduling

Key control activities may include:
- Use formal procedures for planning and scheduling processing activities. Gain authorization for the initial schedules and changes to these schedules.
- Schedule authorized changes to systems, applications, configurations, networks and procedures with consideration of maintenance windows, minimizing impact on business operations, urgency of change.
- Ensure that the scheduling of batch jobs takes into consideration business requirements, priorities, conflicts between jobs and workload balancing. Put procedures in place to identify, investigate and approve departures from standard job schedules.

Key Control Activities Related to IT Infrastructure Monitoring	Key control activities may include: • Define and implement a procedure to resolve and correct job failures, including balancing or abnormal condition controls. • Implement automated tools and processes to immediately detect, notify and rectify critical processing failures. • Define and implement a process for event logging that identifies the level of information to be recorded based on a consideration of risk and performance. • Identify and maintain a list of infrastructure assets that need to be monitored based on service criticality, and the relationship between configuration items and services that depend on them. • Define and implement rules that identify and record threshold breaches and event conditions. Find a balance between generating spurious minor events and significant events so event logs are not overloaded with unnecessary information. • Produce event logs and retain them for an appropriate period to assist in future investigations. • Ensure that incident tickets are created in a timely manner when monitoring activities identify deviations.
Key Control Activities Related to Sensitive Documents and Output Devices	Sensitive documents and output devices activities may include: • Establish procedures for monitoring event logs and conduct regular reviews. • Establish procedures to govern the receipt, storage, use, removal and disposal of special forms and output devices into, within and outside of the enterprise. • Assign access privileges to sensitive documents and output devices based on the least privilege principle, balancing risk and business requirements. • Establish an inventory of sensitive documents and output devices, and conduct regular reconciliations. • Establish appropriate physical safeguards over special forms and sensitive devices. • Ensure that sensitive documents and media containing sensitive information are labeled, stored and transported securely. • Define and implement a process for destroying sensitive information and output devices (e.g., degaussing of electronic media, physical destruction of memory devices, making shredders or locked paper baskets available to destroy special forms and other confidential papers).
Key Control Activities Related to Preventive Maintenance for Hardware	Preventive maintenance for hardware activities may include: • Establish a preventive maintenance plan for all hardware and software, considering cost/benefit analysis, vendor recommendations and support, risk of outage, qualified personnel and other relevant factors. • Review all activity logs on a regular basis to identify critical hardware and software components that require preventive maintenance, and update the maintenance plan accordingly. • Establish maintenance agreements involving third-party access to the enterprise's IT facilities for onsite and offsite activities (e.g., outsourcing). Establish formal service contracts containing or referring to all necessary security conditions, including access authorization procedures, to ensure compliance with the enterprise's security policies and standards. • In a timely manner, communicate to affected customers and users the expected impact (e.g., performance restrictions or system unavailability) of maintenance activities. • Ensure that ports, services, user profiles or other means used for maintenance or diagnosis are active only when required. • Incorporate planned downtime in an overall production schedule, and schedule the maintenance activities to minimize the adverse impact on business processes.

Metrics for Process Monitoring

The following table provides examples of metrics for monitoring the IT operations management process.

Metrics for Monitoring the Process to Manage IT Operations					
Attribute	Target	Current Period	Prior Period	Prior Period −1	Prior Period −2
Number of nonstandard operational procedures executed					
Number of incidents caused by operational problems					
Ratio of events compared to the number of incidents					
Percent of critical operational event types covered by automatic detection systems					

G. THE PRACTITIONER'S PERSPECTIVE

Introduction	IT operations is the production arm of IT. Its primary role is the execution and delivery of scheduled and unscheduled IT services that support business processes. The following section provides an overview of how the five CRISC domains (listed below) relate to IT operations management in practice: • Domain 1—Risk Identification, Assessment and Evaluation • Domain 2—Risk Response • Domain 3—Risk Monitoring • Domain 4—Information Systems Control Design and Implementation • Domain 5—Information Systems Control Monitoring and Maintenance
Domain 1—Risk Identification, Assessment and Evaluation	During systems development and/or acquisition, the risk identified with the delivery of processes should be included as part of the overall design and selection process. Once this has been established, the primary risk identification is focused on those incidents or issues that could affect the delivery of IT production services. Distributed systems, in which many operational activities are performed by the business unit, create additional risk since they are generally not managed by IT professionals, and their job function includes other operational duties. Additional risk is introduced where applications or operations are outsourced.
Domain 2—Risk Response	Risk response is based on the effective mitigation, resolution and notification of system or equipment failures and ensuring that all failures are rapidly identified, analyzed and contained. The problem/incident tracking system should provide the notification and tracking of the response. Most operations risk is predictable and responses should be included in the operations procedures.
Domain 3—Risk Monitoring	The key risk indicators (KRIs) described previously provide a monitoring process for key issues that would normally be experienced by an IT operations organization. They provide a good view of trends and will highlight problems requiring management attention.
Domain 4—Information Systems Control Design and Implementation	The controls identified in the IS controls and objectives provide a resource of best practices for an IT operations unit. The risk practitioner will be concerned with the operations of IS controls being conducted with effective selection, configuration, implementation and maintenance of IS controls.

Domain 5—Information Systems Control Monitoring and Maintenance	IT operations management monitoring activities include:
	• Monthly operations management monitoring described previously provides a monitoring benchmark. This report should be provided to IT management and business stakeholders on a monthly basis and summarized on an annual basis.
	  – Timely reporting using control metrics defined previously should be distributed to stakeholders within IT and the business units affected.
	  – Annual evaluation of the maturity of controls provides a barometer of controls in their current state, comparison to previous periods, and the target maturity level. Target maturity levels should be agreed on by stakeholders and evaluated by senior management as part of the IT annual review.
	• Maturity assessment can be performed as a self-assessment, with oversight by an objective third party, peer review, or independent assessment (internal audit or external provider).
	• Monitoring related to IT operations should be reported and evaluated routinely.

H. SUGGESTED RESOURCES FOR FURTHER STUDY

Suggested Resources for Further Study

In addition to the resources cited throughout this manual, the following resources are suggested for further study:
- ISACA:
  - COBIT 5, 2012
  - *Enterprise Value: Governance of IT Investments, The Val IT Framework 2.0*, 2008
  - *Implementing and Continually Improving IT Governance*, 2010
  - *The Risk IT Framework*, 2009
  - *The Risk IT Practitioner Guide*, 2009

Study Questions, Answers and Explanations

Note: For more practice questions, you may also want to obtain a copy of the *CRISC® Review Questions, Answers & Explanations Manual 2013*, which consists of 200 multiple-choice study questions, answers and explanations, and the *CRISC® Review Questions, Answers & Explanations Manual 2013 Supplement*, which consists of 100 new multiple-choice study questions, answers and explanations.

Study Questions

Introduction — Study questions are grouped by domain. Answers and explanations are provided following the 25 questions.

DOMAIN 1—RISK IDENTIFICATION, ASSESSMENT AND EVALUATION

Question 1 The **MOST** significant drawback of using quantitative analysis instead of qualitative risk analysis is the:

A. higher cost.
B. lower objectivity.
C. higher reliance on skilled personnel.
D. lower management buy-in.

Question 2 Which of the following business requirements **MOST** relates to the need for resilient business and information systems processes?

A. Effectiveness
B. Confidentiality
C. Integrity
D. Availability

Question 3 An enterprise that chooses not to engage in e-commerce is demonstrating a form of:

A. risk avoidance.
B. risk transfer.
C. risk treatment.
D. risk acceptance.

Question 4 The risk to an information system that supports a critical business process is owned by:

A. the IT director.
B. senior management.
C. the risk management department.
D. the system users.

Question 5 The **FIRST** step in the risk assessment process is the identification of:

A. assets.
B. threats.
C. vulnerabilities.
D. threat sources.

DOMAIN 2—RISK RESPONSE

Question 6 The preparation of a risk register begins in which risk management process?

A. Risk response planning
B. Risk monitoring and control
C. Risk identification
D. Risk management strategy planning

Question 7 To address the risk of the failure of operations staff to perform the daily backup, management requires that the systems administrator sign off on the daily backup. This is an example of:

A. risk transference.
B. risk avoidance.
C. risk mitigation.
D. risk acceptance.

Question 8 When a risk cannot be sufficiently mitigated through manual or automatic controls, which of the following options will **BEST** protect the enterprise from the potential financial impact of the risk?

A. Insuring against the risk
B. Updating the IT risk registry
C. Improving staff training in the risk area
D. Outsourcing the related business process to a third party

Question 9 When responding to an identified risk event, the **MOST** important stakeholders involved in reviewing risk response options to an IT risk are the:

A. information security managers.
B. internal auditors.
C. incident response team members.
D. business managers.

Question 10 What risk elements **MUST** be known in order to accurately calculate residual risk?

A. Threats and vulnerabilities
B. Inherent risk and control risk
C. Compliance risk and reputation
D. Risk governance and risk response

DOMAIN 3—RISK MONITORING

Question 11 The **MOST** important reason to maintain key risk indicators (KRIs) is because:

A. complex metrics require fine-tuning.
B. threats and vulnerabilities change over time.
C. risk reports need to be timely.
D. they help to avoid risk.

Question 12 To be effective, risk mitigation **MUST**:

A. minimize the residual risk.
B. minimize the inherent risk.
C. reduce the frequency of a threat.
D. reduce the impact of a threat.

Question 13 Which of the following is the **BEST** measure of the operational effectiveness of risk management process capabilities?

A. Key performance indicators (KPIs)
B. Key risk indicators (KRIs)
C. Base practices
D. Metric thresholds

Question 14 During a data extraction process the total number of transactions per year was forecasted by multiplying the monthly average by twelve. This is considered:

A. a controls total.
B. simplistic and ineffective.
C. a duplicates test.
D. a reasonableness test.

Question 15 The **PRIMARY** objective difference between an internal and an external risk management assessment would be the reviewer's:

A. professionalism.
B. quality of work.
C. independence.
D. ease of access.

DOMAIN 4—INFORMATION SYSTEMS CONTROL DESIGN AND IMPLEMENTATION

Question 16 The **BEST** control to prevent unauthorized access to an enterprise's information is user:

A. accountability.
B. authentication.
C. identification.
D. access rules.

Question 17 Which of the following should be considered **FIRST** when designing information systems controls?

A. The organizational strategic plan
B. The existing IT environment
C. The present IT budget
D. The IT strategic plan

Question 18 Which of the following controls **BEST** protects an enterprise from unauthorized individuals gaining access to sensitive information?

A. Using a challenge response system
B. Forcing periodic password changes
C. Monitoring and recording unsuccessful logon attempts
D. Providing access on a need-to-know basis

Question 19 A poor choice of passwords and transmission over unprotected communications lines are examples of:

A. vulnerabilities.
B. threats.
C. probabilities.
D. impacts.

Question 20 Which of the following is the **BEST** defense against successful phishing attacks?

A. An intrusion detection system (IDS)
B. Spam filters
C. End-user awareness
D. Application hardening

DOMAIN 5—INFORMATION SYSTEMS MONITORING AND MAINTENANCE

Question 21 An enterprise has implemented a tool that correlates information from multiple systems. This is an example of a monitoring tool that focuses on:

A. transaction data.
B. configuration settings.
C. system changes.
D. process integrity.

Question 22 The **BEST** test for confirming the effectiveness of the system access management process is to map:

A. access requests to user accounts.
B. user accounts to access requests.
C. user accounts to human resources (HR) records.
D. the vendor database to user accounts.

Question 23 Which of the following provides the **BEST** assurance that a firewall is configured in compliance with an enterprise's security policy?

A. Review the actual procedures.
B. Interview the firewall administrator.
C. Review the parameter settings.
D. Review the device's log file for recent attacks.

Question 24 One way to verify control effectiveness is by determining:

A. its reliability.
B. whether it is preventive or detective.
C. the capability of providing notification of failure.
D. the test results of intended objectives.

Question 25 Which of the following is the **MOST** effective way to ensure that outsourced service providers comply with the enterprise's information security policy?

A. Periodic audits
B. Security awareness training
C. Penetration testing
D. Service level monitoring

Study Questions, Answers and Explanations

ANSWERS AND EXPLANATIONS

Introduction	Answers and explanations are provided for the 25 study questions.
Answer Key	The following table: • Lists and explains the correct answers to the study questions • Explains why the other answer choices are incorrect *ISACA JOURNAL* VOLUME 1, 2012

Domain 1—Risk Identification, Assessment and Evaluation				
Answers and Explanations			
Question No.	Correct Answer	Incorrect Choice	Explanation
1	A		Quantitative risk analysis is generally more complex and thus more costly than qualitative risk analysis.
		B	Neither of the two risk analysis methods is fully objective; while the qualitative method subjectively assigns High, Medium and Low frequency and impact categories to a specific risk, subjectivity within the quantitative method is often expressed in mathematical "weights."
		C	To be effective, both processes require personnel who have a good understanding of the business.
		D	Quantitative analysis generally has a better buy-in than qualitative analysis to the point where it can cause overreliance on the results.
2		A	Confidentiality deals with the protection of sensitive information from unauthorized disclosure. While the lack of system resilience can in some cases affect data confidentiality, resilience is more closely linked to the business information requirement of availability.
		B	Integrity relates to the accuracy and completeness of information as well as to its validity in accordance with business values and expectations. While the lack of system resilience can in some cases affect data integrity, resilience is more closely linked to the business information requirement of availability.
		C	Effectiveness deals with information being relevant and pertinent to the business process as well as being delivered in a timely, correct, consistent and usable manner. While the lack of system resilience can in some cases affect effectiveness, resilience is more closely linked to the business information requirement of availability.
	D		Availability relates to information being available when required by the business process—now and in the future. Resilience is the ability to provide and maintain an acceptable level of service during disasters or when facing operational challenges.

Study Questions, Answers and Explanations

Domain 1—Risk Identification, Assessment and Evaluation (cont.)
Answers and Explanations

Question No.	Correct Answer	Incorrect Choice	Explanation
3	A		Each business process involves inherent risk. Not engaging in any activity avoids the inherent risk associated with the activity.
		B	Risk transfer/sharing means reducing either risk frequency or impact by transferring or otherwise sharing a portion of the risk. Common techniques include insurance and outsourcing. These techniques do not relieve an enterprise of a risk, but can involve the skills of another party in managing the risk and reducing the financial consequence if an adverse event occurs.
		C	Risk treatment means that action is taken to reduce the frequency and/or impact of a risk.
		D	Acceptance means that no action is taken relative to a particular risk, and loss is accepted when/if it occurs. This is different from being ignorant of risk; accepting risk assumes that the risk is known, i.e., an informed decision has been made by management to accept it as such.
4		A	The IT director manages the IT systems on behalf of the business owners.
	B		Senior management is responsible for the acceptance and mitigation of all risk.
		C	The risk management department determines and reports on level of risk, but does not own the risk.
		D	The system users are responsible for utilizing the system properly and following procedures, but they do not own the risk.
5	A		Asset identification is the most crucial and first step in the risk assessment process. Risk identification, assessment and evaluation (analysis) should always be clearly aligned to assets. Assets can be people, processes, infrastructure, information or applications.
		B	While threats tie into the risk assessment process, they are not relevant unless they can be related to specific assets.
		C	While vulnerabilities tie into the risk assessment process, they are not relevant unless they can be related to specific assets.
		D	While threat sources tie into the risk assessment process, they are not relevant unless they can be related to specific assets.

Domain 2—Risk Response
Answers and Explanations

Question No.	Correct Answer	Incorrect Choice	Explanation
6		A	In the risk response planning process, appropriate responses are chosen, agreed on, and included in the risk register.
		B	Risk monitoring and control often requires identification of new risks and reassessment of risks. Outcomes of risk reassessments, risk audits and periodic risk reviews trigger updates to the risk register.
	C		While the risk register details all identified risks, including description, category, cause, probability of occurring, impact(s) on objectives, proposed responses, owners, and current status, the primary outputs from risk identification are the initial entries into the risk register.
		D	Risk management strategy planning describes how risk management will be structured and performed.
7		A	The stem does not describe the sharing of risk. Transference is the strategy that provides for sharing risk with partners or taking insurance coverage.
		B	The stem does not describe risk avoidance. Avoidance is a strategy that provides for not implementing certain activities or processes that would incur risk.
	C		Mitigation is the strategy that provides for the definition and implementation of controls to address the risk described.
		D	The stem does not describe risk acceptance. Acceptance is a strategy that provides for formal acknowledgment of the existence of a risk and the monitoring of that risk.
8	A		An insurance policy can compensate the enterprise up to 100 percent by transferring the risk to another company.
		B	Updating the risk registry (with lower values for impact and probability) will not actually change the risk, only management's perception of it.
		C	Staff capacity to detect or mitigate the risk may potentially reduce the financial impact, but insurance allows for the risk to be completely mitigated.
		D	Outsourcing the process containing the risk does not necessarily remove or change the risk.
9		A	Information security managers may best understand the technical tactical situation, but business managers are accountable for managing the associated risk and will determine what actions to take based on the information provided by others, which includes collaboration with, and support from, IT security managers.
		B	This is not internal audit's function. Business managers set priorities, possibly consulting with other parties, which may include internal audit.
		C	The incident response team must ensure open communication to management and stakeholders to ensure that business managers/leaders understand the associated risk and are provided enough information to make informed risk-based decisions.
	D		Business managers are accountable for managing the associated risk and will determine what actions to take based on the information provided by others.

Study Questions, Answers and Explanations

	Domain 2—Risk Response (cont.)		
	Answers and Explanations		
Question No.	Correct Answer	Incorrect Choice	Explanation
10		A	Threats and vulnerabilities are elements of inherent risk. They do not accurately calculate residual risk.
	B		Inherent risk (threats $\times$ vulnerabilities) multiplied by control risk is the formula to calculate residual risk.
		C	Compliance risk is the current and prospective risk to earnings or capital arising from violations of, or nonconformance with, laws, rules, regulations, prescribed practices, internal policies and procedures, or ethical standards. Compliance risk can lead to reputational damage.
		D	Risk governance and risk response are risk domains, not risk elements, to calculate residual risk.

	Domain 3—Risk Monitoring		
	Answers and Explanations		
Question No.	Correct Answer	Incorrect Choice	Explanation
11		A	While most key risk indicator (KRI) metrics need to be optimized in respect to their sensitivity, the most important objective of KRI maintenance is to ensure that KRIs continue to effectively capture the changes in threats and vulnerabilities over time.
	B		Threats and vulnerabilities change over time and KRI maintenance ensures that KRIs continue to effectively capture these changes.
		C	Risk reporting timeliness is a business requirement, but is not a driver for KRI maintenance.
		D	Risk avoidance is one possible risk response. Risk responses are based on KRI reporting.
12	A		The objective of risk reduction is to reduce the residual risk to levels below the enterprise's risk tolerance level.
		B	The inherent risk of a process is a given and cannot be affected by risk reduction/risk mitigation efforts.
		C	Risk reduction efforts can focus on either avoiding the frequency of the risk or reducing the impact of a risk.
		D	Risk reduction efforts can focus on either avoiding the frequency of the risk or reducing the impact of a risk.
13	A		Key performance indicators (KPIs) are assessment indicators that support the judgment of the process performance of a specific process.
		B	Key risk Indicators (KRIs) only provide insights into potential risks that may exist or be realized within a concept or capability that they monitor.
		C	Base practices are activities that, when consistently performed, contribute to achieving a specific process purpose. However, base practices need to be complemented by work products to provide reliable evidence of the performance about a specific process.
		D	Metric thresholds are decision or action points that are enacted when a KPI or KRI reports a specific value or set of values.

Domain 3—Risk Monitoring (cont.)
Answers and Explanations

Question No.	Correct Answer	Incorrect Choice	Explanation
14		A	The described test does not ensure that all transactions have been extracted.
		B	While simplistic, the reasonableness test is a valid foundation for more elaborate data validation tests.
		C	The described test does not identify duplicate transactions.
	D		Reasonableness tests make certain assumptions about the information as the basis for more elaborate data validation tests.
15		A	The level of professionalism is not dependent on whether a risk assessment is conducted by external or internal resources.
		B	The level of quality is not dependent on whether a risk assessment is conducted by external or internal resources.
	C		Independence is the freedom from conflict of interest and undue influence. By the mere fact that the external auditors are from a different entity, their independence level is higher than if the reviewer were from inside the entity for which they are performing a review. Independence is directly linked to objectivity.
		D	The ease of access is not dependent on whether a risk assessment is conducted by external or internal resources.

Domain 4—Information Systems Control Design and Implementation
Answers and Explanations

Question No.	Correct Answer	Incorrect Choice	Explanation
16		A	User accountability does not prevent unauthorized access; it maps a given activity or event back to the responsible party.
	B		Authentication verifies the user's identity and the right to access information according to the access rules.
		C	User identification without authentication does not grant access.
		D	Access rules without identification and authentication do not grant access.
17	A		Review of the enterprise's strategic plan is the first step in designing effective IS controls that would fit the enterprise's long-term plans.
		B	Review of the existing IT environment, although useful and necessary, is not the first task that needs to be undertaken.
		C	The present IT budget is just one of the components of the strategic plan.
			The IT strategic plan exists to support the enterprise's strategic plan.
18		A	Verifying the user's identification through a challenge response does not completely address the issue of access risk if access was not appropriately designed in the first place.
		B	Forcing users to change their passwords does not guarantee that access control is appropriately assigned.
		C	Logon and monitoring unsuccessful access attempts does not address the risk of appropriate access rights.
	D		Physical or logical system access should be assigned on a need-to-know basis, where there is a legitimate business requirement based on least privilege and segregation of duties.

Study Questions, Answers and Explanations

Domain 4—Information Systems Control Design and Implementation (cont.)
Answers and Explanations

Question No.	Correct Answer	Incorrect Choice	Explanation
19	A		Vulnerabilities represent characteristics of information resources that may be exploited by a threat. The stem describes such a situation.
		B	Threats are circumstances or events with the potential to cause harm to information resources. The stem does not describe a threat.
		C	Probabilities represent the likelihood of the occurrence of a threat. The stem does not describe a probability.
		D	Impacts represent the outcome or result of a threat exploiting a vulnerability. The stem does not describe an impact.
20		A	An intrusion detection system (IDS) does not protect against phishing attacks since phishing attacks usually do not have the same patterns or unique signatures.
		B	While certain highly specialized spam filters can reduce the number of phishing e-mails that reach their addressees' "in" boxes, they are not as effective in addressing phishing attacks as end-user awareness.
	C		Phishing attacks are a type of to social engineering attack and are best defended by end-user awareness training.
		D	Application hardening does not protect against phishing attacks since phishing attacks generally use e-mail as the attack vector, with the end-user as the vulnerable point, not the application.

Domain 5—Information Systems Control Monitoring and Maintenance
Answers and Explanations

Question No.	Correct Answer	Incorrect Choice	Explanation
21	A		Monitoring tools focusing on transaction data generally correlate information from one system to another, such as employee data from the human resources (HR) system with spending information from the expense system or the payroll system.
		B	Configuration settings are generally compared against predefined values and not based on the correlation between systems.
		C	System changes are compared from a previous state to the current state.
		D	Process integrity is confirmed within the system.
22		A	Tying access requests to user accounts confirms that all access requests have been processed; however, the test does not consider user accounts that have been established without the supporting access request.
	B		Tying user accounts to access requests confirms that all existing accounts have been approved.
		C	Tying user accounts to human resources (HR) records confirms whether user accounts are uniquely tied to employees.
		D	Tying vendor records to user accounts may confirm valid accounts on an e-commerce application, but is similarly flawed as choices A and C since it does not consider user accounts that have been established without the supporting access request.

Domain 5—Information Systems Control Monitoring and Maintenance (cont.)

Answers and Explanations

Question No.	Correct Answer	Incorrect Choice	Explanation
23		A	While procedures may provide a good understanding of how the firewall is supposed to be managed, they do not reliably confirm that the firewall configuration complies with the enterprise's security policy.
		B	While interviewing the firewall administrator may provide a good process overview, it does not reliably confirm that the firewall configuration complies with the enterprise's security policy.
	C		A review of the parameter settings will provide a good basis for comparison of the actual configuration to the security policy and will provide reliable audit evidence documentation.
		D	While reviewing the device's log file for recent attacks may provide indirect evidence about the fact that logging is enabled, it does not reliably confirm that the firewall configuration complies with the enterprise's security policy.
24		A	Reliability is not an indication of control strength; weak controls can be highly reliable, even if they do not meet the control objective.
		B	The type of control (preventive or detective) does not help determine control effectiveness.
		C	Notification of failure does not determine control strength.
	D		Control effectiveness requires a process to verify that the control process worked as intended and meets the intended control objectives.
25	A		Regular audits can spot gaps in information security compliance.
		B	Training can increase user awareness of the information security policy, but is not more effective than auditing.
		C	Penetration testing can identify security vulnerability, but cannot ensure information compliance.
		D	Service level monitoring can only pinpoint operational issues in the enterprise's operational environment.

Glossary

Term	CRISC Definition
Access control	The processes, rules and deployment mechanisms which control access to information systems, resources and physical access to premises
Access rights	The permission or privileges granted to users, programs or workstations to create, change, delete or view data and files within a system, as defined by rules established by data owners and the information security policy
Accountability	The ability to map a given activity or event back to the responsible party
Application control	The policies, procedures and activities designed to provide reasonable assurance that objectives relevant to a given automated solution (application) are achieved
Assessment indicator	Used to assess whether process attributes have been achieved There are two types of assessment indicators: • Process capability indicators, which apply to capability levels 1 to 5 • Process performance indicators, which apply exclusively to capability level 1
Asset	Something of either tangible or intangible value worth protecting, including people, information, infrastructure, finances and reputation
Asset management	Accounts for all IT assets and optimizes the value provided by these assets
Attribute indicator	An assessment indicator that supports the judgment of the extent of achievement of a specific process attribute
Authentication	The act of verifying identity, i.e., user, system **Scope note:** Can also refer to the verification of the correctness of a piece of data
Availability	Ensuring timely and reliable access to and use of information
Balanced scorecard (BSC)	Developed by Robert S. Kaplan and David P. Norton as a coherent set of performance measures organized into four categories that includes traditional financial measures, but adds customer, internal business process, and learning and growth perspectives
Base practice	An activity that, when consistently performed, contributes to achieving a specific process purpose (ISO/IEC 15504:1, 3.17)
Business case	Documentation of the rationale for making a business investment, used both to support a business decision on whether to proceed with the investment and as an operational tool to support management of the investment through its full economic life cycle
Business continuity plan (BCP)	A plan used by an enterprise to respond to disruption of critical business processes Depends on the contingency plan for restoration of critical systems
Business goal	The translation of the enterprise's mission from a statement of intention into performance targets and results
Business impact	The net effect, positive or negative, on the achievement of business objectives
Business impact analysis (BIA)	A process to determine the impact of losing the support of any resource **Scope note:** The BIA assessment study will establish the escalation of that loss over time. It is predicated on the fact that senior management, when provided reliable data to document the potential impact of a lost resource, can make the appropriate decision.

Glossary

Term	CRISC Definition
Business impact analysis/ assessment (BIA)	Evaluating the criticality and sensitivity of information assets An exercise that determines the impact of losing the support of any resource to an organization, establishes the escalation of that loss over time, identifies the minimum resources needed to recover, and prioritizes the recovery of processes and supporting system **Scope note:** This process also includes addressing: • Income loss • Unexpected expense • Legal issues (regulatory compliance or contractual) • Interdependent processes • Loss of public reputation or public confidence
Business objective	A further development of the business goals into tactical targets and desired results and outcomes
Business process owner	The individual responsible for identifying process requirements, approving process design and managing process performance **Scope note:** Must be at an appropriately high level in the enterprise and have authority to commit resources to process-specific risk management activities.
Business risk	A probable situation with uncertain frequency and magnitude of loss (or gain)
Capability	An aptitude, competency or resource that an enterprise may possess or require at an enterprise, business function or individual level that has the potential or is required to contribute to a business outcome and to create value
Capability dimension	The set of elements in a process assessment model explicitly related to the Measurement Framework for Process Capability (ISO/IEC 15504:1, 3.18)
Capability indicator	An assessment indicator that supports the judgment of the process capability of a specific process (ISO/IEC 15504:1, 3.19)
Capability maturity model (CMM)	1. Contains the essential elements of effective processes for one or more disciplines It also describes an evolutionary improvement path from *ad hoc*, immature processes to disciplined, mature processes with improved quality and effectiveness. 2. CMM for software, from the Software Engineering Institute (SEI), is a model used by many enterprises to identify best practices useful in helping them assess and increase the maturity of their software development processes. **Scope note:** CMM ranks software development organizations according to a hierarchy of five process maturity levels. Each level ranks the development environment according to its capability of producing quality software. A set of standards is associated with each of the five levels. The standards for level one describe the most immature or chaotic processes and the standards for level five describe the most mature or quality processes. A maturity model that indicates the degree of reliability or dependency the business can place on a process achieving the desired goals or objectives A collection of instructions that an enterprise can follow to gain better control over its software development process

Glossary

Term	CRISC Definition
Change management	Effective change management—within the IT service domain—helps ensure that changes to the IT infrastructure are applied using standardized methods and procedures to minimize the number and impact of any related incidents upon service. Change management comprises that aspect of the overall IT governance framework, which is enterprise sensitive, dealing with IT configuration, release and problem management issues. **Scope note:** Changes in the IT infrastructure may arise as part of project, program or service improvement initiatives, based on IT services-related problems or in response to changing legislative requirements.
Compensating control	An internal control that reduces the risk of an existing or potential control weakness resulting in errors and omissions
Computer emergency response team (CERT)	A group of people integrated at the enterprise with clear lines of reporting and responsibilities for standby support in case of an information systems emergency This group will act as an efficient corrective control, and should also act as a single point of contact for all incidents and issues related to information systems.
Confidentiality	Preserving authorized restrictions on access and disclosure, including means for protecting privacy and proprietary information
Configuration item (CI)	Component of an infrastructure—or an item, such as a request for change, associated with an infrastructure—which is (or is to be) under the control of configuration management
Configuration management	The control of changes to a set of configuration items over a system life cycle Configuration management focuses on the processes relating to the establishment and maintenance of hardware and software baselines, hardware versions and models, software versions (commercial off-the shelf software [COTS]) and release baselines (changes to internally developed systems) and related service agreements
Control	The means of managing risk, including policies, procedures, guidelines, practices or organizational structures, which can be of an administrative, technical, management, or legal nature
Data classification	The assignment of a level of sensitivity to data (or information) that results in the specification of controls for each level of classification Levels of sensitivity of data are assigned according to predefined categories as data are created, amended, enhanced, stored or transmitted. The classification level is an indication of the value or importance of the data to the enterprise.
Data classification scheme	An enterprise scheme for classifying data by factors such as criticality, sensitivity and ownership
Data custodian	The individual(s) and department(s) responsible for the storage and safeguarding of computerized data
Data owner	The individual(s), normally a manager or director, who has responsibility for the integrity, accurate reporting and use of computerized data
Database management system (DBMS)	A software system that controls the organization, storage and retrieval of data in a database
Defense in depth	The practice of layering defenses to provide added protection Defense in depth increases security by raising the effort needed in an attack. This strategy places multiple barriers between an attacker and an enterprise's computing and information resources.
Detective controls	Exists to detect and report when errors, omissions and unauthorized uses or entries occur

Term	CRISC Definition
Direct information	The process in which the risk practitioner directly gathers the data needed to monitor system operations or controls without the assistance of another individual This is the most trustworthy data since the analyst has direct access to the data source and thereby prevents any alteration or deletion by another person.
Disaster recovery plan (DRP)	A set of human, physical, technical and procedural resources to recover, within a defined time and cost, an activity interrupted by an emergency or disaster
Drag time	The calculation of the impact that a delay in one activity will have on the completion date of the project The drag of an activity is equal to its duration if it does not have another activity in parallel; however, if it has parallel activities, its drag is the lesser of either the activity duration or the total float of the parallel activity.
Enterprise risk management (ERM)	The discipline by which an enterprise in any industry assesses, controls, exploits, finances and monitors risks from all sources for the purpose of increasing the enterprise's short- and long-term value to its stakeholders
ERP (enterprise resource planning) system	A packaged business software system that allows an enterprise to automate and integrate the majority of its business processes, share common data and practices across the entire enterprise, and produce and access information in a real-time environment
Event	Something that happens at a specific place and/or time
Event type	For the purpose of IT risk management, one of three possible sorts of events: threat event, loss event and vulnerability event. **Scope note:** Being able to consistently and effectively differentiate the different types of events that contribute to risk is a critical element in developing good risk-related metrics and well-informed decisions. Unless these categorical differences are recognized and applied, any resulting metrics lose meaning and, as a result, decisions based on those metrics are far more likely to be flawed.
Evidence	1. Information that proves or disproves a stated issue 2. Information an auditor gathers in the course of performing an IS audit; relevant if it pertains to the audit objectives and has a logical relationship to the findings and conclusions it is used to support
Fallback procedures	A plan of action or set of procedures to be performed if a system implementation, upgrade or modification does not work as intended **Scope note:** May involve restoring the system to its state prior to the implementation or change. Fallback procedures are needed to ensure that normal business processes continue in the event of failure and should always be considered in system migration or implementation.
Feasibility study	A phase of a system development life cycle (SDLC) methodology that researches the feasibility and adequacy of resources for the development or acquisition of a system solution to a user need
Framework	A generally accepted, business-process-oriented structure that establishes a common language and enables repeatable business processes **Note:** This term may be defined differently in different disciplines. This definition suits the purposes of this manual.
Frequency	A measure of the rate by which events occur over a certain period of time
Generic practice	An activity that, when consistently performed, contributes to the achievement of a specific process attribute (ISO/IEC 15504:1, 3.22)

Glossary

Term	CRISC Definition
Governance	Ensures that stakeholder needs, conditions and options are evaluated to determine balanced, agreed-on enterprise objectives to be achieved; setting direction through prioritization and decision making; and monitoring performance and compliance against agreed-on direction and objectives Scope note: Conditions can include the cost of capital, foreign exchange rates, etc. Options can include shifting manufacturing to other locations, subcontracting portions of the enterprise to third-parties, selecting a product mix from many available choices, etc.
Governance of enterprise IT	A governance view that ensures that information and related technology support and enable the enterprise strategy and the achievement of enterprise objectives; this also includes the functional governance of IT, i.e., ensuring that IT capabilities are provided efficiently and effectively.
Impact	The magnitude of harm that could be caused by a threat's exploitation of a vulnerability
Impact analysis	A study to prioritize the criticality of information resources for the enterprise based on costs (or consequences) of adverse events In an impact analysis, threats to assets are identified and potential business losses determined for different time periods. This assessment is used to justify the extent of safeguards that are required and recovery time frames. This analysis is the basis for establishing the recovery strategy.
Incident	Any event that is not part of the standard operation of a service and that causes, or may cause, an interruption to, or a reduction in, the quality of that service
Indirect information	The process in which the risk practitioner relies on another individual or process to provide data This could lead to alteration of the data prior to submission to the analyst.
Information security	Ensures that within the enterprise, information is protected against disclosure to unauthorized users (confidentiality), improper modification (integrity), and non-access when required (availability)
Information systems (IS)	The combination of strategic, managerial and operational activities involved in the gathering, processing, storing, distributing and use of information, and its related technologies. Scope note: Information systems are distinct from information technology (IT) in that an information system has an IT component that interacts with the process components.
Information technology (IT)	The hardware, software, communication and other facilities used to input, store, process, transmit and output data in whatever form
Infrastructure as a Service (IaaS)	Offers the capability to provision processing, storage, networks and other fundamental computing resources, enabling the customer to deploy and run arbitrary software, which can include operating systems (OSs) and applications
Inherent risk	1. The risk level or exposure without taking into account the actions that management has taken or might take (e.g., implementing controls) 2. The risk that a material error could occur, assuming that there are no related internal controls to prevent or detect the error
Integrity	Guarding against improper information modification or destruction, and includes ensuring information non-repudiation and authenticity
Internal control	The policies, procedures, practices and organizational structures designed to provide reasonable assurance that the business objectives will be achieved and undesired events will be prevented or detected and corrected
IT architecture	Description of the fundamental underlying design of the IT components of the business, the relationships among them, and the manner in which they support the enterprise's objectives

Term	CRISC Definition
IT infrastructure	The set of hardware, software and facilities that integrates an enterprise's IT assets **Scope note:** Specifically, the equipment (including servers, routers, switches, and ca!bling), software, services and products used in storing, processing, transmitting and displaying all forms of information for the organization's users
IT-related incident	An IT-related event that causes an operational, developmental and/or strategic business impact
IT risk	The business risk associated with the use, ownership, operation, involvement, influence and adoption of IT within an enterprise
IT risk issue	1. An instance of an IT risk 2. A combination of control, value and threat conditions that impose a noteworthy level of IT risk
IT risk profile	A description of the overall (identified) IT risk to which the enterprise is exposed
IT risk register	A repository of the key attributes of potential and known IT risk issues Attributes may include name, description, owner, expected/actual frequency, potential/actual magnitude, potential/actual business impact and disposition.
IT risk scenario	The description of an IT-related event that can lead to a business impact
IT strategic plan	A long-term plan (i.e., three- to five-year horizon) in which business and IT management cooperatively describe how IT resources will contribute to the enterprise's strategic objectives (goals)
IT tactical plan	A medium-term plan (i.e., six- to 18-month horizon) that translates the IT strategic plan direction into required initiatives, resource requirements and ways in which resources and benefits will be monitored and managed
Key control	The controls that provide the best gauge of the maturity and effectiveness of the control environment and the risk management effort A key control is one which, if breached, would have the largest impact on the business, often by affecting more than one system, department or product line.
Key performance indicator (KPI)	A measure that determines how well the process is performing in enabling the goal to be reached **Scope note:** A lead indicator of whether a goal will likely be reached, and a good indicator of capabilities, practices and skills. It measures an activity goal, which is an action the process owner must take to achieve effective process performance.
Key risk indicator (KRI)	A subset of risk indicators that are highly relevant and possess a high probability of predicting or indicating important risk **Scope note:** See *risk indicator*.
Leading practice	A frequent or usual action performed as an application of knowledge
Likelihood	In nontechnical terms, "likelihood" is usually a synonym for "probability;" but in statistical usage, a clear technical distinction is made. It would be improper to switch "likelihood" and "probability" in the two following sentences: • "If I were to flip a fair coin 100 times, what is the *probability* of it landing heads-up every time?" • "Given that I have flipped a coin 100 times and it has landed heads-up 100 times, what is the *likelihood* that the coin is fair?" To determine the likelihood of a future adverse event, enterprises must analyze threats to a system, potential vulnerabilities and the controls in place

Glossary

Term	CRISC Definition
Loss event	Any event where a threat event results in loss
Magnitude	A measure of the potential severity of loss or the potential gain from realized events/scenarios
Management	Plans, builds, runs and monitors activities in alignment with the direction set by the governance body to achieve the enterprise objectives
Nondisclosure agreement (NDA)	A legal contract between at least two parties that outlines confidential materials that the parties wish to share with one another for certain purposes, but wish to restrict from generalized use; a contract through which the parties agree not to disclose information covered by the agreement
Nonrepudiation	The assurance that a party cannot later deny originating data; provision of proof of the integrity and origin of the data and that can be verified by a third party
Objectivity	The ability to exercise judgment, express opinions and present recommendations with impartiality
Ongoing monitoring	The gathering of control data as part of normal processes where control data may be supplied to the analyst on a routine scheduled basis
Operational level agreement (OLA)	An internal agreement covering the delivery of services that support the IT organization in its delivery of services
Performance indicator	A set of metrics designed to measure the extent to which performance objectives are being achieved on an ongoing basis **Scope note:** Performance indicators can include service level agreements (SLAs), critical success factors (CSFs), customer satisfaction ratings, internal or external benchmarks, industry best practices and international standards.
Platform as a Service (PaaS)	Offers the capability to deploy onto the cloud infrastructure customer-created or -acquired applications that are created using programming languages and tools supported by the provider
Portfolio	The set of projects owned by a company **Scope note:** It usually includes the main guidelines relative to each project, including objectives, costs, timelines and other information specific to the project.
Privilege	The level of trust with which a system object is imbued
Problem	In information technology (IT), the unknown underlying cause of one or more incidents
Problem escalation procedure	The process of escalating a problem up from junior to senior support staff, and ultimately to higher levels of management
Problem management	Systematic assessment and resolution through a defined set of IT change and control activities, of technical and procedural issues arising from the use of IT systems in a production environment
Preventive control	An internal control that is used to avoid undesirable events, errors and other occurrences that an enterprise has determined could have a negative material effect on a process or end product
Process assessment model	A model suitable for the purpose of assessing process capability, based on one or more process reference models (ISO/IEC 15504:1, 3.33)
Process attribute	A measurable characteristic of process capability applicable to any process (ISO/IEC 15504:1, 3.31)
Process attribute rating	A judgment of the degree of achievement of the process attribute for the assessed process (ISO/IEC 15504:1, 3.32)
Process capability	A characterization of the ability of a process to meet current or projected business goals (ISO/IEC 15504:1, 3.33)
Process capability level	A point on the six-point ordinal scale (of process capability) that represents the capability of the process, each level building on the capability of the level below (ISO/IEC 15504:1, 3.36)

Term	CRISC Definition
Process capability level rating	A representation of the achieved process capability level derived from the process attribute ratings for an assessed process (ISO/IEC 15504:1, 3.37)
Process outcome	An observable result of a process (ISO/IEC 15504:1, 3.44) **Scope note:** An outcome is an artefact, a significant change of state or the meeting of specified constraints.
Process performance indicator	Process performance indicators (base practices and work products) are specific for each process and are used to determine whether a process is at capability level 1. These are based on material in COBIT 4.1. The process capability indicators are generic for each process attribute for capability levels 1 to 5.
Process purpose	The high-level measurable objectives of performing the process and the likely outcomes of effective implementation of the process (ISO/IEC 15504:1, 3.47)
Process reference model	A model composed of definitions of processes in a life cycle described in terms of process purpose and outcomes, together with an architecture describing the relationships among the processes (ISO/IEC 15504:1, 3.48)
Program	A structured grouping of interdependent projects that is both necessary and sufficient to achieve a desired business outcome and create value These projects could include, but are not limited to, changes in the nature of the business, business processes and the work performed by people as well as the competencies required to carry out the work, the enabling technology, and the organizational structure.
Project	A structured set of activities concerned with delivering a defined capability (that is necessary but not sufficient, to achieve a required business outcome) to the enterprise based on an agreed-on schedule and budget
Qualitative risk analysis	Defines risk using a scale or comparative values (i.e., defining risk factors in terms of high/medium/low or on a numeric scale from 1 to 10) It is based on judgment, intuition and experience rather than on financial values.
Quantitative risk analysis	The use of numerical and statistical techniques to calculate likelihood and impact of risk It uses financial data, percentages and ratios to provide an approximate measure of the magnitude of impact in financial terms.
Recovery point objective (RPO)	Determined based on the acceptable data loss in case of a disruption of operations It indicates the earliest point in time to which it is acceptable to recover the data. The RPO effectively quantifies the permissible amount of data loss in case of interruption.
Recovery strategy	An approach by an enterprise that will ensure its recovery and continuity in the face of a disaster or other major outage
Recovery testing	A test to check the system's ability to recover after a software or hardware failure
Recovery time objective (RTO)	The amount of time allowed for the recovery of a business function or resource after a disaster occurs
Release management	Focuses on the software release life cycle processes, ranging from the initial development of a piece of software to its eventual release, and creating updated versions of the released version to help improve the software or to fix existing software bug(s) This terminology can also be applied to the hardware acquisition, deployment, upgrade and retirement processes.
Residual risk	The remaining risk after management has implemented risk response

Glossary

Term	CRISC Definition
Resilience	The ability of a system or network to resist failure or to recover quickly from any disruption, usually with minimal recognizable effect
Risk	The combination of the probability of an event and its consequence (ISO/IEC 73)
Risk aggregation	The process of integrating risk assessments at a corporate level to obtain a complete view on the overall risk for the enterprise
Risk analysis	1. A process by which frequency and magnitude of IT risk scenarios are estimated 2. The initial steps of risk management: analyzing the value of assets to the business, identifying threats to those assets and evaluating how vulnerable each asset is to those threats **Scope note:** It often involves an evaluation of the probable frequency of a particular event, as well as the probable impact of that event.
Risk appetite	The amount of risk, on a broad level, that an entity is willing to accept in pursuit of its mission
Risk assessment	A process used to identify and evaluate risk and its potential effects **Scope note:** Includes assessing the critical functions necessary for an enterprise to continue business operations, defining the controls in place to reduce enterprise exposure and evaluating the cost for such controls. Risk analysis often involves an evaluation of the probabilities of a particular event.
Risk culture	The set of shared values and beliefs that governs attitudes toward risk-taking, care and integrity, and determines how openly risks and losses are reported and discussed
Risk evaluation	The process of comparing the estimated risk against given risk criteria to determine the significance of the risk (ISO/IEC Guide 73:2002)
Risk factor	A condition that can influence the frequency and/or magnitude and, ultimately, the business impact of IT-related events/scenarios
Risk governance	Risk governance is a strategic business function that helps ensure that: • Risk management activities align with the enterprise's opportunity and loss capacity and leadership's subjective tolerance of it. • The risk management strategy is aligned with the overall business strategy. Enterprise decisions consider the full range of (risk) opportunities and consequences.
Risk Identification	The process of determining and documenting the risk that an enterprise faces The identification of risk is based on the recognition of threats, vulnerabilities, assets and controls in the enterprise's operational environment.
Risk indicator	A metric capable of showing that the enterprise is subject to, or has a high probability of being subject to, a risk that exceeds the defined risk appetite
Risk management	1. The coordinated activities to direct and control an enterprise with regard to risk **Scope note:** In the International Standard, the term "control" is used as a synonym for "measure" (ISO/IEC Guide 73:2002). 2. One of the governance objectives. Entails recognizing risk; assessing the impact and likelihood of that risk; and developing strategies, such as avoiding the risk, reducing the negative effect of the risk and/or transferring the risk, to manage it within the context of the enterprise's risk appetite.
Risk map	A (graphic) tool for ranking and displaying risks by defined ranges for frequency and magnitude

Glossary

Term	CRISC Definition
Risk portfolio view	1. A method to identify interdependencies and interconnections among risk, as well as the effect of risk responses on multiple types of risk 2. A method to estimate the aggregate impact of multiple risks (e.g., cascading and coincidental threat types/scenarios, risk concentration/correlation across silos) and the potential effect of risk response across multiple risks
Risk scenario	A description of an event that can lead to a business impact
Risk scenario analysis	Risk scenario analysis is a technique used to: • Describe risk in a more concrete and tangible manner • Allow for proper risk assessment and analysis
Risk tolerance	The acceptable level of variation that management is willing to allow for any particular risk as the enterprise pursues its objectives
Scope creep	Also called requirement creep, this refers to uncontrolled changes in a project's scope.
Separate evaluation	The gathering of data through a separate or on demand process—where the control data is not supplied automatically or as a part of normal scheduled business processes
Service level agreement (SLA)	An agreement, preferably documented, between a service provider and the customer(s)/user(s) that defines minimum performance targets for a service and how they will be measured
Slack time (float)	Time in the project schedule, the use of which does not affect the project's critical path; the minimum time to complete the project based on the estimated time for each project segment and their relationships **Scope note:** Slack time is commonly referred to as "float" and generally is not "owned" by either party to the transaction.
Statement of work (SOW)	A formal document that captures and defines the work activities, deliverables, and time line a vendor must execute in performance of specified work for a client The SOW usually includes detailed requirements and pricing, with standard regulatory and governance terms and conditions
Software as a Service (SaaS)	Offers the capability to use the provider's applications running on cloud infrastructure The applications are accessible from various client devices through a thin client interface such as a web browser (e.g., web-based email).
Standard	A mandatory requirement, code of practice or specification approved by a recognized external standards organization, such as International Organization for Standardization (ISO)
Strategic planning	The process of deciding on the enterprise's objectives, on changes in these objectives, and the policies to govern their acquisition and use
System development life cycle (SDLC)	The phases deployed in the development or acquisition of a software system **Scope note:** SDLC is an approach used to plan, design, develop, test and implement an application system or a major modification to an application system. Typical phases of SDLC include the feasibility study, requirements study, requirements definition, detailed design, programming, testing, installation and postimplementation review, but not the service delivery or benefits realization activities.
Threat	Anything (e.g., object, substance, human) that is capable of acting against an asset in a manner that can result in harm **Scope note:** A potential cause of an unwanted incident (ISO/IEC 13335)

Term	CRISC Definition
Threat event	Any event where a threat element/actor acts against an asset in a manner that has the potential to directly result in harm
Vulnerability	A weakness in the design, implementation, operation or internal control of a process that could expose the system to adverse threats from threat events
Vulnerability event	Any event where a material increase in vulnerability results. Note that this increase in vulnerability can result from changes in control conditions or from changes in threat capability/force.
Work product	An artefact associated with the execution of a process (ISO/IEC 15504:1,3.55)

Page intentionally left blank

Suggested Resources for Further Study

As candidates read through this manual and encounter topics that are new to them or ones for which they feel their knowledge and experience are limited, additional references should be sought. Suggested resources for further study are provided in each of the chapters in Parts I and II of this manual. Also presented, below, is a comprehensive alphabetical list that includes all of the references provided in Parts I and II. Publications in **boldface** are available through the ISACA Bookstore.

Alberts, Christopher; Audrey Dorofee; *Managing Information Security Risks: The OCTAVE^{SM} Approach*, Addison-Wesley Professional, USA, 2002

Barnier, Brian; *The Operational Risk Handbook for Financial Companies: A Guide to the New World of Performance-oriented Operational Risk*, Harriman House Ltd, UK, 2011

Blokdijk, Gerard; Claire Engle; Jackie Brewster; *IT Risk Management Guide—Risk Management Implementation Guide: Presentations, Blueprints, Templates; Complete Risk Management Toolkit Guide for Information Technology Processes and Systems*, Emereo Publishing, Australia, 2008

British Standards Institution (BSI), BS 25999, *A Code of Practice for Business Continuity Management*, UK, 2007

Business Continuity Institute (BCI), *Good Practice Guidelines 2010*, UK, 2010

Committee of Sponsoring Organizations of the Treadway Commission (COSO), *Internal Control—Integrated Framework: Guidance on Monitoring Internal Control Systems*, USA, 2009

European Commission, "European Union (EU) Data Protection Directive," *http://ec.europa.eu/justice/data-protection/index_en.htm*

IEEE Computer Society, Standard 1074-2006 for Developing a Software Project Lifecycle Process, USA, 2006, *www.ieee.org*

International Organization for Standardization (ISO), ISO/IEC 22301:2012, *Societal security—Business continuity management systems—Requirements, www.iso.org/iso/catalogue_detail?csnumber=50038*

ISO, ISO 27001, *Information security management—Specification with guidance for use*, Switzerland, 2005. This is the replacement for BS7799-2. It is intended to provide the foundation for third-party audit and is harmonized with other management standards, such as ISO/IEC 9001 and 14001.

ISO, ISO/IEC 27001:2005, *Information technology—Security techniques—Information security management systems—Requirements*, Switzerland, 2005

ISO, ISO/IEC 27002:2005, *Information technology—Security techniques—Code of practice for information security management*, Appendix 1, Switzerland, 2005

ISO, ISO/IEC 27004:2009, *Information technology—Security techniques—Information security management—Measurement*, Switzerland, 2009

ISO/IEC 27005:2009, *Information technology—Security techniques—Information security management—Measurement*, Switzerland, 2009

ISO, ISO/IEC 27005:2011, *Information technology—Security techniques—Information security risk management*, Switzerland, 2011

ISO, ISO/IEC 38500:2010 *Corporate governance of information technology*, Switzerland, 2010

International Project Management Association (IPMA), *IPMA Competence Baseline (ICB), Version 3.0*, The Netherlands, 2006

Note: Publications in bold are stocked in the ISACA Bookstore.

Suggested Resources for Further Study

ISACA
- **Frameworks and Related Publications:**
  - *COBIT 5*, 2012 *(A Business Framework for the Governance and Management of Enterprise IT)*
  - *COBIT and Application Controls: A Management Guide*, 2009
  - *COBIT 5: Enabling Processes*, 2012
  - *COBIT 5 Implementation*, 2012
  - *Enterprise Value: Governance of IT Investments, The Val IT Framework 2.0*, 2008
  - *Implementing and Continually Improving IT Governance*, 2010
  - *ITAF: A Professional Practices Framework for IT Assurance*, 2008
  - *IT Assurance Guide: Using COBIT*, 2007
  - *The Business Model for Information Security (BMIS)*, 2010
  - *The Risk IT Framework*, 2009
  - *The Risk IT Practitioner Guide*, 2009
  - *Value Management Guidance for Assurance Professionals: Using Val IT 2.0*, 2010
- **Executive and Management Guidance:**
  - *Information Security Governance: Guidance for Boards of Directors and Executive Management*, 2nd Edition, 2006
- **Practitioner Guidance:**
  - *Change Management Audit/Assurance Program*, 2009
  - *Monitoring Internal Control Systems and IT: A Primer for Business Executives, Managers and Auditors on How to Embrace and Advance Best Practices*, 2010
  - *Systems Development and Project Management Audit/Assurance Program*, 2009
- **Certification Publications:**
  - *CISM Review Manual 2012*, 2011

ITIL Service Operations, Her Majesty's Government Cabinet Office, UK 2011

Jones, J., *An Introduction to Factor Analysis of Information Risk (FAIR)*, Risk Management Insight LLC, USA, November 2006

Kouns, Jake; Daniel Minoli; *Information Technology Risk Management in Enterprise Environments: A Review of Industry Practices and a Practical Guide to Risk Management Teams*, **Wiley-Interscience, USA, 2010**

Kovacich, Gerald L.; Edward Halibozek; *The Manager's Handbook for Corporate Security: Establishing and Managing a Successful Assets Protection Program*, Butterworth-Heinemann, USA, 2003

Krutz, Ronald L.; Russell Dean Vines; *The CISM Prep Guide: Mastering the Five Domains of Information Security Management*, Wiley, USA, 2003

Lientz, Bennet P.; Lee Larssen; *Risk Management for IT Projects: How to Deal With Over 150 Issues and Risks*, Butterworth-Heinemann, USA, 2006

Maizlish, Bryan; Robert Handler; *IT Portfolio Management Step-by-Step: Unlocking the Business Value of Technology*, **John Wiley & Sons, USA, 2005**

National Institute of Standards and Technology (NIST), *Guide for Assessing the Security Controls in Federal Information Systems and Organizations, Building Effective Security Assessment Plans*, Special Publication (SP) 800-53A, Revision 1, USA, 2010

NIST, *Guide for Mapping Types of Information and Information Systems to Security Categories*, Special Publication (SP) 800-60 USA, 2008

NIST, *Guide for Security-Focused Configuration Management of Information Systems*, Special Publication (SP) 800-128, USA, 2011, *http://csrc.nist.gov/publications/nistpubs/800-128/sp800-128.pdf*

NIST, *Risk Management Guide for Information Technology Systems*, Special Publication (SP) 800-30, *www.csrc.nist.gov*

NIST, *Security Controls in External Environments*, Special Publication (SP) 800-53, Revision 3, Section 2.4, Appendix 2, USA, 2009

Note: Publications in bold are stocked in the ISACA Bookstore.

Suggested Resources for Further Study

NIST, Special Publication (SP) 800-53, Revision 3, Online Database, USA, *http://web.nvd.nist.gov/view/800-53/home*

Office of Government Commerce (OGC), *ITIL: IT Service Management*, Version 3, UK, 2009, *http://itil.osiatis.es/ITIL_course/it_service_management/change_management/overview_change_management/overview_change_management.php*

OGC, *Projects in Controlled Environments 2 (PRINCE2): Directing Successful Projects With PRINCE2*, UK, 2009

OGC, *Projects in Controlled Environments 2 (PRINCE2): Managing Successful Projects With PRINCE2*, UK, 2009

Parliament of Australia, "Privacy Amendment (Enhancing Privacy Protection) Bill 2012," *www.aph-gov.au/Parliamentary_Business/Bills_Legislation*

Peltier, Thomas A.; *Information Security Risk Analysis, 3rd Edition*, Auerbach Publications, USA, 2010

Project Management Institute (PMI), *A Guide to the Project Management Body of Knowledge (PMBOK), 4th Edition*, USA, 2008

Sherwood, John; Andrew Clark; David Lynas; *Enterprise Security Architecture: A Business-Driven Approach*; CMP Books, USA, 2005

Westerman, George; Richard Hunter; *IT Risk: Turning Business Threats Into Competitive Advantage*, **Harvard Business School Press, USA, 2007**

Page intentionally left blank

General CRISC Information

Introduction

The CRISC certification is designed to meet the growing demand for professionals who can integrate enterprise risk management (ERM) with discrete IS control skills. The technical skills and practices the CRISC certification promotes and evaluates are the building blocks of success in this growing field, and the CRISC designation demonstrates proficiency in this role.

Requirements for Certification

To earn the CRISC designation, the following requirements must be met:
1. Pass the CRISC exam
2. Submit an application (within five years of the passing date) with verified evidence of a minimum of at least three years of cumulative work experience performing the tasks of a CRISC professional across at least three CRISC domains. There will be no substitutions or experience waivers. A processing fee of US $50 must accompany all applications.
3. Adhere to the ISACA Code of Professional Ethics
4. Agree to comply with the CRISC continuing education policy

Please note that certification application decisions are not final as there is an appeal process for certification application denials. Inquiries regarding denials of certification can be sent to *certification@isaca.org*.

Work Experience

Work experience must be gained within the 10-year period preceding the application for certification or within five years from the date of initially passing the exam. An application for certification must be submitted within five years from the passing date of the CRISC exam. All experience must be verified independently with employers.

> **Note:** A CRISC candidate may choose to take the CRISC exam prior to meeting the experience requirements.

Description of the Exam

The CRISC Certification Committee oversees the development of the exam and ensures the currency of its content.

The exam consists of 200 multiple-choice questions that cover the CRISC job practice domains.

The job practice was developed and validated using prominent industry leaders, subject matter experts and industry practitioners.

General CRISC Information

Exam Scope

There are five domains in the CRISC job practice. Each domain is accompanied by tasks and knowledge statements that depict the tasks performed by CRISCs and the knowledge required to perform these tasks. Exam candidates will be tested based on their practical knowledge associated with performing the tasks in the percentages listed in the following table.

Domain No.	Domain Name	Percentage
1	Risk Identification, Assessment and Evaluation	31%
2	Risk Response	17%
3	Risk Monitoring	17%
4	Information Systems Control Design and Implementation	17%
5	Information Systems Control Monitoring and Maintenance	18%

Note: The percentages listed with the domains indicate the emphasis or percentage of questions that will appear on the exam from each domain. For a description of each domain's task and knowledge statements, visit *www.isaca.org/criscjobpractice*.

Types of Questions

CRISC exam questions are developed with the intent of measuring and testing practical knowledge and the application of general concepts and standards. All questions are designed with one best answer. The candidate is asked to choose the correct or best answer from the options.

The exam consists of 200 multiple-choice questions that relate to the tasks and knowledge statements in each of the five domains. Questions are designed in a manner which refrains from using negative logic.

Every CRISC question has a stem (question) and four choices (possible answers).

Question Component	Description
Stem	The stem may be in the form of a question or incomplete statement. In some instances, a scenario may also be included. The questions with a scenario normally include a description of a situation and require the candidate to answer two or more questions based on the information provided.
Answer choices	Each question has four answer choices. The candidate has to select the single most appropriate answer. Some questions may require the candidate to choose the appropriate answer based on a qualifier, such as **MOST** or **BEST**.

The candidate is required to read the question carefully, eliminate known incorrect answers and then make the best choice possible.

Exam Schedule

The CRISC exam will be administered twice in 2013:
- Saturday, 8 June 2013
- Saturday, 14 December 2013

Any changes to this schedule will be specified in the *CRISC Bulletin of Information* (*www.isaca.org/criscboi*).

General CRISC Information

Exam Registration	Refer to the *CRISC™ Bulletin of Information* at www.isaca.org/criscboi for specific registration deadlines and registration forms. Registration for the exam can be completed online at www.isaca.org/examreg. The *Candidate's Guide to the CRISC™ Exam and Certification* will be sent to candidates upon receipt and recording of their exam registration and payment.
Exam Administration	ISACA has contracted with an internationally recognized testing agency. This not-for-profit corporation engages in the development and administration of credentialing exams for certification and licensing purposes. It assists ISACA in the construction, administration and scoring of the CRISC exam.
Exam Admission Requirements	All of the following are required to be admitted to take the exam: **Timely arrival**—Report to the testing site at the report time indicated on the admission ticket. NO CANDIDATE WILL BE ADMITTED TO THE TEST CENTER ONCE THE CHIEF EXAMINER BEGINS READING THE ORAL INSTRUCTIONS. > **Note:** Candidates who arrive after the oral instructions have begun will not be allowed to sit for the exam and will forfeit their registration fee. **Admission ticket**—To be admitted into the test site, candidates must bring the e-mail printout OR hard copy admission ticket. Candidates can use their admission tickets only at the designated test center on the admission ticket. **Government issued-identification**—To be admitted to the test site, candidates must bring an acceptable form of photo identification (ID) such as a driver's license, passport or other government ID. This ID must be a current and original government-issued identification that is not handwritten and that contains both the candidate's name as it appears on the admission ticket and the candidate's photograph. > **Note:** Candidates who do not provide an acceptable form of identification will not be allowed to sit for the exam and will forfeit their registration fee.
Exam Guidelines: Logistics	DO: • Become familiar with the exact location of, and the best travel route to, the exam site prior to the date of the exam. • Arrive at the exam testing site at the time indicated on your admission ticket. This will give time for candidates to be seated and get acclimated.

General CRISC Information

Exam Guidelines: Tools at the Exam Center	DO: • Bring several no. 2 pencils since pencils will not be provided at the exam site. DO NOT bring any of the following to the exam site: • Study materials (including notes, paper, books or study guides) • Scratch paper • Notepads • Communication devices (e.g., cellular phone, PDA, BlackBerry®, etc.) • Reference materials, including language dictionaries • Calculators • Food or beverages (without advanced authorization) **Note:** Candidates are not allowed to bring any type of communication device (e.g., cellular phone, PDA, BlackBerry, etc.) into the test center. If candidates are viewed with any such device during the exam administration, their exams will be voided and they will be asked to immediately leave the exam site. For further details regarding what personal belongings can (and cannot) be brought into the test site, please visit *www.isaca.org/criscbelongings*.
Exam Guidelines: Your Exam ID	The chief examiner or designate at each test center will read aloud the instructions for entering information on the answer sheet. It is imperative that candidates include their exam identification number as it appears on their admission ticket and any other requested information on their exam answer sheet. Failure to do so may result in a delay or errors.
Exam Guidelines: Question Analysis	The following instructions may help the candidate to find the correct answer more efficiently: • Read the provided instructions carefully before attempting to answer questions. Skipping over these directions or reading them too quickly could result in missing important information and possibly losing credit points. • Mark the appropriate area when indicating responses on the answer sheet. When correcting a previously answered question, fully erase a wrong answer before writing in the new one. • Remember to answer all questions since there is no penalty for wrong answers. Grading is based solely on the number of questions answered correctly. Do not leave any question blank. • Identify key words or phrases in the question (**MOST**, **BEST**, **FIRST** …) before selecting and recording the answer.
Exam Guidelines: Budget Time	The exam is administered over a four-hour period. This allows for a little over one minute per question. Therefore, it is advisable that candidates pace themselves to complete the entire exam. In order to do so, candidates should complete an average of 50 questions per hour. Candidates are asked to sign the answer sheet to protect the security of the exam and maintain the validity of the scores. **Note:** Candidates are urged to record their answers on their answer sheet. No additional time will be allowed after the exam time has elapsed to transfer or record answers should candidates mark their answers in the question booklet.

| **Exam Misconduct** | Candidates who are discovered engaging in any kind of misconduct—such as giving or receiving help; using notes, papers or other aids; attempting to take the exam for someone else; using any type of communication device, including cell phones, during the exam administration; or removing the exam booklet, answer sheet or notes from the testing room—will be disqualified and may face legal action. Candidates who leave the testing area without authorization or accompaniment by a test proctor will not be allowed to return to the testing room and will be subject to disqualification. The testing agency will report such irregularities to ISACA's CRISC Certification Committee.

Candidates may not take the exam question booklet after completion of the exam.

The following are reasons for dismissal or disqualification:
• Unauthorized admission to the test center
• Candidate creates a disturbance or gives or receives help
• Candidate attempts to remove test materials or notes from the test center
• Candidate impersonates another candidate
• Candidate brings items into the test center that are not permitted
• Candidate possession of any communication device (i.e., cell phone, PDA, BlackBerry) during the exam administration
• Candidate unauthorized leave of the test area |
|---|---|
| **Earning the CRISC Certification** | Passing the exam does not grant the CRISC designation.

To become a CRISC, each candidate must complete all requirements, including submitting an application for certification.

In order to become CRISC-certified, candidates must pass the exam and must complete and submit an application for certification (and must receive confirmation from ISACA that the application is approved). A processing fee of US $50 must accompany all applications. The application will be available on the ISACA web site at *www.isaca.org/criscapp*. Once the application is approved, the applicant will be sent confirmation of the approval. The candidate is not CRISC-certified, and cannot use the CRISC designation, until the candidate's application is approved. |
| **Exam Grading Procedure** | The exam consists of 200 items:
• Candidate scores are reported as a scaled score.

A scaled score is a conversion of a candidate's raw score on an exam to a common scale:
• ISACA uses and reports scores on a common scale from 200 to 800.
• A candidate must receive a score of 450 or higher to pass the exam.

A score of 450 represents a minimum consistent standard of knowledge as established by ISACA's CRISC Certification Committee:
• A candidate receiving a score less than 450 is not successful and can retake the exam by registering and paying the appropriate exam fee for any future exam administration. To assist with future study, the result letter each candidate receives includes a score analysis by content area. There are no limits to the number of times a candidate can take the exam.

The CRISC exam contains some questions that are included for research and analysis purposes only. These questions are not separately identified and are not used to calculate your final score. |

General CRISC Information

Exam Result Distribution	**Approximately eight weeks after the test date, the official exam results will be mailed via the post to candidates.** Additionally, with the candidate's consent on the registration form, an e-mail containing the candidates pass/fail status and score will be sent to paid candidates. This e-mail notification will only be sent to the address listed in the candidate's profile at the time of the initial release of the results. To ensure the confidentiality of scores, exam results will not be reported by telephone or fax. To prevent the e-mail notification from being sent to the candidate's spam folder, the candidate should add *exam@isaca.org* to his/her address book, whitelist or safe senders list.
Failed Score Details	For those candidates not passing the exam, the score report contains a subscore for each job practice domain. Subscores help identify those areas in which further study may be needed before retaking the exam. **Note:** Taking either a simple or weighted average of the subscores does not derive the total scaled score.

Rescoring	As all scores are subjected to several quality control checks before they are reported, rescores are unlikely to result in a score change.
	
	Rescoring request process—The rescoring procedure involves a person hand scoring the answer sheet to ensure that no stray marks, multiple responses or other conditions interfered with computer scoring.
	
	Rescoring request submission—Candidates receiving a failing score on the exam may request a rescoring of their answer sheet. The requests must:
	• Be submitted in writing to the ISACA certification department
	• Be submitted within 90 days following the release of the exam results
	• Include:
	  – The candidates's name
	  – The candidate's identification number
	  – The candidate's mailing address
	• Be accompanied by a payment of the US $75 processing fee
	
	Note: Requests for a hand score after the deadline date will not be processed.

Submission of an Application	Once a candidate passes the CRISC exam, the individual has five years from the date of the exam to apply for certification. To be certified, successful candidates must complete the application and submit it for approval. All work experience must be verified using the appropriate forms included in the application. The form can be downloaded at *www.isaca.org/criscapp*. A processing fee of US $50 must accompany all applications. Once certified, the new CRISC will receive a CRISC certificate and a pin. At the time of application, individuals must also acknowledge that ISACA reserves the right, but is not obligated, to publish or otherwise disclose their CRISC status. Please note that certification application decisions are not final as there is an appeal process for certification application denials. Inquiries regarding denials of certification can be sent to *certification@isaca.org*.

General CRISC Information

Experience in Risk Management and IS Control	Certification is granted initially to individuals who have completed the CRISC exam successfully and meet the following work experience requirements: 1. A minimum of at least three years of cumulative work experience performing the tasks of a CRISC professional across at least three CRISC domains. There will be no substitutions or experience waivers. 2. Experience must have been gained within the 10-year period preceding the application date for certification or within five years from the date of initially passing the exam. 3. All experience must be verified independently with employers.
Adherence to the ISACA Code of Professional Ethics	Members of ISACA and/or holders of the CRISC certification agree to the ISACA Code of Professional Ethics. Failure to adhere to this Code of Professional Ethics can result in an investigation into a member's and/or certification holder's conduct and, ultimately, in disciplinary measures. The ISACA Code of Professional Ethics can be viewed online at *www.isaca.org/ethics*.
Adherence to the CRISC CPE Policy	To maintain certification, the CRISC must: • Pay an annual maintenance fee • Attain and report an annual minimum of 20 continuing professional education (CPE) hours per cycle year • Attain and report a minimum of 120 CPE hours for the three-year reporting period The objectives of the continuing professional education program are to: • Maintain an individual's competency by requiring the update of existing knowledge and skills in the areas of risk and IS control. • Provide a means to differentiate between qualified CRISCs and those who have not met the requirements for continuation of their certification. • Provide a mechanism for monitoring risk and IS control professionals' maintenance of their competency. • Aid management in developing sound risk and IS control functions by providing criteria for personnel selection and development. The CRISC CPE policy can be viewed at *www.isaca.org/crisccpepolicy*.
Revocation of CRISC Certification	The CRISC Certification Committee may, at its discretion and after due and thorough consideration, revoke an individual's CRISC certification for any of the following reasons: • Failing to comply with the CRISC CPE policy • Violating any provision of the ISACA Code of Professional Ethics • Falsifying or deliberately failing to provide relevant information • Intentionally misstating a material fact • Engaging or assisting others in dishonest, unauthorized or inappropriate behavior at any time in connection with the CRISC exam or the certification process • As of January 2013, all appeals resulting in reinstatement related to revocations more than 60 days old will require a US $50 reinstatement fee to be processed.

List of Exhibits

Starting Page No.

Introduction
- Exhibit B.1: High-level Relationship Between CRISC Domains ... ix
- Exhibit B.2: Process-specific Knowledge Statements Addressed in Part II ... x

Part I—Risk Management and Information Systems Control Theory and Concepts

Domain 1—Risk Identification, Assessment and Evaluation
- Exhibit 1.1: Risk Map Indicating Risk Appetite Bands ... 9
- Exhibit 1.2: Elements of a Risk Culture ... 12
- Exhibit 1.3: IT Risk in the Risk Hierarchy ... 23
- Exhibit 1.4: IT Risk Categories ... 23
- Exhibit 1.5: Domain 1 High-level Process Phases ... 24
- Exhibit 1.6: Risk Scenario Development ... 26
- Exhibit 1.7: Risk Scenario Components ... 29
- Exhibit 1.8: Risk Register Entry ... 33
- Exhibit 1.9: Risk Factors in Detail ... 34
- Exhibit 1.10: Expressing IT Risk in Business Terms ... 65
- Exhibit 1.11: Risk Communication Components ... 67

Domain 2—Risk Response
- Exhibit 2.1: Risk Response Process ... 75
- Exhibit 2.2: Risk Response Options and Parameters ... 79
- Exhibit 2.3: Risk Response Prioritization Options ... 81

Domain 3—Risk Monitoring
- Exhibit 3.1: Process Capability Levels and Related Performance Attributes ... 109
- Exhibit 3.2: Process Capability Model ... 110

Domain 4—Information Systems Control Design and Implementation
- Exhibit 4.1: Control Category Interdependencies ... 124
- Exhibit 4.2: The Control Life Cycle ... 125
- Exhibit 4.3: System Development Life Cycle ... 131
- Exhibit 4.4: Parallel Changeover ... 147
- Exhibit 4.5: Phased Changeover ... 148
- Exhibit 4.6: Abrupt Changeover ... 149
- Exhibit 4.7: Relationship Among Portfolio, Program and Project Management ... 153
- Exhibit 4.8: Project Management Iron Triangle ... 153
- Exhibit 4.9: Critical Path Method (CPM) ... 156
- Exhibit 4.10: Relationships Between Project Management Elements ... 160
- Exhibit 4.11: Sample Gantt Chart ... 162
- Exhibit 4.12: PERT Network-based Chart ... 164

Domain 5—Information Systems Control Monitoring and Maintenance
- Exhibit 5.1: Process Dimension Using COBIT 4.1 ... 173
- Exhibit 5.2: Process Capability Levels ... 174
- Exhibit 5.3: PA1.1 Process Performance ... 174
- Exhibit 5.4: Generic Work Products and Relation to Capability Level ... 175
- Exhibit 5.5: Process Dimension and Process Performance Indicators ... 176
- Exhibit 5.6: The Control Life Cycle ... 177
- Exhibit 5.7: Direct and Indirect Information ... 183
- Exhibit 5.8: Continuous Auditing Activities ... 188
- Exhibit 5.9: Continuous Monitoring Tools—Advantages and Disadvantages ... 197
- Exhibit 5.10: Cause-and-effect Diagram Based on Software Change Failure ... 203

List of Exhibits

Part II—Risk Management and Information Systems Control in Practice

2. Portfolio, Program amd Project Management
    Exhibit 2.1:   Relationship Among Portfolio, Program and Project Management...229

3. Change Management
    Exhibit 3.1:   Relationship Among Change Management, Release Management,
                       Problem Management and Configuration Management 255

7. Configuration Management
    Exhibit 7.1:   Relationship Among Change Management, Configuration Management,
                       Release Management and Problem Management..322

8. Problem Management
    Exhibit 8.1:   Relationship Between Events, Incidents and Problems..336

Page intentionally left blank

Index

A slash (/) indicates that the terms are synonymous within this manual.

"See also" indicates that the terms are related or relevant to one another.

A

Acceptance Testing/Final Acceptance Testing, 142
Accountability, 16 (See also Responsibility, RACI Chart)
Accreditation, 143
Automated Tool/Tools
    Monitoring, 188-194
    Segregation of Duties (SoD), 62
Availability, 53 (See also Resilience)

B

Balanced Scorecard (BSC), 65
Benefit Risk/Business Risk, 132
Business Impact/Impact, 25
Business Impact Analysis (BIA), 50
Business Opportunity, 87
Business Requirements for Information, 65
Business Risk/Benefit Risk, 132

C

Change Management, 253
Changeover Technique/Go-live Technique, 146
    Abrupt Changeover, 149
    Parallel Changeover, 147
    Phased Changeover, 148
Confidentiality, 53 (See also Accountability, Reliability)
Configuration Management, 62, 319
Continuous Monitoring, 195
Continuity Management, 281
Control, 122
    Design, 128, 131
    Element, 184-185
    Implementation, 92
    Inventory, 89
    Maintenance, 206
    Monitoring, 183

D

Data
    Access, 104
    Aggregation, 98
    Analysis, 104, 106
    Conversion, 145-146 (See also Data Migration)
    Extraction, 104
    Management, 349 (see Knowledge Management)
    Migration, 145 (See also Data Conversion)
    Validation, 104
Delivery Risk/Project Risk, 132 (See also Business Risk)
Design (Phase), 137

E

Error Management, 194
Error Rate, 100, 150

F

Failure Modes and Effects Analysis (FMEA), 46
Fallback/Rollback, 145 (See also System Development Life Cycle [SDLC])
Feasibility Study, 134
Framework, 16, 64 (See also Practice [Leading], Standard)
Frequency/Likelihood, 9, 198 (See also Impact)

G

Go-live Technique/Changeover Technique, 146

I

Impact/Magnitude of Impact, 40 (See also Frequency, Likelihood)
Incident Response, 93-94
Information
    Direct, 163, 167
    Indirect, 183
Information Security Management, 299
Information Systems (IS) Control Metrics for:
    Change Management, 263
    Configuration Management, 328
    Continuity Management, 294
    Information Security Program Management, 314
    Information Security Services Management, 315
    IT Operations, 380
    Knowledge Management, 359
    Managing the IT Strategy, 223
    Portfolio Management, 247
    Problem Management, 344
    Program and Project Management, 248
    Third-party Service Management, 276
Integrated Test Facility (ITF), 196
Integration Testing/Interface Testing, 141
Integrity, 53 (See also Availability, Confidentiality)
Interface Testing/Integration Testing, 141
International Electronic Commission (IEC), 19 (See also International Organization for Standardization [ISO])
International Organization for Standardization (ISO), 19 (See also International Electronic Commission [IEC])
IS Control
    Maintenance, 180-181
    Monitoring, 180-181, 183, 185
IT Capability, 36
IT Operations Management, 369

Index

K

Key Control Activities, Process-specific to:
    Change Management, 261-263
    Configuration Management, 327-328
    Continuity Management, 291-294
    Information Security Management, 309-313
    IT Operations, 378-379
    Knowledge Management, 357-359
    Managing the IT Strategy, 221-222
    Portfolio Management, 244-246
    Problem Management, 342-344
    Program and Project Management, 244-246
    Third-party Service Management, 275-276
Key Performance Indicator (KPI), 100
Key Risk Indicator (KRI), 44, 100-103 (See also Key Performance Indicator [KPI])
Key Risk Indicators (KRIs) for:
    Change Management, 260
    Continuity Management, 290
    Configuration Management, 326
    Information Security Management, 308
    IT Operations, 377
    Knowledge Management, 356
    Managing the IT Strategy, 220
    Portfolio Management, 243
    Problem Management, 341
    Program and Project Management, 243
    Third-party Service Management, 274
Knowledge Management, 349

L

Likelihood/Frequency, 45 (See also Impact)
Loss Event, 30

M

Magnitude of Impact/Impact, 9
Management Practices, Process-specific to:
    Change Management, 256
    Configuration Management, 323
    Continuity Management, 284
    Information Security Management, 302
    IT Operations, 372
    Knowledge Management, 352
    Managing the IT Strategy, 214
    Portfolio Management, 230
    Problem Management, 337
    Program and Project Management, 233
    Third-party Service Management, 270
Metrics, Performance for:
    Change Management, 263
    Configuration Management, 328
    Continuity Management, 294
    Information Security Program Management, 314
    Information Security Services Management, 315
    IT Operations, 380
    Knowledge Management, 359
    Managing the IT Strategy, 223
    Portfolio Management, 247
    Problem Management, 344
    Program and Project Management, 248
    Third-party Service Management, 276
Monitoring
    Frequency, 198
    Tools, 194

O

Opportunity/Risk Opportunity, 52, 91
Opportunity Management, 7

P

Parallel Changeover/Changeover, 147
Phased Changeover/Changeover, 148
Physical Environment Management, 327
Pilot Testing, 92 (See also Test Types)
Postimplementation Review, 150
Practice (Leading), 18, 137 (See also Framework, Standard)
Problem Management, 333
Process Capability Model, 108, 110
Process Goals for:
    Change Management, 256
    Configuration Management, 323
    Continuity Management, 284
    Information Security Management, 302
    IT Operations, 372
    Knowledge Management, 352
    Portfolio Management, 230
    Problem Management, 337
    Program and Project Management, 233
    Third-party Service Management, 270
Process Monitoring for:
    Change Management, 263
    Configuration Management, 328
    Continuity Management, 294
    Information Security Program Management, 314
    Information Security Services Management, 315
    IT Operations, 380
    Knowledge Management, 352
    Managing the IT Strategy, 223
    Portfolio Management, 247
    Problem Management, 344
    Program and Project Management, 248
    Third-party Service Management, 276
Project Closeout, 150 (See also System Development Life Cycle [SDLC])
Project Controlling/Project Management, 227
Project Initiation, 134
Project Management/Project Controlling, 227
Project Risk/Delivery Risk, 132, 133
Project Scope Management/Scope Management, 155

Q

Quality Assurance (QA) Testing, 142 (See also Test Types)
Qualitative Risk Analysis, 43-44
Quantitative Risk Analysis, 46

Index

R

RACI Chart, 16
Reliability, 53 (See also Availability, Confidentiality, Integrity)
Reporting, 106, 194
Requirements Definition, 136
Resilience, 14 (See also Availability)
Responsibility, 16, 39 (See also Accountability)
Risk, 55-58
    Inherent, 52
    IT-related, 54
    Residual, 55
Risk Acceptance, 78
Risk Analysis, 24
    Qualitative, 43-44
    Quantitative, 46
Risk Appetite, 9, 11 (See also Risk Tolerance)
Risk Articulation, 84 (See also Risk Communication)
Risk Assessment, 21-22
Risk Avoidance, 77
Risk Awareness, 66
Risk Categories, 23
Risk Communication, 66-68, 112 (See also Risk Articulation)
Risk Components, 68
Risk Culture, 13
Risk Evaluation, 21-22 (See also Risk Assessment)
Risk Event, 95 (See also Loss Event, Threat Event, Vulnerability Event)
Risk Factors, 37, 55, 241, 273, 289, 306, 325, 340, 355, 376
    External, 35
    Internal, 37
Risk Factors Related to:
    Change Management, 259
    Configuration Management, 325
    Continuity Management, 289
    Information Security Management, 306
    IT Operations, 376
    Knowledge Management, 355
    Managing the IT Strategy, 218
    Portfolio Management, 241
    Problem Management, 340
    Program and Project Management, 241
    Third-party Service Management, 273
Risk Governance, 6 (See also Risk Management)
Risk Governance Objectives, 8
Risk Hierarchy, 23
Risk Identification, 21-22
Risk Indicator, 100, 102 (See also Key Risk Indicator [KRI])
Risk Management, 87 (See also Risk Governance)
    Guiding Principles for, 5
Risk Management Capability, 36
Risk Management, Considerations for:
    Change Management, 259
    Continuity Management, 289
    Configuration Management, 325
    Information Security Management, 306
    IT Operations, 376
    Knowledge Management, 355
    Managing the IT Strategy, 218
    Portfolio Management, 241
    Problem Management, 340
    Program and Project Management, 241
    Third-party Service Management, 273
Risk Mitigation, 77
Risk Monitoring, 106-107
Risk Opportunity/Opportunity, 91
Risk Reporting, 92, 113-115
    Channels, 113-115
    Criteria, 112
    Input, 107-109
    Output, 107-109
    Stakeholders, 107-109
    Tools, 115
Risk Response, 70-72, 74, 84-85
    Guidelines, 83
    Integration, 83
    Parameters, 79
    Plan, 82, 92
    Prioritization, 81
    Tracking, 83
    Tracking/Monitoring, 83
Risk Scenario, 22, 24-25, 28
    Development, 26-28, 30-31
Risk Scenarios Related to:
    Change Management, 260
    Continuity Management, 289
    Configuration Management, 326
    Information Security Management, 307
    IT Operations, 377
    Knowledge Management, 356
    Managing the IT Strategy, 220
    Portfolio Management, 243
    Problem Management, 341
    Program and Project Management, 243
    Third-party Service Management, 273
Risk Sharing/Risk Transfer, 77
Risk Tolerance, 7-8 (See also Risk Appetite)
Risk Tolerance Threshold, 90
Risk Transfer/Risk Sharing, 77
Rollback/Fallback, 145 (See also System Development Life Cycle [SDLC])
Root Cause (Analysis), 132

S

Scope Management/Project Scope Management, 155
Sensitivity Analysis, 49
Service Level Agreement (SLA), 269
Software Acquisition, 136
Software Development, 132
Stakeholders, 113-115
Standard, 18 (See also Framework, Practice [Leading])
System Design, 138
System Development, 139
System Development Life Cycle (SDLC), 132
System Testing, 141 (See also Software Testing, Test Types)

T

Test Approaches/Test Types, 141
Test Types, 141
Third-party Service Management, 235
Threat, 110

Threats Affecting:
- Change Management, 259
- Configuration Management, 325
- Continuity Management, 289
- Information Security Management, 307
- IT Operations, 376
- Knowledge Management, 355
- Managing the IT Strategy, 218
- Portfolio Management, 241
- Problem Management, 341
- Program and Project Management, 241
- Third-party Service Management, 273

Threat Analysis, 110-111
Threat Event, 30, 405
Tools/Monitoring Tools, 190-193
- Tools for Monitoring Changes, 192
- Tools for Monitoring Conditions, 191-192
- Tools for Monitoring Process Integrity, 193
- Tools for Monitoring Reporting, 194
- Tools for Monitoring Transaction Data, 180

U

Unit Testing, 141 (See also Test Types)
User Acceptance Testing (UAT), 142 (See also Test Types)

V

Vulnerability Event, 30, 405
Vulnerabilities Common to:
- Change Management, 259
- Continuity Management, 289
- Configuration Management, 325
- Information Security Management, 307
- IT Operations, 376
- Knowledge Management, 355
- Managing the IT Strategy, 218
- Portfolio Management, 241
- Problem Management, 341
- Program and Project Management, 241
- Third-party Service Management, 273

Your Evaluation of the CRISC™ Review Manual

ISACA continuously monitors the swift and profound professional, technological and environmental advances affecting risk and IS control professionals. Recognizing these rapid advances, the *CRISC™ Review Manual* will be updated annually.

To assist ISACA in keeping abreast of these advances, please take a moment to evaluate the *CRISC™ Review Manual 2013*. Such feedback is valuable to fully serve the profession and future CRISC exam registrants.

To complete the evaluation on the web site, please go to *www.isaca.org/studyaidsevaluation*.

Thank you for your support and assistance.

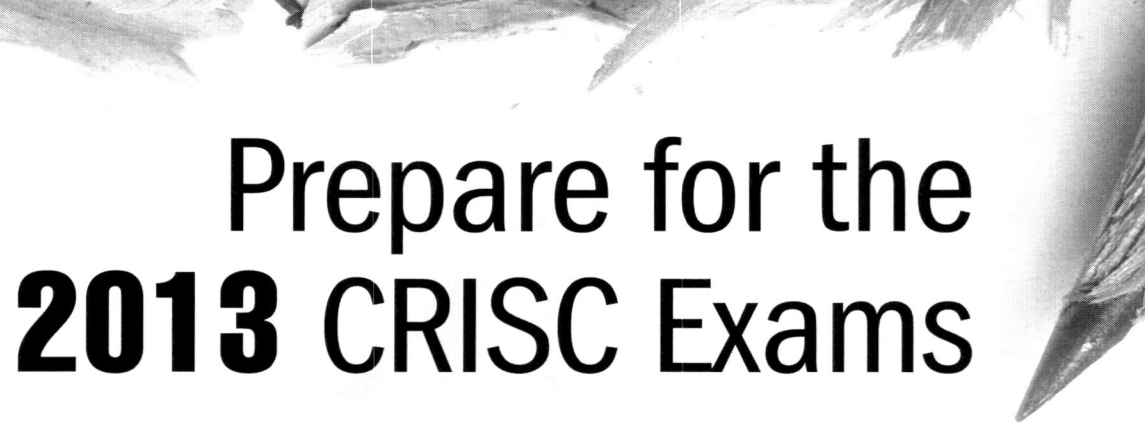

Prepare for the 2013 CRISC Exams

2013 CRISC Review Resources for Exam Preparation and Professional Development

Successful Certified in Risk and Information Systems Control™ (CRISC™) exam candidates have an organized plan of study. To assist individuals with the development of a successful study plan, ISACA® offers several study aids and review courses to exam candidates. These include:

Study Aids

- *CRISC™ Review Manual 2013*
- *CRISC™ Review Questions, Answers & Explanations Manual 2013*
- *CRISC™ Review Questions, Answers & Explanations Manual 2013 Supplement*
- *CRISC™ Exam Self-study Subscription—6 months*

To order, visit *www.isaca.org/criscbooks*.

Review Courses

- Chapter-sponsored review cources *(www.isaca.org/criscreview)*